Python® for Data & Analytics

A Business-Oriented Approach

Daniel H. Groner
Fordham University

Prospect Press

Founded in 2014, Prospect Press serves the academic discipline of Information Systems by publishing essential and innovative textbooks across the curriculum including introductory, emerging, and upper-level courses. Based in Burlington, Vermont, Prospect Press distributes titles worldwide. We welcome new authors to send proposals or inquiries to Beth.golub@prospectpressvt.com.

Editor: Beth Lang Golub
Production Management: Scribe Inc.
Cover Design: Annie Clark

eTextbook:
Edition 1.0
ISBN: 978-1-943153-98-5
Available from Redshelf and VitalSource

Printed Paperback:
Edition 1.0
ISBN: 978-1-943153-99-2
Available from Redshelf

For more information, visit https://www.prospectpressvt.com/textbooks/groner-python

To my family:
Susan, Stefanie, Adam, and Rebecca

Contents

Preface

About the Text

Welcome to *Python for Data & Analytics: A Business-Oriented Approach*. This book introduces essential programming concepts using the Python language as well as how Python can be used in conjunction with packages to access data from various sources and perform different types of analyses. The book uses examples from finance, logistics, marketing, accounting, and health.

Python books typically fall into one of two categories:

- an introductory book with a computer science–style presentation
- a more advanced book on a specific topic—such as finance, data analytics, or a particular Python package—with little introduction to the Python language

This requires potentially using two books or finding/creating supplemental material for teaching Python in a business-oriented programming course. In contrast, this book covers an introduction to the Python language as well as business-oriented uses of Python and key packages for data and analytics.

Target Audience

The book is designed for Python programming and data science courses at either the undergraduate or graduate level within a business or information systems curriculum. Basic math is the only prerequisite for the book. The book can be used in a number of ways:

- for a one-semester course on programming fundamentals and data access
- for a one-semester course on programming fundamentals with additional analytics topics such as visualization, statistics, and/or text analysis
- for a one-semester course on programming fundamentals with selected topics from both data access and analytics
- for a two-semester course sequence on Python programming fundamentals, data access, and analytics

Key Features of the Book

Business, information systems, and data analytics orientation

A key goal of the book is to be a good fit for students studying business, information systems, and/or data science. To this end, the text is conversational but still precise and concise. In Part I, the examples used are not complex (typically relying on basic math) and are described so as to minimize the need for any prerequisite course. In Parts II and III, real and realistic data examples of small and moderate sizes are presented and used. These include stock prices and stock index lists, electric car usage by country, health club membership data, sales transactions, product reviews, and customer demographics. Conversely, the book omits some topics commonly covered in a computer science–oriented text, such as sorting algorithms or recursion.

Animated example program videos

Selected programs presented in the book are animated and narrated in CodeInMotion videos, which visually connect the key aspects of programming: code, flow, variables, input, and output. This allows students to see programs as "living and breathing" code as opposed to just reading static code on the page, thereby providing a better understanding of the dynamics of how programs work. Programs with accompanying videos are marked with this icon: ▷.

Spreadsheets vs. programming

Students in an introductory programming course may have prior experience with spreadsheets. Some spreadsheet concepts can help students learn programming, while others can cause confusion. This book includes sections to highlight certain analogies between spreadsheets and Python programming. These help leverage students' spreadsheet background while also highlighting the differences between spreadsheet modeling and programming. For example, while a spreadsheet's calculation engine automatically determines the order of calculations, a programmer needs to think in recipes (algorithms). As another example, while spreadsheets allow repetition of calculations using copying, programming allows repetition of logic using explicit looping statements. Inclusion of compare/contrast between spreadsheets and programming can help students draw analogies as well as understand differences.

Programs and notebooks

Python solutions are generally developed either in scripts (.py files), often using an integrated development environment (IDE), or using a more interactive incremental approach using a notebook-style tool. In learning and professional practice, there is value in using both types of tools, and the book supports both approaches. For Part I, the book presents small complete programs that are suitable for running in an IDE or in a single cell of a notebook. In most of the later chapters, there are sequences of notebook cells to evolve the concepts as well as self-contained programs to tie things together.

Programs, notebooks, and data files presented in the textbook are provided to instructors in a .zip file. Instructors can provide these to students, who can then view/edit/run programs and notebooks. Further, users of the eTextbook can open a notebook in Google Colab by clicking on this icon: CO. (Note: running notebooks in Colab requires a Google account.)

Modular content

Instructors may prefer a particular order for certain topics. The book is designed to allow some flexibility in topic sequence. For example, Chapter 5 on "Defining Functions" can be taught before Chapter 3 on "Decisions" and Chapter 4 on "Repetition" if desired. Later chapters in Part I cover files, strings, and dictionaries, and these can be covered in any order. For example, this allows dictionaries to be covered right after Chapter 6 on "Lists." Further, the different types of analytics covered in Part III assume only coverage of Chapter 10 on pandas; they do not depend on any of the other chapters in Part II.

Python for data access and preparation

The textbook provides added breadth of coverage for data-oriented Python packages relative to other books. Data acquisition and preparation can be a significant part of an analysis or a system. The textbook covers key packages and approaches related to data, including how pandas and other packages can be used to read web pages, access web APIs, query databases, and access spreadsheet data. Standards related to these are likely new to students, so these are covered in brief in the corresponding chapters: HTML (web pages), JSON (web services), and SQL (databases).

Chapter Elements

At the beginning of each chapter, there are:

- Chapter Objectives: Each chapter opens with the primary learning objectives.
- Chapter Outlines: The sections of the chapter are outlined after the chapter objectives.

Within chapters there are these special items:

- Video: Selected example programs are available in short animated CodeInMotion videos.

Colab: Python .ipynb notebooks are available to open, edit, and run; users of the eTextbook can click on this icon to open a notebook in Google Colab.

Spreadsheet Compare/Contrast: For a given concept, similarities and differences between programs and spreadsheets are discussed.

Tip: For certain concepts, an additional note or recommendation is provided.

At the end of chapter sections are:

Section Summary: A recap of the major concepts is given at the end of sections.

Test Your Understanding: At the end of most sections, multiple choice and true/false questions are provided related to the section.

At the end of each chapter are:

Chapter Review: A summary of the concepts and techniques presented in the chapter is provided.

Key Terms: For the chapter, key terms are defined.

Programming Problems: A number of programming problems that focus on the concepts and techniques introduced in the chapter are included at the end of most chapters.

Overview of Chapters and Appendixes

The book is divided into three major sections:

I: Introduction to Python Programming
II: Python for Data
III: Python for Analytics

Part I focuses on the Python language. The first two chapters introduce basic programming concepts, including simple types (integer, float, string), basic statements, and several Python built-in functions.

The middle chapters of Part I relate to program flow. Chapter 3 ("Decisions") introduces the `if` statement for modeling choices. Chapter 4 ("Repetition") introduces the `while` and `for` statements. Chapter 5 ("Defining Functions") shows variations for creating and using functions. Chapter 5 has no dependency on Chapters 3 and 4, so can be covered earlier if desired.

The last chapters of Part I focus on data. Chapter 6 shows how to use Python lists to store an ordered collection of items. Chapter 7

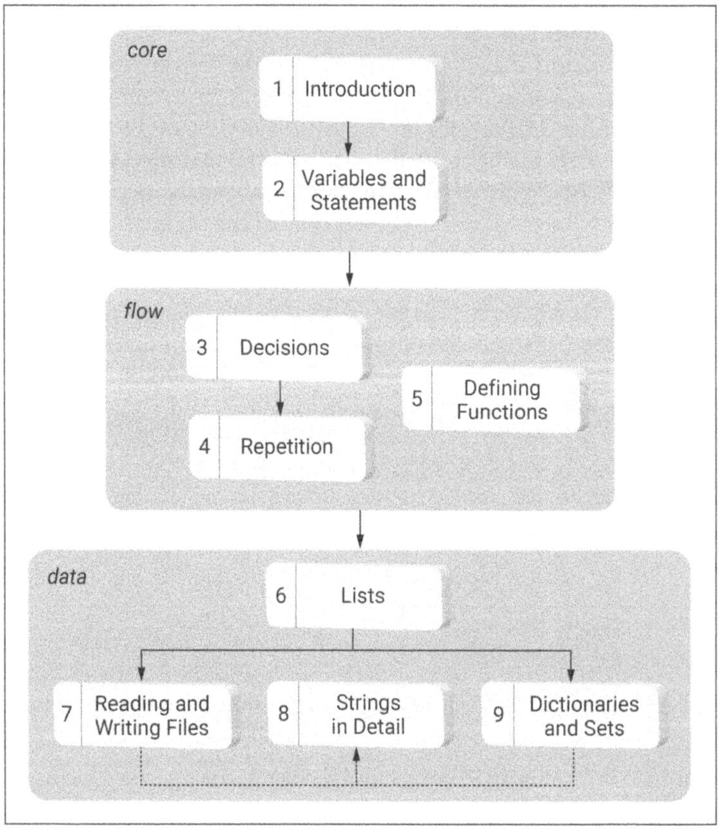

Figure P-1: Part I chapter dependencies

covers reading and writing text data files. Strings are covered in depth in Chapter 8 (and introduced in Chapter 2). Chapter 9 covers dictionaries and sets, which are other data structures for storing collections of items. Note that Chapters 7, 8, and 9 can be covered in any order; for example, dictionaries can be covered right after lists or strings can be covered before files.

Part II covers how Python can be used with selected packages to organize and access data. The pandas package, one of the most popular Python packages, is introduced in Chapter 10 and used throughout the rest of the book. Chapter 11 covers approaches for handling data quality issues, as well as how to transform the organization of pandas data.

The remaining chapters in Part II cover how to access different types of data sources. Accessing web pages using pandas for tabular data, and Beautiful Soup for other cases, is covered in Chapter 12. This chapter also provides a brief introduction to HTML. Web APIs, an alternative to web pages for accessing data and calculations available on the web, are covered in Chapter 13. Chapter 14 covers querying data from relational databases with SQL and pandas. This chapter includes a very brief introduction to SQL queries. Chapter 15 covers accessing data from Excel spreadsheets using pandas for tabular data and the openpyxl package for other data layouts. Note that Chapters 12, 13, 14, and 15 can be covered in any order.

Part III covers how Python can be used with selected packages for analysis. Chapter 16 covers charting with the pandas, Matplotlib, and seaborn packages. A wide variety of statistical topics are covered in Chapter 17. An introduction to text analysis is covered in Chapter 18. The last chapter is a statistical case study on providing incentives to retain customers of a hypothetical telecom company.

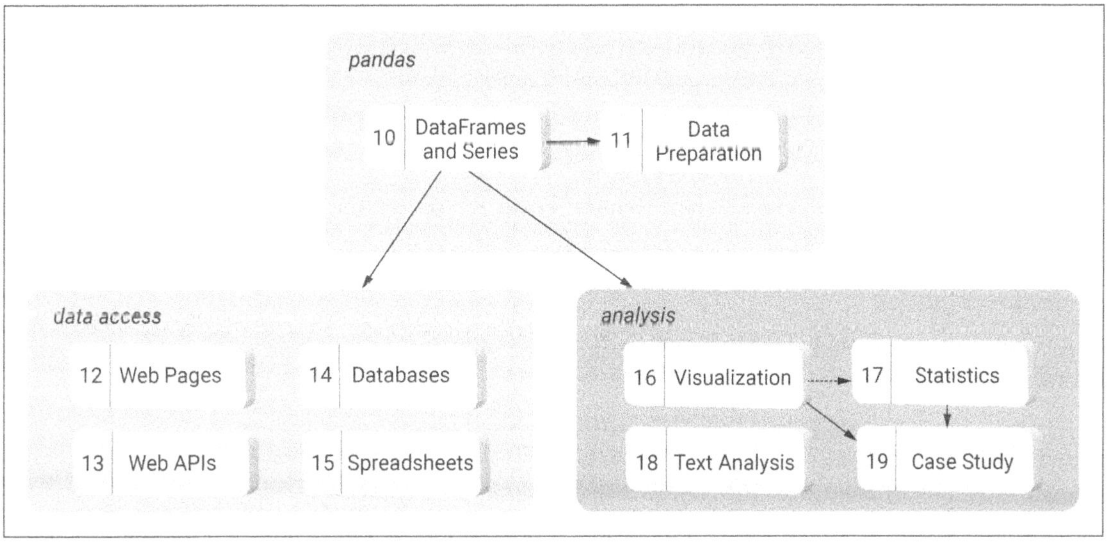

Figure P-2: Part II and III chapter dependencies

Appendixes

Appendixes A1 through A5 provide information for getting started with particular Python tools: Jupyter Notebook, Google Colab, Thonny, Spyder, and PyCharm.

Anaconda is a popular distribution that bundles Python, tools, and packages. This is introduced in Appendix A6. How to install Python packages is covered in Appendix A7.

In order to use packages, you need to import them into your program, and Appendix B discusses this.

Formatting numbers is covered briefly in Chapter 2. Appendix C provides more detail about formatting numbers, strings, and dates.

The exception framework is Python's approach to handling errors. This is covered briefly in Chapter 7 and in more detail in Appendix D.

Python provides extensibility by allowing programmers to define their own classes (also known as types). This is covered in Appendix E.

Chapter 7 covers reading and writing data files in detail. For instructors who would like to cover reading data files in a targeted way, Appendix F provides a brief recap of reading files. This can also provide background for late sections in Chapters 8 and 9, which use file reading in their examples.

Development Tools

There are a variety of development tools available for Python, and instructors may choose a particular integrated development environment (IDE) and/or notebook tool to use in conjunction with this book. To help students get started with particular development tools, Appendix A introduces selected notebook tools (Jupyter Notebook, Google Colab) and IDEs (Thonny, Spyder, PyCharm). (Note: running notebooks in Colab requires a Google account.)

Instructor Resources

Instructor resources include:

- example programs, notebooks, and data files in a .zip file that can be provided to students
- PowerPoint slides for all chapters, including links to notebooks for easy access during lectures
- a test bank
- an instructor guide with answers to programming problems

To access the Instructor Resources:

- Visit https://www.prospectpressvt.com/textbooks/groner-python.
- Scroll to the red horizontal menu bar.
- Click on **Instructor Resources**.
- Click on the yellow box with **Resources Login**.
- Click on **Instructor Resources Access Request Form** and fill out the form. Only verified faculty will receive access information.

Student Resources

As described earlier, selected code examples are presented in CodeInMotion video animations, accessed as follows:

- Users of the eTextbook can click on the video icon ▷.
- Users of the print version can access links to the videos using the instructions below:
  - Visit https://www.prospectpressvt.com/textbooks/groner-python.
  - Scroll to the red horizontal menu bar.
  - Click on **Videos** to find the video animations, organized by chapter.

Acknowledgments

I would like to thank Robert Chiang, Associate Professor at Fordham University and Area Chair of the Information, Technology, and Operations (ITO) Area in the Gabelli School of Business, for suggesting and supporting advanced Python courses in the ITO Area. I would also like to thank Yilu Zhou, Associate Professor at Fordham University and program director for MS Business Analytics and MS Information Systems at the Gabelli School of Business, for providing materials, guidance, and review for the chapter on text analysis.

Aspects of the book grew out of courses at Fordham in introductory and advanced Python. Several graduate assistants helped develop content for these courses, including Elif Aydin, Daryl Chen, Ting Ding, Kriti Dogra, Daniel Li, Jie Lu, and Sonya Song.

Albert Madansky, H. G. B. Alexander Professor Emeritus of Business Administration at the University of Chicago's Booth School of Business, provided significant feedback on the manuscript, as well as technical guidance for the chapter on statistics. Beth Moore, Senior Director of Operations at the Gabelli School of Business, provided specific feedback on many of the chapters in Part I.

The videos accompanying the book use CodeInMotion, which had significant conceptual, design, and development input from Adam Groner.

I was fortunate to work with Beth Lang Golub, president and editor of Prospect Press, since the inception of the book. Her guidance has been key throughout the development of the book. I'd also like to thank Annie Clark, artist at Prospect Press, for her work on figures and icons in the book.

I would like to thank the following reviewers for their valuable feedback on the manuscript:

Deepak Agrawal, Saint Peter's University
Dr. Solomon Antony, Murray State University
Peter Arsenault, Bentley University
Sarbani Banerjee, SUNY Buffalo State College
George Chen, Carnegie Mellon University
Amy J. Connolly, James Madison University
Susan S. Conrad, PhD, Marymount University
John D. Delano, PhD, Cedarville University
Zachary Dodds, Harvey Mudd College
Mark Frydenberg, Bentley University
Xinyu Fu, Joseph M. Katz Graduate School of Business, University of Pittsburgh
Adrian Gardiner, Georgia Institute of Technology
Ashish Mohan Gulati, Arizona State University
Mark Hwang, Central Michigan University
Neil Hwang, Bronx Community College, City University of New York
Nanda Kumar, CUNY Baruch College
Dr. Theodore Larson, University of North Texas Dallas
Dr. Mary Lebens, Metropolitan State University
Prescott Joseph Lerch, Augusta Technical College
Dr. Morteza Mashayekhi, Business Analytics Department, Rider University
Dr. Jacqueline Masloff, Bentley University
Kevin Mentzer, Nichols College
Neba Nfonsang, University of Denver
Susan Palocsay, James Madison University
Hilde Patron, University of West Georgia
Mike Pavlak, Moraine Valley Community College
Elham Rastegari, Creighton University
Michele Samorani, Santa Clara University
Karl R. B. Schmitt, Trinity Christian College
William D. Senn, PhD, Tarleton State University
Anthony Varghese, University of Wisconsin–River Falls
Dr. Jiin Wang, Professor of Computer Information Systems, Alabama State University
Siamak Zadeh, Golden Gate University
Hongwei Zhu, UMass Lowell

About the Author

Daniel H. Groner is a Clinical Associate Professor in the Information, Technology, and Operations Area at Fordham University's Gabelli School of Business. He teaches introductory and advanced classes in Python at the undergraduate and graduate levels. He also teaches courses in Business Analytics, Database Systems, and Optimization Models.

Dan has over 35 years of experience in business analytics systems spanning finance, logistics, and marketing, with a focus on optimization-based models. He is the coauthor of books on the Java programming language as well as the founder of Rose River Software, which develops interactive visualization tools for teaching programming languages.

PART I

Introduction to Python Programming

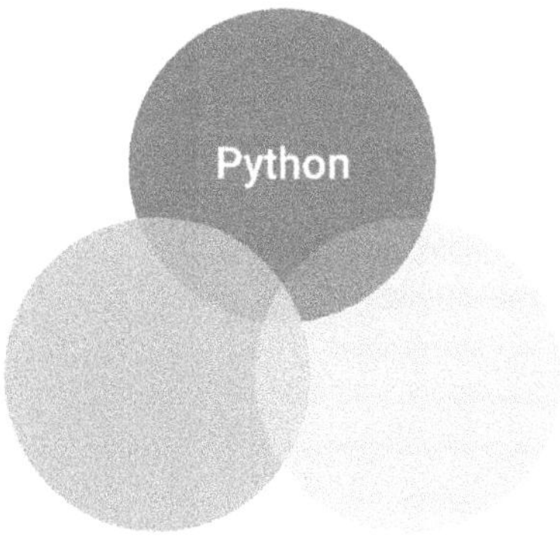

Chapter 1 provides an introduction to programming in general as well as Python in particular. The core elements of programming—variables and statements—are covered in Chapter 2. The middle chapters of Part I are about program flow: how to handle decisions, define repeating logic, and create and use a set of logic (functions). The last chapters of Part I focus on data: lists of data, how to read from and write to data files, strings in more detail, and other data structures (dictionaries and sets).

Chapter 1

Introduction to Programming and Python

Chapter Objectives

In this chapter, you'll learn:

- what a computer program is
- about the input/processing/output model of programs
- some analogies between programming and cooking that may be useful as you learn to program
- different types of problems that Python is used to solve
- how Python compares with some other popular programming languages
- different types of tools for using Python
- some key Python features

Chapter Outline

1.1 About Programming

Why programming?

Most books about computer programming start with the question "What is a computer program?" We'll answer that in a moment, but let's first answer the question "Why are so many people learning computer programming?"

The interest in and value of computer programming have grown due to the intersection of several factors:

- *Data availability and accessibility*: The tremendous growth of publicly available data on the internet, as well as organizations' continually increasing ability to capture and store internal data, has provided a wealth of information to access and analyze using computer programs.
- *Analysis of data*: Visualization, statistics, machine learning, and other techniques have been developed that allow for data to be interpreted, helping people arrive at recommendations and potentially make better decisions.
- *Programming languages* allow for access to data from a variety of sources and provide ways to transform and analyze the data. These languages, including Python, have increasingly incorporated great tools and libraries that allow programmers to be incredibly productive. For example, in Python, one line of code can access a detailed table of data on the internet, another line can summarize this into a concise table, and a third line can display the result in a chart.

This book's title and its coverage reflect these three intersecting areas—programming, data, and analysis. Knowing a programming language like Python allows you to access and analyze data as well as build systems that can automate the accessing, processing, and analysis of data.

Some jobs may require you to focus on one aspect: for example, data collection, data analysis, or systems building. For other roles, you may draw on experience in all these aspects. In other types of positions—for example, technical support, project management, or quality assurance—you may not develop programs but instead work with others who are developers. But regardless of your focus, knowledge of a programming language is becoming a valuable skill for an expanding set of jobs and careers across a wide range of functional areas.

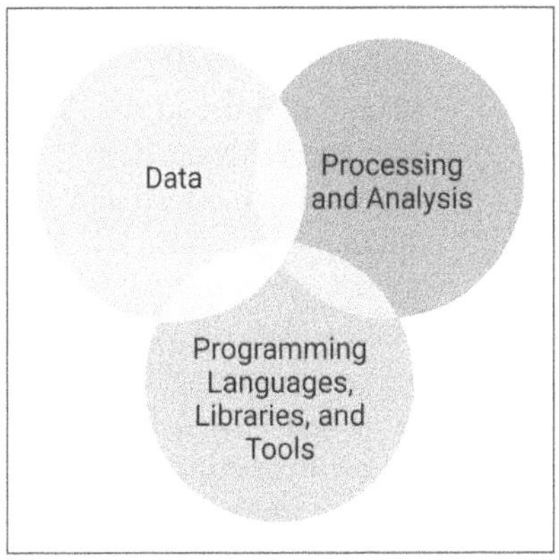

Figure 1-1: Programming, data, processing, and analysis

What is a computer program?

Having looked at why programming is useful to learn, let's look at what a computer program is. A computer program takes input, processes it in a desired way, and provides output. The program is written in a particular language, which provides ways to store and manipulate data. Your goal as a programmer is to decide how best to arrange and manipulate the data to achieve the desired output. There are of course a variety of ways you can do this, and many of these are covered in the book.

To further describe the above, programs:

- *contain data*: Data may be simple—such as numbers, letters, and words—or data may be a composite and organized in some way: for example, in a list of numbers.
- *operate on the data*: The operations on the data are defined in statements and follow a sequence to achieve the desired results. Although programs can do very complicated things, fortunately at their heart there are only a few types of statements. These are what you'll learn in the first few chapters of the book.
- *are written in a computer-interpretable language*: Programs are written in a particular language, and each language has its own particular syntax. For example, in some languages, using uppercase vs. lowercase often matters. Further, certain punctuation like commas and colons need to be in the right place. With practice you are less likely to make mistakes, but following the syntax of a language can be frustrating when you start coding. Fortunately the tools you'll use to program often point out the source of errors to help you correct your programs to get them to work.

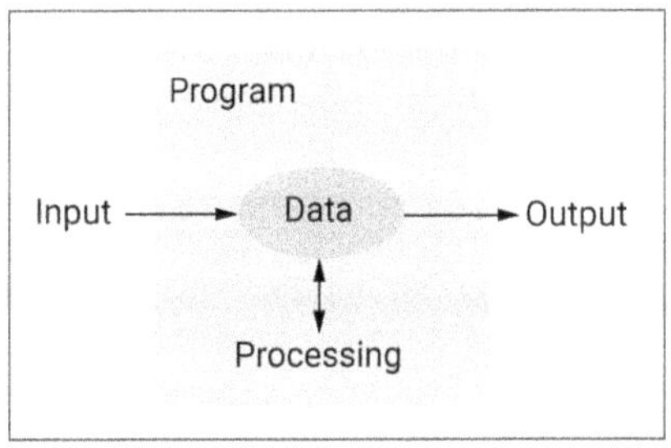

Figure 1-2: Input/processing/output (IPO)

Input/processing/output (IPO)

In general, programs collect input, process the data, and provide output. This pattern is referred to as input/processing/output (IPO), and is shown in Figure 1-2. For example, consider an online application for ordering takeout food from a restaurant. Here the inputs include the items available on the menu and their prices, as well as which items a user orders. The processing includes calculating the pre-tax cost of the order, the amount of tax, and the after-tax cost of the order. The output would include what was ordered, the costs, and when the order will be available for pickup.

Your goal in programming is to decide on the inputs, how to best organize the data, what processing is needed, and how to output the results.

More specifically there are a variety of forms that input and output may take. Figure 1-3 shows some examples of these.

As Figure 1-3 shows, your program's input may come from different types of physical sources. Similarly, a program's output may be sent to different types of physical destinations. For example, the

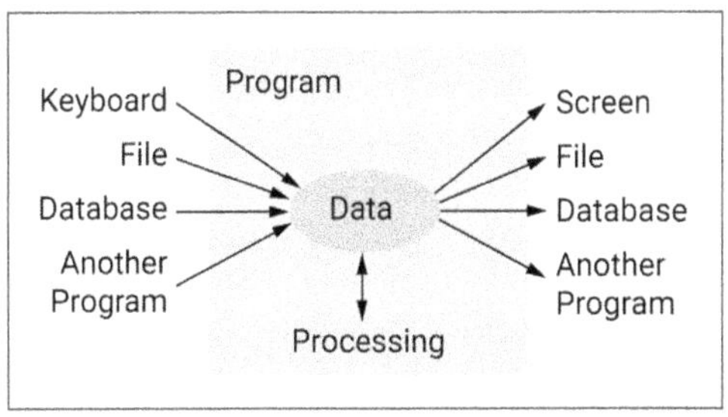

Figure 1-3: Some types of input and output

simplest arrangement is to have input received from the keyboard in response to prompts from the program, and after processing, the program can provide its results to your screen. This is what our initial chapters will assume in order to simplify the input and output and allow us to focus on the different types of processing your programs can do.

Once input is read into a program, the data are stored in the computer's memory. This is known as RAM, for random access memory. Access to data in RAM is fast as compared to that of other types of storage such as accessing files on your hard drive. However, data in RAM are only retained during the running of the program, whereas data on a hard drive are retained after the program ends.

In subsequent chapters, we'll look beyond the keyboard and screen. For example, we'll see how you can read data from a file and write results to a file. Another source of input can be a database, and in Part II, we'll look at how to query a database.

More advanced cases involve programs communicating with other programs. For example, if you would like to access data from a web page, your program can communicate with another program running on a server on the internet. In Part II, we'll see some alternatives for doing this. Beyond web pages, there are other ways to work with features available on the internet, and Part II also covers this in the chapter about web APIs.

There are other potential sources of input for a program. For example, the touchscreen of a phone or tablet is another way a program can receive input. Other devices for providing input to programs include a mouse, touchpad, or microphone. For output, a program can send sound to the speakers of a phone or computer. Although these are beyond the scope of the book, they are further examples of ways you can interact with a computer program by giving input and getting output.

The art and science of programming

Much of this book covers the science of programming in Python. For example, this includes the syntax of the code you need to write so your programs will operate the way you want. However, there is a difference between writing adequate programs and writing good (or great!) programs, and this is found in the art of programming. In Figure 1-1, there is an oval for data, but you have a lot of latitude for how to organize the oval. Knowing when to use what approach is part of the art of programming and comes with experience. As another example, the processing logic of the program can often be coded in different ways. But some may be more efficient, others may be more readable, and yet other approaches may be based on the work of other programmers' code. Deciding what approach to take for logic is another aspect of the art of programming. As part of this book, there are tips that may help you decide between alternative approaches, though there is no substitute for experience in order to become a great programmer.

1.2 Programs as Recipes

In this chapter, we've referred to processing as a key aspect of programming, and in subsequent chapters, we'll look at what kinds of processing your program can do. Essentially programs are recipes for processing data, and so it may be helpful to draw some analogies between cooking recipes and some aspects of programming. For our analogies, we'll use the example of cooking a spinach and mushroom omelet.

In cooking, recipes start with a list of <u>ingredients</u> you need to gather in order to make the dish. For our omelet we'll need eggs, spinach, mushrooms, and other ingredients. The ingredients in a recipe are somewhat similar to the data inputs in your program. Further, the cooking ingredients can come from different sources (pantry, refrigerator, freezer). This is analogous to the diversity of data input sources, such as a keyboard, files, databases, and other input alternatives.

As there are different <u>types of ingredients</u> for cooking, there are different types of data available for you to use when constructing your program, such as numbers and strings. Also, as cooking ingredients are stored in containers, data in programs are stored in containers called variables, which your code can access and update.

Further, in cooking there are <u>steps</u> indicated for working with the ingredients: combining, stirring, frying, sautéing. The steps in a recipe are analogous to the statements you use that direct the program to perform its processing. In Chapter 2, we'll cover some basic statements as well as some data types for writing simple programs.

In cooking, you may have alternatives—for example, choosing between butter and margarine. Similarly, programs may need to choose between alternatives, such as in response to a user's input choice. The selection between alternatives is covered in Chapter 3.

Repeating an action or step may be part of a recipe. For example, when making an omelet, you need to crack open each egg. Further, you need to keep whisking the eggs until the whites and yolks are blended. Similarly, programs often need to repeat steps. For example, a program may need to read each line of a file. Reading the file's first line is a first step, and reading the file's subsequent lines are subsequent repeating steps. Repetition is covered in Chapter 4.

Certain dishes are made from parts that are assembled. For example, an omelet would have the eggs cooked separately from the filling (for example, sautéing spinach and mushrooms), and then these would be put together. Similarly, programs are commonly made from parts called functions and methods, each of which represents a particular set of logic. Using functions is covered in Chapter 2 and making functions is covered in Chapter 5.

You need to decide on how to serve the omelet (paper plate, fancy china, etc.). Analogously, programs can provide their results to different forms of output: for example, to the screen, a file, a database, or other physical options. Also, the cooked dish may be plated to look a particular way. Similarly, programs have the ability to format the results in alternative ways.

Cooking	Cooking Example	Programming	Coverage
storage	refrigerator	input source	Chapters 2, 7, 14
ingredient	¼ pound mushrooms	variable	Chapter 2
direction	cut up mushrooms	statement	Chapter 2
alternatives	butter or margarine	decisions	Chapter 3
repetition	break each egg; whisk eggs until blended	repetition	Chapter 4
parts	cook eggs; sauté mushrooms	functions	Chapters 2, 5
serving	plates	output destination	Chapters 2, 7, 14
plating	arrange on plate	formatting	Appendix C
problem	pieces of shell in eggs	exception	Appendix D
cookbook	breakfast recipes	packages	Parts II, III

Table 1-1: Cooking and programming analogies

As a last analogy, a set of recipes may be collected into a cookbook. For example, an omelet recipe might be one of several recipes in a breakfast cookbook. In programming, a collection of items may be combined in a library (also called a package), and as a programmer, you may use some of these items when building your program. We'll cover several packages in Parts II and III.

Spreadsheet Compare/Contrast
Programming Steps vs. Spreadsheet Recalculation

When developing a program, we need to state the processing logic's order explicitly. This is in contrast to spreadsheets, which have calculation engines that determine the order of calculations for you. This is a key difference between building a spreadsheet and writing a program. Considering the right sequence of steps is part of your goal when developing programs and can be challenging as you write some of your initial programs.

1.3 About Python

Having taken a look at programming in general, let's now introduce Python. Python is one of the most popular programming languages. You can see this by searching for "Top Programming Languages" in a browser. Python has become widely adopted not just because it is easy to learn but also because it can tackle complicated problems.

Some uses of Python

Python is a general-purpose language and a great choice for several types of programs, including:

- *applications*, from small apps to large-scale enterprise-wide systems
- *data science* for seeking patterns in data
- *automation utilities* to streamline certain tasks
- *prototypes* to verify a certain technical approach will work
- *model development* to build algorithms for solving particular problems

Program Type	Example
Application	A university's course registration system keeps track of all courses offered, including their descriptions and prerequisites, information about enrolled students, and which students are registered for which courses.
Data Science	A marketing analyst has detailed transaction data from the company's sales database and wishes to find patterns in recent sales to help decide where best to spend the marketing budget in the period ahead.
Automation	A planning analyst needs to consolidate forecast data in spreadsheets from different business areas in order to create a company-wide projected income statement.
Prototype	A sales engineer for a health technology company needs to confirm that a client's new system can provide data that can be imported into the company's primary analytics application.
Model Development	An analytics specialist is developing a model-based approach to optimize production planning at a consumer goods company.

Table 1-2: Some example uses of Python

Python compared with some other languages

Although there are hundreds of programming languages, it may be useful to compare Python to some of the most popular languages. Python is considered to be a general-purpose language; other examples are C++ and Java. Python can be easier to learn and write than those languages, and so it can be said that it is fast at "develop time." Conversely, C++ can be harder to learn, but it is designed to run on a computer more rapidly—in other words, it is fast at "run time." Java has been in use for many years for large-scale projects, so it is often a good choice for large-scale systems.

For web-based applications, the languages web browsers understand are: HTML to represent page content, CSS to define the page layout, and JavaScript to handle a web page's activity. There are other approaches to web development, but notably, web browsers by default cannot run Python. Behind the scenes on web servers, Python may be a good choice for implementing the back-end logic of a web-based system. Also, in some cases you may want to read a web page's data with a program; some approaches for doing this with Python are covered in Chapter 12, which also has a brief introduction to HTML.

Certain languages are designed with a data-oriented focus. For example, Structured Query Language (SQL) is the most commonly used language for working with relational databases. In Chapter 14, we'll see some ways Python and SQL can work together for relational database queries. Other data-oriented languages are XML and JSON, which are designed to represent data structures (as opposed to performing any processing). XML or JSON are often used when two programs communicate with each other. We'll look at JSON as part of Chapter 13, which covers web APIs.

Python version history

Python was designed and initially implemented by Guido van Rossum with several goals, such as ease of use, readability, and extensibility. Its initial public release was in 1991, and subsequent major releases were version 1.0 in 1994, 2.0 in 2000, and 3.0 in 2008. Notably, version 3 is not backward compatible with version 2, and in this book, we use Python version 3. Due to differences between versions 2 and 3, if you search for help with Python on the web, you should indicate "Python 3" as part of your search.

As of this writing, the latest Python version is 3.10.3, released in March 2022. In recent years, new Python versions have been released once a year, with micro releases several times a year. As a result, Python continues to improve over time.

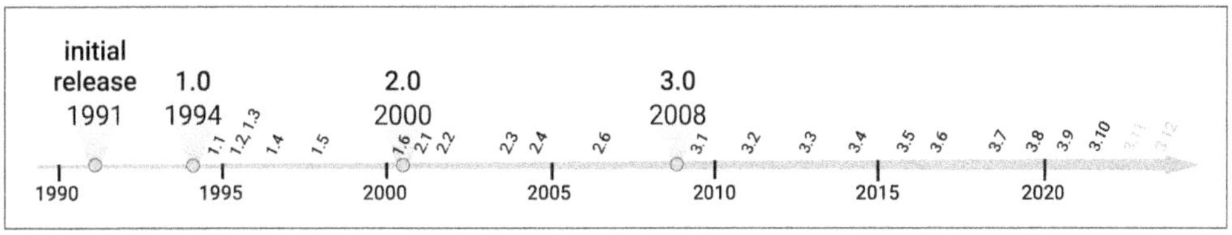

Figure 1-4: Key Python versions

1.4 Programming with Python

There are a number of ways to run Python (as described in the next section), as well as a number of different tools for developing with Python. Several of these tools are described in Appendix A of this book.

Alternative ways to run Python

You can use Python in a variety of ways, depending on your needs. For example, you can use Python interactively, trying one line of code at a time. This is very helpful when learning a new feature or seeing how certain aspects work together. The standard interactive prompt for Python is shown as:

```
>>>
```

In contrast to interactive use, you can write a few or many lines of code and save these in a file in order to run multiple lines of related code together. In this case, you save your program in a file with the .py extension: for example, myprogram.py, which is referred to as a script. (Python may convert your program to a .pyc file, which is more efficient for it to run.) When creating program files, you will typically use an integrated development environment (IDE), which combines editing, running, and debugging features into one tool. IDEs also let you try code lines interactively. Several IDEs are introduced in Appendix A, including Thonny, Spyder, and PyCharm.

A third way you can use Python is with a notebook tool, such as Jupyter Notebook or Google Colab. This is a middle ground between using Python interactively and via scripts. In the notebook approach, you write one or several lines of code in a code cell, which you can run and see the results. Data from one cell are available to subsequent code cells. This approach is very good for exploring and experimenting with data. Notebook tools also allow you to add documentation cells so that notebook steps are easy to understand. In this book, we show notebook cells like this:

```
[1]: language = 'Python'
     print('Welcome to', language, '!')
```

```
Welcome to Python !
```

Some key Python features

As we'll see starting in Chapter 3, Python requires programs to use indenting for certain cases. In particular, code is written in Python like this:

```
parent statement:
    child statement 1
    child statement 2
    ...
```

The above layout can be easier to write and read as compared with other programming languages that use curly braces { } to group their child statements, as here:

```
parent statement {
        child statement 1
        child statement 2
        ...
   }
```

Another way Python can be more readable than other program languages is in its use of English keywords instead of symbols. For example, in some languages, to combine logic, you would use && or ||, but in Python, you use words (and, or).

Python also provides a good set of data types for storing related sets of data together. These have very concise features for working with the data, allowing you to do more with less code. We'll cover these in Chapter 6 (Lists) and Chapter 9 (Dictionaries and Sets). Relatedly, Strings, covered in detail in Chapter 8, have concise and comprehensive features.

Certain programming languages require you to be explicit about the type of data involved. In contrast, Python determines and keeps track of data types for you. Further, some languages require you to include code to handle the use of your computer's memory, but Python handles this for you. Implicit data typing and memory management are examples of ways Python can allow you to focus more on the problem you're solving and a little less on the technical details.

Another important feature of Python is that it runs on different operating systems, such as Windows, macOS, Linux, and others. This allows you to develop a program on one type of computer and have it run on another type of computer.

Python with packages

The Python language is only part of the reason so many people have adopted it. There are also a large number of high-quality and featureful libraries that have propelled Python adoption for many uses. Because Python has an open architecture, programmers have developed and published libraries of code (called packages) that you can use to be very productive. For example, with Python and a package, your program can gather a large amount of data, create a summary report, and display the results in a chart—all in a few lines of code. Further, these libraries are available in a central location, allowing you to find, download, and start using them very quickly. This book covers particular packages related to data organization and data access in Part II, and analysis-oriented packages are covered in Part III.

Python documentation

The official documentation for the Python language is at:

docs.python.org/3

While this book will give you a solid introduction to Python, the official site has detailed coverage that you may find of interest if you want to dive further into a specific aspect of the language.

Chapter Review

Programs collect input and process it to create desired output. The statements that process the data are written in a particular programming language that has a particular syntax. Input to a program may come from different types of sources: for example, your keyboard, files on a disk, a database, or other programs. As well, output may go to different destinations, including your screen, a file, a database, or another program.

Python has evolved over 30 years to become one of the most-used program languages. Its popularity is due in part to Python's ease of use, as well as the wide variety of powerful packages you can leverage. You can also run Python in a few different ways: interactively, in script programs, and in notebooks. Python's ease of use, flexibility, and power make it a great tool for many uses, including application development, data science, automation, prototyping, and model development.

This book will show you how to use Python and several key packages. In Part I, we'll see fundamental aspects of the language and fundamental data types. Part II focuses further on data—how you can use the pandas package to work with data series and matrices, as well as how to access data from different sources. In Part III, different types of analysis are covered, including visualization, statistics, and text analysis.

Key Terms

database	Storage of data in a specific organized way.
function	A named set of logic.
IDE	Integrated development environment, used for writing, running, and debugging a program.
input	Data that are taken into a program (or into a part of a program).
interactive	Working with a program with one or few lines of logic at a time.
keyword	A particular word used by a programming language in a particular way.
library	A set of code and/or data types that can be used when writing programs.
memory	Space within a computer that can store data for a computer program to work with.
notebook	A set of programming logic and documentation, divided into cells.
output	Results from a program or part of a program.
processing	A series of steps for getting, transforming and/or saving data.
PyPi	The central repository for Python packages.
Python package	A software library that can be used to facilitate writing programs.
Python script	A file with Python code that can be run as a program.
relational database	A type of database that stores its information in tables.
statement	A singular step in a program.
variable	A named location for storing data in a program.
web page	A document that can be displayed in a web browser.

Chapter 2

Variables and Statements

Chapter Objectives

In this chapter, you'll learn:

- how you can use the basic building blocks—constants, variables, and statements—to write your first programs
- how your programs can display information on the screen using the `print()` function
- how you can make your programs interactive using the `input()` function
- about Python's basic types for representing numbers and text: `int`, `float`, and `str`
- how to write mathematical statements to calculate results in your programs
- how you can use key math functions that allow rounding and other operations to expand your mathematical vocabulary in Python
- about a few string operations for working with text data

Chapter Outline

2.1 About Constants, Variables, and Statements

In all programming languages, ***constants***, ***variables***, and ***statements*** are the essential ingredients. Constants represent specific values, such as numbers or text. Variables are named places in computer memory that will contain these values. Basic statements provide your program with the ability to access and update variable values. We use constants, variables, and statements to solve a problem by writing a program that combines them.

In this chapter, we'll start by looking at how you can create a numeric variable and set it to a value. You'll see how variable values can be set in several ways, including requesting keyboard input from the program's user to get values. Since variables have names, we'll cover the rules and suggestions for variable names. In addition to a name, each Python variable has a ***type***—we'll initially look at the basic types for numbers and text.

Later in the chapter, we'll look at some ways you can write mathematical statements, including some useful mathematical functions Python provides.

2.2 Variables

Creating a variable and assigning it a value

Variables are names that point to values. For example, the following Python statement creates a variable named `sales`, which points to the value 132:

```
sales = 132
```

You can think of the above statement as a name that points to a value, as shown in Figure 2-1.

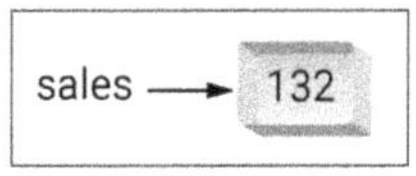

Figure 2-1: A variable includes a name and pointer to a value

> **Spreadsheet Compare/Contrast**
> **Python Variables and Spreadsheet Cells**
>
> In some ways, a variable in Python is like a cell in a spreadsheet. A spreadsheet cell can contain a value such as a number or text. Similarly, a Python variable can point to a value such as a number or string (and other types that we'll see later). And as you can change the value of a spreadsheet cell, your program can also change the value associated with a variable. However, a spreadsheet cell may contain either an entered value or a formula, whereas programs have a sharper distinction between variables and statements. And as spreadsheet formulas are typically hidden and cells display formula results, your program statements are explicitly shown, and their results are displayed only when explicitly requested.

Displaying a variable's value

In order to see a variable's value, your program needs a statement that explicitly displays the value. The simplest Python statement for this uses the `print()` function. (Although it sounds like `print()` will send something to your printer, by default, it will display results to your screen.) For example, Program 2-1 creates and assigns the variable `sales` and then uses the `print()` function to display the value of `sales` to the screen.

Program 2-1

```
1 sales = 132
2 print(sales)
```

Program 2-1 Output

```
132
```

Note that in Program 2-1, line 2 has

- the word `print`, indicating the name of a function
- a set of parentheses
- something inside the parentheses—in this case, the name of the variable: sales

`print()` is one of nearly 70 functions that Python provides—these are called ***built-in functions***. Functions are building blocks you can use to concisely do particular work in your programs. We'll see a few more functions in this chapter, for getting keyboard input as well as some mathematical functions.

Let's step through what happens when Program 2-1 is run. First, a variable named `sales` is created and assigned the value of 132. Then the `print()` function looks up the value assigned to the variable `sales` and displays that value on the screen. Importantly, note that the assignment statement in line 1 is "silent"—it associates the value with the variable but does *not* display it to your computer's screen. The second statement actively displays the value to the screen.

Let's augment our program to provide a label for our value. You can provide multiple items to the `print()` function by separating them with commas. For example, Program 2-2 provides the `print()` function with a label as well as the name of the variable of interest.

Program 2-2 Video animation available.*

```
sales = 132
print('Sales =', sales)
```

Program 2-2 Output

```
Sales = 132
```

In Program 2-2, we provide the `print()` function with two items: the text label `'Sales ='`, which is a Python ***string***, as well as the variable `sales`, which contains the value 132. The `print()` function displays both the text label and the variable's value on the same line on the screen, as shown in Program 2-2's output.

Reassigning a variable

Variables are so named because their values can vary during the life of the program's run. Depending on the logic of your program, some variables might be set to a value, and the value may remain unchanged during the program's run. Other variables may have an initial value, then later in the program's run, these variables will have a different value as the program executes. Deciding which variables should be unchanging (static) vs. which should be changing (dynamic) is a key part of your effort in developing a program. Program 2-3 shows a simple case where a variable's value is changed, and then redisplayed.

Program 2-3 Video animation available.

```
1 # The variable sales is created, initialized to 132, and the value is displayed
2 sales = 132
3 print('Sales =', sales)
4
5 # The variable sales is set to a new value, and the new value is displayed
6 sales = 143
7 print('Sales is now =', sales)
```

Program 2-3 Output

```
Sales = 132
Sales is now = 143
```

In Program 2-3, line 2 creates the variable `sales` and sets its initial value to 132. Later in line 6, the program changes the variable's value to 143.

Program 2-3 also includes the use of ***comments*** in Python. Comments are indicated by the hashtag character (#), as shown in lines 1 and 5. Comments are useful for describing a program and its parts to help clarify what different sections of a program are doing. The text you include after a comment does not affect the logic of the program; in fact, when the program is executing, comments are ignored. Comments can also be placed on the same line as a statement. For example:

```
sales = 132  # create a variable sales and initialize its value
```

Setting a variable's value from user input

So far, our examples involved values that are hard-coded in the program logic. In programs, some variables may be set this way, while other variables are set by user entry via the keyboard, data read from a file, or data

---

* Go to the Videos page at www.prospectpressvt.com/textbooks/groner-python.

accessed from a database. In this chapter, and other early chapters, we'll focus on user input from the keyboard; in later chapters we'll look at other ways to read data into programs.

In Python, you can use the `input()` function to ask the user for input and get their typed reply into your program. `input()` can be seen as a related function to `print()`. Whereas `print()` provides a way to *display* information to the user, `input()` provides a way to *get* information from the user. The `input()` function can take a string within its parenthesis, which the program displays on the screen to the user in order to prompt for the information being requested.

To demonstrate the use of the `input()` function, assume you wanted to write a program that asks the user for their name and then displays a greeting to the user. The recipe for this is as follows:

1) Display a prompt to the user. For example, `Please enter your name:`
2) Allow the user to type their name using the keyboard.
3) Assign the user's entry to a variable, which might be called `name`.
4) Display the greeting to the user on the screen, including the name they entered.

Program 2-4 shows the use of both the `print()` and `input()` functions in order to achieve the above steps in Python.

Program 2-4 Video animation available.

```
name = input('Please enter your name: ')
print('Hi', name)
```

Program 2-4 Example Run

```
Please enter your name: David
Hi David
```

When Program 2-4 is run, the `input()` function first displays the phrase `Please enter your name:` on the screen; then the program waits for the user to type an answer and hit the return key. In the above example run, the user types **`David`**. (In this book, we'll use **`bold`** to indicate what the user types; your Python tool may show user input in color.) Once the return key is hit, the program assigns the user's entry to the variable called `name`. The second line of the program then displays the item `'Hi'` as well as the value of the variable `name`. Note that line 1 does several things: displays a prompt, waits for and receives the keystrokes the user types, creates a variable called `name`, and assigns to it the characters the user typed as the value for the variable `name`.

In contrast with the `print()` function, the `input()` function is typically used on the right side of an equals sign so that its result (what the user entered) can be assigned to a variable. In this example, the program assigns the value provided by `input()` to the variable `name`. The value provided by `input()` is called the ***returned result***, because this value is returned to the program from the `input()` function for the program's further use.

Rules for valid variable names

So far, we've chosen words for variable names (`sales`, `name`). In general, words can be used as variable names, but there are a few rules you need to follow when choosing variable names.

Rule 1: Variable names can start with a letter (A–Z, a–z) or the underscore character (_). Note that variable names cannot start with a digit or other special characters.

Rule 2: Variable names can contain letters, underscores, or digits (0–9) after the first character of the name. For example, `sales2020` is a valid variable name, but `2020sales` is not.

Rule 3: Python has *keywords* that cannot be used as variable names. For example, the word `yield` is a keyword. Table 2-1 contains the list of Python keywords as of Python version 3.8.5.

Rule 4: Upper- vs. lowercase matters. For example, `Sales` is a different variable than `sales`.

If you try to use a variable name that breaks rules 1, 2, or 3, you will get an error message that includes the phrase `SyntaxError: invalid syntax`. If you unintentionally capitalize a variable name that should be lowercase, your program will not run correctly or may generate an error (or incorrect results) when it executes.

and	continue	finally	is	raise
as	def	for	lambda	return
assert	del	from	None	True
async	elif	global	nonlocal	try
await	else	if	not	while
break	except	import	or	with
class	False	in	pass	yield

Table 2-1: Python keywords

Suggestions for good variable names

The previous section indicates the rules for valid variable names—in other words, names you can use for variables. Here are suggested guidelines for naming your variables.

Guideline 1: Choose variable names that help make your program readable.

Guideline 2: Use words that represent the value. For example, `sales` is a better variable name than `s`.

Guideline 3: Start variable names with a lowercase letter. For example, use `sales` instead of `Sales`.

Guideline 4: In certain cases, consider using compound words, either with uppercase letters within (for example, `salesForecast`) or separated by underscores (like `sales_forecast`).

Guideline 5: Use abbreviations to avoid variable names that are too long. There is no firm rule for how long is too long, but names longer than 20 characters might be considered long.

Spreadsheet Compare/Contrast

Variable Names

As described earlier, spreadsheet cells have some similarities to program variables. To continue this analogy, spreadsheet cells have names, each of which corresponds to its row and column: for example, cell A1. In this sense, spreadsheet variable names are preassigned with a nondescript label. In contrast, in Python, you need to give each variable a name, which gives you the opportunity to choose a good variable name. Choosing good variable names is one key aspect to making your program more readable to you and others.

Section Summary

Variables are named locations for data. Statements are actions that can create, access, and update variable values. Variable names need to follow certain rules. The `print()` function is used to display information on the screen. The `input()` function is used to prompt for and get information from the user via the keyboard.

Test Your Understanding

1) How can statements interact with variables?
   a) statements can create variables
   b) statements can get variable values
   c) statements can change variable variables
   d) all of the above

2) What function can be used to display information on the screen?
   a) `print()`
   b) `input()`
   c) `display()`
   d) (a) and (b)

3) Which is not a valid Python variable name?
   a) `Sales2020`
   b) `2020Sales`
   c) `sales_2020`
   d) `SALES`

2.3 Python's Basic Types

As we saw in the previous section, variables have names by which a variable's value can be set, accessed, or updated. In addition to a name and a value, Python values and variables also have a ***type***, which provides information to Python about what can be done with the value. For example, Python can subtract numeric variable values, but it can't subtract text values. Conversely, Python can find the first character of a text variable, but it can't find the first character of a numeric variable.

In this section, we'll look closer at the basic variable types for Python numbers and text.

integer (`int`)

In the previous section, the initial examples used integer numbers. Python refers to this as the `int` type. Integer constants are written as digits, possibly with a negative sign. You should not include commas in your numbers. For example, instead of using:

```
thous = 1,000
```

you should write:

```
thous = 1000
```

In the above statement, Python determines that 1000 is an `int` and so tags the variable as an `int` variable.

Note that in some programming languages, there is a minimum and maximum allowable value for an integer, but Python does not have a preassigned integer value limit.

`float`

In the previous sections, all numbers have been integers, but of course Python allows for numbers with decimal places. The standard type for this in Python is ***float***. Constant values can be written with digits and a decimal point, and with a leading minus sign for negative numbers. So for example:

```
sales = 143.56
```

will cause Python to associate the `float` type to the variable sales, as well as assigning the value of 143.56 to sales.

Large float values can be represented using scientific notation with the letter *e*. For example, 1.2e5 in Python represents 1.2×10^5, which is 120,000. Similarly, small values can be represented—for example, 1.2e-5 is 1.2×10^{-5}, which is 0.000012.

Unlike integers, Python `float` constants and variables have minimum and maximum values. For example, the largest Python `float` value is approximately 1.8e+308. The smallest value above 0 is approximately 2.2e-308. In nonscientific applications, these limits are usually not a problem.

In Python and many other programming languages, there is a limitation on the accuracy of float values, due to how numbers are represented internally as a sequence of bits in the computer. At first glance, this may seem surprising and of concern, but in practice, this is not usually an issue. As an example, if you enter this expression in the Python interpreter:

```
>>> 140-124.8
```

the result is given as:

```
15.200000000000003
```

The zeros and trailing 3 are a symptom of the inexact representation of the result of 140 – 124.8. In practice, we can round the result, as we'll see later in this chapter. Though beyond the scope of this book, Python also provides the `Decimal` type for correct rounding of floating point arithmetic.

string (`str`)

Python strings are used to represent text, as a sequence of letters, digits, and other characters. String constants begin and end with either single quotes (apostrophes) or double quotes. Simple examples of string statements are:

```
label = "Sales to Date:"
header = 'Income'
```

If the text within a string contains an apostrophe, it is simplest to use double quotes to surround the text. For example:

```
"ABC's inventory"
```

There are some special characters that you can include in strings. These are all indicated by using the backslash \, as shown in Table 2-2. You can enter each example statement in the Python interpreter to see its result.

characters	meaning	Example statement
\\	A single backslash	print('over\\under')
\'	Single quote	print('Acme\'s Income')
\"	Double quote	print("see \"note section\"")
\t	Tab (1 or more spaces)	print('Actual\tProjected')
\n	New line	print('Division', '\n', 'Footwear')

Table 2-2: Special characters in strings

More about types

In many programming languages, each variable's type needs to be declared in the program. For example, in Java, a variable can be defined like:

```
int count = 1;
```

This indicates that `count` is a variable with an initial value of 1, and it is an integer variable.

Python uses a different approach—it determines the type of variable for you. For example, in the Python statement:

```
count = 1
```

Python sees the value 1, and determines that the variable `count` will at that point be an integer.

This smart typing aspect of Python means that a variable's type is "behind the scenes." As you develop programs, though you don't need to indicate to Python the type, in general, it can be helpful for you to think about the types your variables will have. When working interactively, you may find the `type()` function helpful to tell you about the type of a particular variable. For example, if we enter the statements interactively:

```
>>> count = 1
>>> type(count)
```

we'll see the result as:

```
<class 'int'>
```

indicating that Python has determined that the variable `count` has type `int`. Though this may be obvious here, in more involved statements, the `type()` function can clarify for us the type of a variable.

Another aspect of Python's typing is that it is dynamic. As a result, a variable can initially point to one type of data (for example, a string) and, at a later point in the program, point to another type (for example, a `float`). This flexibility can be useful in more advanced programs, but it can be confusing as well. In general in this book, we'll avoid switching a variable's type. If you find that your code is switching a variable's type, you may want to introduce a new variable so that each variable's type remains fixed. For example, instead of writing:

```
# variation 1: variable weight's type changes
weight = input('Enter weight: ') # weight is initially a string
weight = float(weight)           # weight's type switches to float
```

you can use:

```
# variation 2: two separate variables are used
strWeight = input('Enter weight: ') # strWeight is a string
weight = float(strWeight)           # weight is a float
```

Section Summary

Each value and variable in Python has a type; the type allows Python to know what operations on the data can be done. The basic numeric types are `int` (integer) and `float`. Large and small float constants can be specified in scientific notation. Text values can be represented with the `str` (string) type, using either apostrophes or double quotes to surround the text.

Test Your Understanding

1) Variables have which aspects?
   a) name
   b) value
   c) type
   d) all of the above

2) After the following statement executes:
   `income = 12.0`
   What is the type of the variable income?
   a) `int`
   b) `float`
   c) `str`

2.4 More on Input and Output

In Section 2.2, we used the `input()` function to get keyboard input from the user, and assign this to a string variable. In this section, we look at getting numeric input from the user. We'll also look at a refinement in using the `print()` function, to have more control over the output.

More about input: Getting strings vs. getting numbers from the keyboard

An important feature of the `input()` function is that it always returns a string result, since a string can represent anything a user may enter with the keyboard. What if you need a numerical value in order to use it in a mathematical equation? Python provides two functions, `int()` and `float()`, that allow you to convert a string into a number. The `float()` function is more general since it allows decimal places.

As a simple example, we can write a program that prompts the user for two numbers, and calculates their sum.

Program 2-5 Video animation available.

```
1  # Calculate the total of two numbers
2
3  # Prompt for inputs
4  value1 = float(input("Enter a number: "))
5  value2 = float(input("Enter another number: "))
6
7  # Sum the variable's values
8  total = value1 + value2
9
10 # Display the result
11 print(total)
```

Program 2-5 Example Run

```
Enter a number: 15
Enter another number: 23
38.0
```

On line 4, the result of the `input()` function (what the user enters) is further processed by the `float()` function, so that value1 will contain a numeric value. Similarly, line 5 will provide the value2 variable with a float value.

Line 8 adds the numbers stored in value1 and value2, and assigns the results to a new variable named "total." This is the first mathematical statement we have seen, and we'll look further at math statements in the next section. Line 11 displays the value of "total" to the screen.

Note that the `float()` function works if its argument can be converted into a number; otherwise, an error will result when the program is run. For example, if you enter a word in answer to either of the input prompts, Python will generate a ValueError. In Appendix D, we'll see how this can be handled better.

More about output: `print()`'s `sep` and `end` parameters

We've seen that the `print()` function provides an easy way to display text and variable values to the screen. Within the parentheses, you can provide a single item. For example:

```
print(sales)
```

Alternatively, two or more items can be included within the parentheses, separated by commas. In this case, `print()` will by default display each item on the screen with a space in between each. For example, if the variable `sales` is set to 132, the statement:

```
print('Sales =', sales)
```

will result in the output:

```
Sales = 132
```

Note that the output displayed on the screen has a space after the equals sign—this is because by default, the `print()` function places a space in between each of the items it prints.

In some cases, you do not want a space between the displayed items. One way to control this is to provide `print()` with a ***named parameter*** called `sep`, short for separator. Here is an example statement that displays items without any space between them:

```
print('Sales =', sales, sep='')
```

In the above statement, the `sep=''` (note this is two apostrophes as opposed to one double quote) is indicating that the separator between the items to display should not be a space; instead, nothing should be in between items displayed by `print()`. This optional parameter for the `print()` function can in fact be set to other strings. For example:

```
print('Sales', sales, sep='\t')
```

uses the special tab character to separate the displayed values.

Another useful option for the `print()` function is the `end` parameter. This parameter indicates what to do after the `print()` function has displayed all its items. By default, after a `print()` has displayed its items, the next `print()` will start on the next line of the screen. This is a common case. Another common case is to use a series of `print()` statements to build up a displayed line. In this case, you can use `end=''` to indicate that the next `print()` will continue on the same line as the prior `print()`. For example:

```
print('Sales = ', end='')
print(sales)
```

will result in a displayed result like:

```
Sales = 132
```

More about strings: + and `str()`

Another useful way to display multiple items is to join them together with the + operator. If the items are strings, Python will join them together (***concatenate*** them). For example, assuming the variable name is set to the value "John," then:

```
print('Hello ' + name)
```

will create a string that combines `'Hello '` with the value "John," and display:

```
Hello John
```

As used in this way, the + symbol requires strings on both sides. For example, if `sales` is the numeric variable 132, then:

```
print('Sales =' + sales)
```

will result in an error, since `'Sales ='` is a string and the variable `sales` is a number.

To address this, you can use Python's `str()` function, which converts a value into a string. So the statement:

```
print("Sales =" + str(sales))
```

will work.

Tip

Splitting Long Statements into Multiple Lines

If you'd like to display a sentence with many variable values, the `print()` statement to do this can become long. To support long statements, Python allows statements to span multiple lines in a few different ways.

One recommendation is to use a new line for a statement after the comma separating function arguments. For example,

```
print("For a height of", height, "and weight of", weight,
      "the calculated BMI is", bmi)
```

Another recommendation is to keep your code visible—in other words, keep statements (or parts of statements) so they do not flow past the width of the screen.

Section Summary

The `input()` function always returns a string, representing the keyboard entry from the user. The `float()` and `int()` functions can be used to convert strings (for example, as returned by the `input() function`) to numbers, so that mathematical expressions can be performed.

The `print()` function can be controlled using the `sep` (separator) argument, to control what is displayed between items. The `end` argument can also be used to control the behavior of `print()` after it displays its results.

The + sign can be used to join (concatenate) strings. The `str()` function can be used to convert values to a string representation.

Test Your Understanding

1) What function can be used together with `input()` to assign variables to numeric values from keyboard input?
   a) `int()`
   b) `float()`
   c) `print()`
   d) (a) and (b)
2) What optional named parameter can be provided to the `print()` function to control what is displayed between its comma-separated items?
   a) end
   b) sep
   c) space
3) What does the `str()` function do?
   a) Creates a string representation of values
   b) Creates a numeric representation of a string
   c) Provides a way for user input from the keyboard

2.5 Mathematical Statements

Calculations in Python can be performed with the following operators:

+ addition
- subtraction
* multiplication
/ division
** exponentiation
% remainder

The order of operations follows the standard algebraic rules (also known as "PEMDAS")—parentheses, exponentiation, then multiplication and division, then addition and subtraction, with left-to-right order.

Spreadsheet Compare/Contrast
Order of Operations

The order of operations in Python is the same as used in spreadsheets. The symbols are mostly the same as those used by spreadsheets: () * / + -. However, exponentiation in Python is by **, whereas in spreadsheets, ^ is used.

To demonstrate a calculation in Python, let's consider the following health-related example. BMI (body mass index) is a simple measure for checking if a person's weight is in a healthy range relative to their height. Given a height *H* in inches, and weight *W* in pounds, the BMI is calculated as:

$$\text{BMI} = W / H^2 \times 703$$

(A healthy BMI is considered to be in the range between 18.5 and 24.9.)

Program 2-6 calculates BMI given a person's height and weight.

Program 2-6 Video animation available.

```
1 # Calculate BMI, given height and weight
2
3 # Prompt for inputs
4 height = float(input("Enter height (inches): "))
5 weight = float(input("Enter weight (pounds): "))
6
7 # Calculate the BMI result:
8 bmi = weight / (height ** 2) * 703
9
10 # Display the result
11 print("BMI =", bmi)
```

Program 2-6 Example Run

```
Enter height (inches): 68
Enter weight (pounds): 145
BMI = 22.044766435986162
```

Line 8 is the key calculation in Program 2-6. Note that on line 8, the use of parentheses is not needed, as exponentiation is done before multiplication and division. But parentheses are used here to make the statement a little more readable.

The example run's output is displayed to many decimal places. In order to display fewer decimal places, you can use the built-in Python function `round()`. You can provide `round()` with two values within the parentheses, the first value indicating what to round, and the second indicating how many decimals to round to. The second value is optional; if you omit it, the result is rounded to the nearest integer. For example, in Program 2-6, line 11 could be replaced with:

```
print("BMI =", round(bmi, 1))
```

in which case the displayed result would be: 22.0.

> **Tip**
> **Organizing Programs Using Input/Processing/Output (IPO)**
>
> As described in Chapter 1, IPO (input/processing/output) is one way to characterize the main aspects of programs. In many of our examples, we will use these three parts to help organize the programs. For example, Program 2-5 collects inputs in lines 4 and 5, processes the data in line 7, and outputs the result in line 11. Dividing programs into input, processing, and output steps can make programs easier to design, develop, and maintain.

As another simple example with calculations, assume you have both actual and projected values—for example, for sales. Program 2-7 calculates the difference and percent difference between actual and projected values. Like Program 2-6, this example prompts for two inputs, but it calculates and displays two results.

Program 2-7

```
# Calculate difference and percent difference between two values

# Prompt for and get inputs
salesHistory = float(input("Please enter last year's sales: "))
salesProjected = float(input("Please enter this year's projected sales: "))

# Calculate results
salesChange = salesProjected - salesHistory
salesChangePct = salesChange / salesHistory * 100

# Display results
print("Projected sales change:", round(salesChange, 1))
print("Projected sales % change:", round(salesChangePct,1))
```

Program 2-7 Example Run

```
Please enter last year's sales: 34.5
Please enter this year's projected sales: 40
Projected sales change: 5.5
Projected sales % change: 15.9
```

Section Summary

Math expressions can be formed using *, /, +, -, **, %, as well as parentheses. The order of operations follows standard algebra convention. The `round()` function can be used to round a value to the nearest integer or to a certain number of decimal places.

Test Your Understanding

For this section, assume variables have these values:

a = 1
b = 3
c = 2
d = 1.6

1) What is the value of `a * b ** c / a`?
   a) 1
   b) 2
   c) 3
   d) 9
2) What is the value of `round(d, a)`?
   a) 1
   b) 2
   c) 1.6

2.6 More about Math

In this last section, we'll look at a few more math-related functions.

Let's consider an example related to inventory management. Assume an item has a steady demand during the year. There is a cost for placing an order with the supplier and a cost for restocking this item. There is also a cost for storing items—for example, in a warehouse. Given these costs and our demand forecast, we'd like to calculate how much of the item to reorder to minimize our annual ordering and storing costs. This amount is referred to as the economic order quantity (EOQ).

Given the following assumptions:

D = annual demand (units per year)
S = reorder cost ($ per order)
H = holding cost ($ per unit per year)

the EOQ is calculated as:

EOQ = sqrt(2 × D × S / H)

For example, if demand is expected to be 100 units/year, reorder cost is $5/order, and holding cost is $1.25/ unit/year, then the EOQ = sqrt(2 × 100 × 5 /1.25) = sqrt(800) = 28.3 units.

Program 2-8 is a Python program for calculating the EOQ based on input assumptions for demand and costs.

Program 2-8

```
# Calculate the economic order quantity (EOQ)

# Prompt for and get inputs
demand = float(input("Enter the projected demand (units/year): "))
reorderCost = float(input("Enter the reorder cost ($/order): "))
holdingCost = float(input("Enter the holding cost ($/year/unit): "))

# Calculate result
eoq = (2 * demand * reorderCost / holdingCost) ** .5

# Display result
print("Economic Order Quantity:", round(eoq))
```

Program 2-8 Example Run

```
Enter the project demand (units/year): 100
Enter the reorder cost ($/order): 5
Enter the holding cost ($/unit/year): 1.25
Economic Order Quantity: 28
```

Let's enhance this program to introduce one additional aspect. Assume there may be a minimum order size that our supplier requires.

In order to ensure a minimum order size, we can use the Python built-in function `max()`, which determines the maximum of two (or more) values. Program 2-9 shows the use of the `max()` function to enhance the order quantity calculator.

Program 2-9

```
# Calculate the EOQ - version 2: minimum order size

# Prompt for and get inputs
demand = float(input("Enter the projected demand (units/year): "))
reorderCost = float(input("Enter the reorder cost ($/order): "))
holdingCost = float(input("Enter the holding cost ($/year/unit): "))
minOrder = float(input("Enter the minimum order size (units/order): "))

# Calculate result
eoq = (2 * demand * reorderCost / holdingCost) ** .5
orderSize = max(eoq, minOrder) # account for the minimum order size

# Display results
print("\nEconomic Order Quantity:", round(eoq))
print("Order Quantity:", orderSize)
```

Program 2-9 Example Run

```
Enter the project demand (units/year): 100
Enter the reorder cost ($/order): 5
Enter the holding cost ($/unit/year): 1.25
Enter the minimum order size (units/order): 30

Economic Order Quantity: 28
Order Quantity: 30.0
```

In the above example on line 11, note the use of the `max()` function, to ensure the order quantity is at least as large as the minimum order size.

In addition to the `max()` function, there is also the Python built-in function `min()`. Both `min()` and `max()` can take two (or more) arguments, and will return a single value as the answer.

Other built-in mathematical functions in Python are `abs()`, which determines the absolute value, and `pow()`, which calculates a value raised to a power. Table 2-3 summarizes the built-in mathematical functions in Python.

Function	**Description**	**Notes**
`abs(x)`	absolute value	
`min(x, y)`	minimum	two or more values can be provided
`max(x, y)`	maximum	
`pow(x, y)`	x^y (*x* raised to the *y*)	alternative to ** operator
`round(x, d)`	Round *x* to *d* decimal places	*d* is an optional argument

Table 2-3: Python built-in mathematical functions

In addition to the functions in Table 2-3, Python provides over 50 mathematical functions in its math package. This includes basic functions like ceiling and floor, functions for trigonometry, power and logarithmic functions, as well as others. These can be found in the Python documentation at:

https://docs.python.org/3/library/math.html

Section Summary

In addition to `round()`, selected other Python built-in functions are `abs()`, `pow()`, `max()`, and `min()`. Table 2-3 summarizes these functions.

Test Your Understanding

For this section, assume variables have these values:

a = -10.44
b = 20
c = 30

1) What is the value of `min(a, max(b, c))`?
   a) -10.44
   b) 20
   c) 30

2) What is the value of `abs(round(a, 1))`?
   a) -10
   b) -10.4
   c) 10
   d) 10.4

2.7 More about Strings

Concatenation

As we saw earlier in the chapter, strings can be defined using apostrophes or double quotes. You can also combine strings using the + sign for concatenation. The result of concatenating strings can be assigned to a string variable; for example, if `s` and `t` are string variables, then

```
r = s + t
```

creates a string by concatenating `s` and `t` and assigning the result to the variable `r`.

In some cases, you may want to append one string to a particular string variable, and store the result back into the particular variable. This can be done as:

```
s = s + t
```

or using the shorthand +=:

```
s += t
```

These variations for string concatenation are shown in Program 2-10.

Program 2-10

```
# String concatenation

message = 'Hi '
name = input('Please enter your name: ')
message += name + '!'
print(message)
```

Program 2-10 Example Run

```
Please enter your name: Adam
Hi Adam!
```

Formatting

When displaying numbers using `print()`, we've seen in some cases we get more decimal places than we want. So far we've addressed this using the `round()` function. However, often we want more control over the way numbers are displayed. For example, we may want commas every three positions, or we may want a fixed number of decimal points. To do this, we can use the built-in `format()` function. Appendix C covers `format()` in detail. In this section, we'll look at its basics.

The `format()` function takes two arguments—what we want to format, and how we want to format it—and returns a string. The second argument is a string and referred to as the ***format specifier***. For example, if we have a float variable `sales` with value 1234.567, and we'd like to display it in 10 positions with two decimals and commas, we can use:

```
print('Sales:', format(sales, '10,.2f'))
```

In the above statement, the format specifier is `'10,.2f'`, which indicates `format()` should use 10 positions, with commas, and two decimal places. The trailing *f* refers to float—if you exclude it you will get a result in scientific notation. The output from the above statement would be:

```
Sales:   1,234.57
```

In this example, the string result of `format()` includes a comma, as well as two leading spaces in order for the result to be the 10 positions requested.

Methods

Python strings have a lot more functionality, which we'll see in Chapter 8. Many of string's features are provided through Python ***methods***. Methods are similar to functions, but are associated with a particular type. Methods can be used by indicating a variable, followed by a period, followed by the method name, and if needed, arguments in parentheses. For example, the string method `capitalize()` will take a string and create a new string with the first letter capitalized. Assuming `name` is a string variable, the syntax to do this is `name.capitalize()`.

Another example string method is `strip()`, which creates a string without any leading or trailing spaces. Program 2-11 is a variation of Program 2-10 that shows the usage of `capitalize()` and `strip()`.

Program 2-11

```
# String concatenation and some methods

message = 'Hi '
name = input('Please enter your name: ')
name = name.strip().capitalize()
message += name + '!'
print(message)
```

Program 2-11 Example Run

```
Please enter your name:   rebecca
Hi Rebecca!
```

In Program 2-11's example run, the user entered several spaces before `rebecca`, as well as several spaces after. These spaces were removed by the `strip()` method.

A lot of functionality in Python is provided via methods, not just for strings, but for other types as well. We cover more methods starting in Chapter 6.

Section Summary

In addition to storing text, strings provide ways to manipulate text. Concatenation is used to join strings together. As well, there are several other string operations, many provided through methods. Methods are similar to functions and provide functionality for strings as well as other types in Python.

Test Your Understanding

For this section, assume variables have these values:

s = `' Hello '`

t = `'there'`

1) What is the value of `s + t`?
   a) `'Hellothere'`
   b) `' Hello there'`
   c) an error
   d) none of the above

2) What is the value of `s.strip()`?
   a) `'Hello'`
   b) `'Hello '`
   c) `' Hello '`
   d) `'hello'`

Chapter Review

Constants, variables, and statements are the primary building blocks of programs. Variables are named locations for data, and statements provide a way to create, access, and update variables. Statements can also get information, such as via the keyboard using `input()`, as well as display information using `print()`. Python provides other ways to access data, such as from files, databases, and the internet—we'll see this in later chapters.

Values and variables in Python have a type. The type allows Python to know what processing can be done with the data. The basic numeric types are `int` (integer) and `float`. Text values can be represented in the `str` (string) type using either apostrophes or double quotes to surround the text. There are many other types in Python, as we'll see in subsequent chapters.

The `input()` function always returns a string, representing the keyboard entry from the user. The `float()` and `int()` methods can be used to convert strings (for example, returned by `input()`) to numbers, so that mathematical expressions can be performed.

Math expressions can be formed using *, /, +, -, **, %, as well as parentheses. The order of operations is the same as standard algebra convention. The `round()` function can be used to round a value to the nearest integer or to a certain number of decimal places. In addition to `round()`, other Python built-in functions are `abs()`, `pow()`, `max()`, and `min()`.

Strings can also be used in expressions. For example, + and += can be used to concatenate strings. Further, strings have methods that provide a variety of ways to work with text data. These are covered in detail in Chapter 8.

Key Terms

argument	An item specified within the parentheses of a function to indicate to the function its input.
built-in function	Logic that is provided with Python that can be used for a specific task: for example, displaying to the screen, getting keyboard input, rounding numbers; functions typically take arguments as their inputs and often provide an output result that is further used in the program.
concatenation	Joining of two strings in Python using the + operator.
constant	A fixed value: for example, a specific number or text.
float	A Python numeric type, which can store numbers with or without fractional parts.
initialize	Create a variable and set it to an initial value.
int	A Python numeric type that can store integer values.
method	Similar to a function, functionality that is used with a particular type.
named parameter	Functions may have optional named parameters that allow passing of additional optional arguments as input to the function.
precedence	The order of operations.
prompt	A displayed message indicating what kind of information the user should enter.
statement	An instruction in a program: for example, to set a variable, display information on the screen, get information from the keyboard, or calculate a value.
str	A Python type that can store text as a sequence of characters.
type	The kind of a Python value or variable: for example, `int`, float, `str` (and many others).
variable	A program location that holds (points to) a value; a variable's value may change during the execution of the program.

Programming Problems

1) Vehicle Routing

An intercampus shuttle can take one of several routes. For each run, a dispatcher checks what the current traffic conditions are to find out the average speed on each route and determine which route the shuttle driver should take.

a) Write a program that:
   - prompts for two inputs: distance in miles, and speed in miles per hour
   - calculates the time the route should take (in hours), and displays this result

b) In addition to displaying the time in hours, enhance your program to display the result in minutes, rounding the result to one decimal.

c) Assume current traffic conditions are those found in the table follows. Run your program three times using the data below to determine the fastest route.

Route type	Distance (miles)	Current speed (miles/hour)
local	30	30
parkway	35	40
highway	48	55

2) Same Price Store

A store charges $2.75 for every product it sells. There is also a sales tax of 6% on all purchases.

a) Write a program that prompts the user for the number of items purchased, and calculates the before-tax cost, sales tax, and total cost. Example run:

```
Number of items: 3
Pretax cost: 8.25
Sales tax:   0.495
Total cost:  8.745
```

b) Refine your program so sales tax is rounded to two decimal places. Example run:

```
Number of items: 3
Pretax cost: 8.25
Sales tax:   0.49
Total cost:  8.74
```

3) Weighted Average Forecast

A marketing analyst needs to provide a product's sales forecast for the coming month. The analyst decides to estimate three scenarios (pessimistic, middle, optimistic) and then calculate the forecast as a weighted average using 0.25 × pessimistic + 0.5 × middle + 0.25 × optimistic. Write a program that prompts the user for three values and then calculates and displays the weighted average forecast. Example run:

```
Enter pessimistic forecast: 100
Enter middle forecast.....: 125
Enter optimistic forecast.: 175
Weighted average forecast = 131
```

4) BMI—Healthy Range

Given a height (in inches) and a weight (in pounds), the body mass index (BMI) is defined as:

$$BMI = weight / height^2 \times 703$$

A "healthy" range for BMI is between 18.5 and 24.9. Given this, the above equation can be rearranged to determine a healthy weight range for a given height:

$$weight\_low = 18.5 \times height^2 / 703$$
$$weight\_high = 24.9 \times height^2 / 703$$

a) Write a program that calculates and displays the healthy range of weight for a given height input by the user.

b) Enhance your program to display a sentence like:

```
For a height of 66 inches, the healthy weight range is between 115 and 154 pounds.
```

5) BMI—Metric

The body mass index (BMI) equation was originally based on metric measures, using meters for height and kilograms for weight. Given a height (in meters) and a weight (in kilograms), BMI is defined as:

BMI = weight / height^2

Write a program that prompts the user for their weight (in pounds) and height (in inches), and converts these to metric measures (height in meters, weight in kilograms), then calculates and displays BMI using the equation above.

6) Investments: Return and Risk

Assume that you have several investment choices. For each investment, there are two equally likely end-of-year forecasts: the optimistic value (up_value), and the pessimistic value (down_value). Assume the expected return and risk of each investment is given by:

expected_end_value = (up_value + down_value) / 2
expected_return = (expected_end_value - initial_investment) / initial_investment
risk = (up_value - down_value) / 2

a) Write a program that for one investment choice, prompts for three inputs:
   - initial investment amount
   - optimistic end value (up_value)
   - pessimistic end value (down_value)

   and calculates the investment's expected return and risk. Here is an example run:

```
Enter start value: 1000
Enter high value: 1100
Enter low value: 975
Average end value = 1037.5
Average return = 3.75
Risk = 62.5
```

b) Run your program three times to calculate the expected return and risk of the following three investment alternatives, assuming an initial investment of $1,000. Determine the investment choice with the lowest risk, and the investment choice with the highest expected return.

	Forecasted ending value	
Alternative	**Pessimistic**	**Optimistic**
Cash fund	$1,000	$1,010
Bond fund	$970	$1,050
Stock fund	$950	$1,100

7) Forecasting—Compound Growth

Based on an advertising campaign, a company is planning higher sales growth in the year ahead, followed by slower sales growth in the following years. Write a program that prompts for:

- last year's sales ($)
- sales growth for the year ahead (%)
- sales growth for subsequent years (%)

and calculates forecasted sales for the next three years. Here is an example run:

```
Enter last year's sales ($): 1015
Enter year-ahead sales growth (%): 9
Enter subsequent years' growth (%): 2
```

```
Forecast Sales:
year 1: 1106.4
year 2: 1128.5
year 3: 1151.0
```

8) Customer Wait Time

A bank has one automated teller machine (ATM) at each of its branches. As the ATMs are old, their processing times are relatively slow, and customers are often waiting in line to use the ATM. The bank can invest in newer ATMs that would process customer requests faster, reducing customer waiting times.

In order to evaluate the benefit of investing in new ATMs, the bank decides to use a queueing model to calculate two measures:

- average customer waiting time in line to use the ATM
- average number of customers in line waiting to use the ATM

The model the bank will use to evaluate the customer waiting time and line length is as follows:

a = average arrival rate of customers (customers per hour)
s = average service rate of ATM (customers per hour)
w = average wait time per customer (in hours) = 1 / (s − a) − 1/s
c = average number of customers in line = a / (s − a) − a/s

a) Write a program that prompts the user for the arrival rate and service rate, and calculates average wait time (in minutes) and average number of customers in line.
b) Run your program to determine average wait times and average number of customers in line for the following ATM model choices. Assume the arrival rate is 30 customers per hour.

ATM model	Service rate (customers / hour)
Old	45
New–fast	55
New–faster	60

Chapter 3

Decisions

Chapter Objectives

In this chapter, you'll learn:

- how you can handle choices between alternatives in your Python programs
- about the three keywords used in Python decision structures: `if`, `elif`, and `else`
- how to use the `if` statement to control whether certain code is executed
- how `if` in combination with `else` allows your program to choose between two alternatives
- how to use `elif` to select between multiple choices
- what nested decisions are, and how they can handle a sequence of decisions

Chapter Outline

3.1 About Decisions in Programs

In the prior chapter, the example programs have a simple flow, from top to bottom. In more involved cases, there may be conditions under which certain statements may or may not execute. These are referred to as ***decision structures***. In this chapter, we'll look at several variations for handling decisions, so you can write programs that select among alternatives. It's a very common requirement for programs to choose among alternatives, and in this chapter, we'll see several examples that use the Python keywords `if`, `else`, and `elif`.

The `if` statement

The main decision-making statement in Python is the `if` statement. The `if` statement contains a ***logical expression*** that evaluates to either True or False. Following the `if` statement are one or more statements that will execute if the logical expression is True. The syntax and pattern of these statements are as follows:

```
if logical_expression:
   statement(s)
```

In other words, the `if` statement involves a question (logical expression), which is followed by a colon. On subsequent <u>indented</u> lines are the statements that will execute if the logical expression is True. Conversely, if the logical expression is False, the subsequent indented lines will *not* execute. Note that Python requires indenting for the lines you want to execute if the logical expression is True.

Let's look at a simple example. In the following program, the user is asked to enter a number. If the number is positive, a message is printed.

Program 3-1

```
1 userNumber = float(input('Please enter a number: '))
2 if userNumber > 0:
3     print('The value you entered was positive')
```

Program 3-1 Example Run

```
Please enter a number: 10
The value you entered was positive
```

In this example, `userNumber > 0` is a logical expression. Depending on the user's input, this expression will evaluate to either True or False. If it is True, the indented `print()` statement will be executed, and the phrase will be displayed on the screen. Note that in the above code, the colon is required at the end of the `if` statement's logical expression; if you omit it, Python will generate an error.

If you run the above program and enter as input a negative value—for example, -1—then the expression `userNumber > 0` will evaluate to False, and the indented print statement will *not* execute. This is the first example we've seen where a statement in a program would not execute.

About the colon and indenting in Python

In Program 3-1, note the colon at the end of line 2, as well as the indented line on line 3. The colon with following indented statements is an important required pattern for if-related logic, as well as several other Python statements we'll see in later chapters.

If you omit the indenting, Python will give you an error. For example, if Program 3-1 were written as:

```
userNumber = float(input('Please enter a number: '))
if userNumber > 0:
print('The value you entered was positive')
```

when you try to run it, you will get:

```
IndentationError: expected an indented block
```

Typical Python style uses three or four spaces for indenting. In addition to being a requirement for certain statements, indenting helps make your programs more readable by offsetting the conditional statements from the related logical statements.

Comparison operators

In the previous example, in the expression `userNumber > 0`, the greater than sign (`>`) is one of several math comparisons that Python provides. Table 3-1 is a list of Python's ***comparison operators***.

Note that in Table 3-1, the comparison operator for equality is == as opposed to a single equal sign. This is because a single equal sign is used to assign a value to a variable, as discussed in the prior chapter. If you incorrectly use a single equal sign in a comparison, Python will generate an error. For example, Program 3-2 has an error in using a single equals sign in line 2.

Operator	Meaning
<	Less than
<=	Less than or equal to
>	Greater than
>=	Greater than or equal to
==	Equal to
!=	Not equal to

Table 3-1: Comparison operators

Program 3-2

```
userNumber = float(input('Please enter a number: '))
if userNumber = 0:
    print('The value you entered was zero')
```

If you try to run this (incorrect) program, Python will generate an error including the following message:

```
    if userNumber = 0:
                  ^
SyntaxError: invalid syntax
```

The fix is simple: use == instead of =, as shown in Program 3-3.

Program 3-3

```
userNumber = float(input('Please enter a number: '))
if userNumber == 0:
    print('The value you entered was zero')
```

In the prior examples, a variable's value is compared to a constant value. You can also use expressions on either side of the logical operator, as shown in Program 3-4.

Program 3-4

```
userNumber1 = float(input('Please enter a number: '))
userNumber2 = float(input('Please enter another number: '))
if userNumber1 * userNumber2 > 0:
    print('The product of your numbers is positive')
```

Program 3-4 Example Run

```
Please enter a number: -3
Please enter another number: -4
The product of your numbers is positive
```

As discussed in the prior chapter, precedence rules apply, including the use of parentheses. Further, math operations such as * / + - have a higher precedence than comparison operators—for example, in the statement:

if a + b < c + d:

the calculations for a + b and c + d are done before the comparison between their results is done.

Section Summary

In programs, it is common to choose whether or not certain statements will execute. The `if` statement is used in Python to test a condition, and if the condition is True, the statement(s) indented under the `if` statement will execute. In Python, the `if` statement ends with a colon, and statements that are conditionally executed are indented under the `if` statement.

Test Your Understanding

1) Assuming *a* and *b* are numeric variables, what is a valid `if` statement?
   a) if a => b:
   b) if a = b:
   c) if a != b:

2) The result of an if's logical expression must be:
   a) True or False
   b) a number
   c) a string

3.2 Multi-way Decisions

Two-way decisions: `if/else`

Another decision programming pattern involves two possible outcomes. If the logical expression is True, a set of one or more statements is executed; if it is False, a different set of one or more statements is executed. To choose between two outcomes, Python provides the `else` keyword. The syntax and pattern of these statements is as follows:

```
if logical_expression:
   statement(s)
else:
   statement(s)
```

Let's look at a simple example using `if` with `else`. In the following program, the user is asked to enter a number. If the entered number is zero, then a message is displayed to indicate that; otherwise, a different message is displayed.

Program 3-5

```
userNumber = float(input('Please enter a number: '))
if userNumber == 0:
    print('The value you entered was zero')
else:
    print('The value you entered was not zero')
```

Program 3-5 Example Run

```
Please enter a number: -1
The value you entered was not zero
```

In the above example, `userNumber == 0` is the logical expression. Depending on the user's input, this expression will evaluate to either True or False. If it is True, the print statement immediately after the `if` statement will be executed. If it is False (the user enters a value other than 0), the print statement after the `else` statement will be executed. In other words, only one of the print statements in the above program will be executed.

Note that although the `if` statement has a condition (in the above example, `userNumber == 0`), the `else` statement does not have a condition. This is because the else section of the code is executed when the `if`'s condition is False.

To further illustrate the use of `if` and `else`, assume we want to write a program to calculate the ending value of a one-year certificate of deposit (CD) investment. The bank offering the one-year CD will pay 3% interest for an investment up to $1,000. For investments of $1,000 or more, the bank will pay 3.25% interest. Program 3-6 computes the ending balance of an investment, where the interest rate depends on the amount of the investment.

Program 3-6 Video animation available.*

```
# Calculate the ending balance of a 1-year certificate of deposit (2 rates)

# Prompt for the initial investment
invest = float(input('Please enter investment amount: '))

# Determine the interest rate, depending on the size of the investment
if invest >= 1000:
    interestRate = 3.25
else:
    interestRate = 3.0

# calculate the ending balance
endBalance = invest * (1 + interestRate/100)

# display the result
print('Interest rate=', interestRate, '%')
print('Ending balance= $', endBalance)
```

Program 3-6 Example Run

```
Please enter investment amount: 1000
Interest rate= 3.25 %
Ending balance= $ 1032.5
```

So far, example programs in this chapter have one statement indented under the `if` and `else`. However, you are not limited to a single indented line if there is more logic needed—more than one statement can be indented. For example, Program 3-6 could be enhanced to add a line under the `if` condition:

```
if invest >= 1000:
   print('Bonus interest earned on investments of $1,000 or more')
   interestRate = 3.25
else:
   interestRate = 3.0
```

In the above case, when the `if` statement is True, the print statement as well as the interest rate assignment statement are executed, since both these statements are indented below the `if` statement.

Spreadsheet Compare/Contrast
Python's if vs. Spreadsheet if

In a spreadsheet, you can use an if function to allow for choices in a formula. The form of the spreadsheet if function is:

```
if(logical_expression, true_expression, false_expression)
```

You can use the if function in a cell, in order to choose between two constants, two equations, or a constant and an equation.

Though Python's `if/else` structure involves multiple lines of code, its logic can be easier to see as compared with an `if()` function in a spreadsheet cell.

Three-way decisions: `if/elif/else`

In the preceding section, we looked at how a choice between two outcomes can be handled by using `if` and `else` in combination. What if you need to handle three different alternative choices and their outcomes? For

* Go to the Videos page at www.prospectpressvt.com/textbooks/groner-python.

example, when you come to an intersection, your choices might be to go straight, turn left, or turn right. In order to handle three-way choices, Python provides the `elif` keyword. The keyword `elif` is short for "else if." With `elif`, your program can check another case, in case the `if` condition is false. The general pattern for using `elif` is:

```
if logical_expression1:
   statement(s)
elif logical_expression2:
   statement(s)
else:
   statements(s)
```

One example for using a three-way choice involves checking whether a number is positive, negative, or zero. Program 3-7 shows how this can be done with `if`, `elif`, and `else`.

Program 3-7

```
1 userNumber = float(input('Please enter a number: '))
2 if userNumber > 0:
3    print('The value you entered was positive')
4 elif userNumber < 0:
5    print('The value you entered was negative')
6 else:
7    print('The value you entered was zero')
```

Program 3-7 Example Run

```
Please enter a number: -1
The value you entered was negative
```

Note that in Program 3-7, though there are three outcomes, there are only two logical conditions tested (on lines 2 and 4). This is because the `else` section of the code covers everything else that is not covered by the `if` and `elif` conditions. Further, note that if the condition on line 2 is True, line 3 will execute, but lines 4 through 7 will not execute. In other words, when the logical expression of the `if` statement is True, the expression of any related `elif` condition is not tested. Conversely, if the logical expression of the `if` statement is False, then the condition of the `elif` is evaluated to see if that condition is True or False. If the logical expression in the `elif` statement is False, then the statement indented under the `else` will execute.

To further illustrate the use of `if`, `elif`, and `else`, let's return to the CD example discussed earlier. Assume now that the bank has decided on a three-rate structure. The bank is offering a one-year CD that will pay 3% interest for an investment up to $1,000. For investments of $1,000 to $9,999, the bank will pay 3.25% interest. For investments of $10,000 or more, the interest rate is 3.5%. Program 3-6 computes the ending balance of an investment where the interest rate depends on the amount of the investment.

Program 3-8 Video animation available.

```
# Calculate the ending balance of a 1-year certificate of deposit (3 rates)

# Prompt for the initial investment
invest = float(input('Please enter investment amount: '))

# Determine the interest rate, depending on the size of the investment
if invest >= 10000:
    interestRate = 3.5
elif invest >= 1000:
    interestRate = 3.25
else:
    interestRate = 3.0

# calculate the ending balance
endBalance = invest * (1 + interestRate/100)

# display the result
print('Interest rate=', interestRate, '%')
print('Ending balance= $', endBalance)
```

Program 3-8 Example Run

```
Please enter investment amount: 1000
Interest rate= 3.25 %
Ending balance= $ 1032.5
```

More decisions: Multiple `elif`

In the previous section, we looked at how three-way choices can be handled with `if`, `elif`, and `else`. In cases where you need to handle more than three choices, you can use multiple `elif` statements. For example, consider CD rates that depend on investment amounts according to the following table:

Investment range	Interest rate
Less than $1,000	3.0%
$1,000–$4,999	3.2%
$5,000–$9,999	3.4%
$10,000 and above	3.6%

Table 3-2: Interest rates for different investment ranges

As this involves four ranges of investments, we can model this using multiple `elif` statements. In particular, the following code can be substituted in Program 3-8:

```
# handle 4 alternative interest rates with multiple elif statements
if invest >= 10000:
   interestRate = 3.6
elif invest >= 5000:
   interestRate = 3.4
elif invest >= 1000:
   interestRate = 3.2
else:
   interestRate = 3.0
```

When to include or omit `else`

In multi-way choices in programs, the `if` statement is always used. Depending on the problem, you may need to include one or more `elif` sections and possibly an `else` section. Sometimes you can solve problems involving selection in different ways.

When should you include or omit an `else` section? In general, if you want to have a section that executes when none of the related `if` or `elif` clauses are True, an `else` section is appropriate. In other words, the `else` section functions as a "none of the above" set of logic in your code. It ensures that one of the `if/elif/else` sections will always execute.

Section Summary

Multi-way decisions in Python can be handled with the keywords `if`, `elif`, and `else`. A two-way choice is modeled using a combination of `if` and `else`. Three-way choices can be handled using `elif` in combination with an `if` and `else`. Cases with four or more alternatives can be handled using multiple `elif` statements in combination with `if` and `else`.

Test Your Understanding

1) In combination with an `if` statement, how many related `elif` statements can there be?
   a) none
   b) one
   c) two or more
   d) all of the above

2) An else statement must always be used in combination with an `if` statement.
   a) True
   b) False

3.3 Nested (Sequential) Decisions

So far, we have looked at small examples that consider choice. In practice, there can be much more complexity. In fact, you can place more choice statements within an `if`, `elif`, or `else` statement. Decisions that follow this structure are referred to as "nested."

Nested decisions are useful for handling multi-dimensional logic. For example, there may be a primary decision and then a secondary decision. Based on the combination of these decisions, appropriate logic would execute.

Let's look at nested decisions by expanding on our CD example. Let's say our bank decides to provide a special interest rate to new customers—in other words, those who don't yet have an account at the bank. Specifically, new customers will earn 3.5% interest on their CD regardless of initial balance. Further, for existing customers, the interest rate is 3.25% for investments of $1,000 or more, or 3% for investments under $1,000. We would like to write a program that calculates the ending CD balance based on these two aspects:

- whether a customer is new or existing
- what the initial investment for the CD is

Another way to show these choices is using a decision tree. In particular, the interest rate for the CD can be determined by the decision tree in Figure 3-1.

The following program shows how to handle these two aspects by considering the customer type (new vs. existing), as well as the amount of the initial investment.

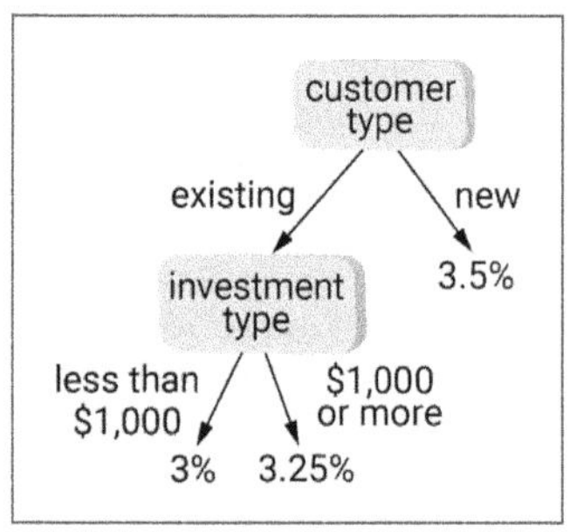

Figure 3-1: Decision tree for interest rate based on customer type and investment amount

Program 3-9 Video animation available.

```
1  # Calculate the ending balance of a 1-year certificate of deposit
2  # Consider new vs existing customers, and 2 rates for existing customers
3
4  # Prompt for type of customer
5  customer = input('Enter n if customer is new, or e for existing: ')
6
7  # Prompt for the initial investment
8  invest = float(input('Please enter investment amount: '))
9
10 if customer == 'e':
11    # Determine the interest rate, depending on the size of the investment
12    if invest >= 1000:
13       interestRate = 3.25
14    else:
15       interestRate = 3.0
16 else:
17    interestRate = 3.5
18
19 # Calculate the ending balance
20 endBalance = invest * (1 + interestRate/100)
21
22 # Display the result
23 print('Interest rate=', interestRate, '%')
24 print('Ending balance= $', endBalance)
```

Program 3-9 Example Run

```
Enter n if customer is new, or e for existing: e
Please enter investment amount: 1000
Interest rate= 3.25 %
Ending balance= $ 1032.5
```

In Program 3-9, note the indenting of lines 12–15, this is where the nesting occurs. Specifically line 12 is nested under line 10, indicating that the investment size check is performed only for existing customers.

For Program 3-9's output, the ending balance number doesn't have a comma or display to two decimal places. To include a comma and two decimal places, we could use for line 24:

```
print('Ending balance= $', format(endBalance, ',.2f'))
```

Section Summary

Nested `if` structures can be used to model multi-part decisions. Within an `if`, `elif`, or `else` block, another `if` structure can be indented to test for additional conditions.

3.4 More about Decisions

We complete the chapter by covering a few more aspects of decisions: combining conditions, comparing strings, and a terse `if/else` statement.

Combining conditions with `and`, `or`, `not`

So far in this chapter, programs have contained simple comparisons between two things, generally a variable and a constant. You can also write more involved comparisons that combine conditions, using these Python keywords:

```
and
or
```

For example, we can rewrite Program 3-10 to make use of the and keyword instead of using nested logic.

Program 3-10

```
1  # Calculate the ending balance of a 1-year certificate of deposit
2  # Consider new vs. existing customers, and 2 rates for existing customers
3
4  # Prompt for type of customer
5  customer = input('Enter n if customer is new, or e for existing: ')
6
7  # Prompt for the initial investment
8  invest = float(input('Please enter investment amount: '))
9
10 if customer == 'e' and invest >= 1000:
11     interestRate = 3.25
12 elif customer == 'e' and invest < 1000:
13     interestRate = 3.0
14 else:
15     interestRate = 3.5
16
17 # Calculate the ending balance
18 endBalance = invest * (1 + interestRate/100)
19
20 # Display the result
21 print('Interest rate=', interestRate, '%')
22 print('Ending balance= $', endBalance)
```

Program 3-10 Example Run

```
Enter n if customer is new, or e for existing: e
Please enter investment amount: 1000
Interest rate= 3.25 %
Ending balance= $ 1032.5
```

The above program demonstrates the use of and in lines 10 and 12. Though this program has fewer lines than Program 3-6, it is less clear since it combines the two decisions (customer type and investment amount) into single statements. However, careful use of the and (and or) keywords allow you to handle more complexity in your programs.

Another Python keyword, not, allows you to test the opposite of a condition. Often, != can be used instead.

Tip

Choosing between Concise vs. Readable Code

Since Python allows for nesting as well as compound conditions, there are often alternative ways to write decision-related code. In general, you should choose a style that is more readable to someone else, as opposed to a variation that is more concise.

Comparing strings

In Program 3-9, the statement:

```
if customer == 'e':
```

compares the string variable `customer` with the string constant `'e'` to check if they are equal. This comparison is similar to how numbers may be compared. Checking for string inequality is also similar to comparing numbers. For example, this statement could have been written as:

```
if customer != 'n':
```

In this case, the variable `customer` is compared with the constant `'n'` to see if they are not equal.

In fact, strings can be compared using the same operators that can be used to compare numbers. However, the meaning of the comparison operators is different when used with strings. For example, the == comparison operator checks whether the sequences of characters between two strings match. Conversely, the != operator checks whether the sequences of characters do not match. Note that using lower- or uppercase matters for string comparisons. For example, if the variable *t* is set to `'rate'` and *s* is set to `'Rate'`, then `s == t` evaluates to False. In practice, this is not a problem, as Python has many features related to strings that allow you to work around this issue, as we'll see in Chapter 8.

In addition to equality testing, Python can compare the alphabetical order of two strings. For these checks, the operators <, >, <=, and >= can be used. The logic used by Python is to check the dictionary order of the strings with the additional rule that uppercase letters come before lowercase letters.

You can interactively check the comparison order in the Python interpreter using constants or variables. For example, entering:

```
'this' < 'that'
```

will return `False`, since `'this'` comes after `'that'` in dictionary order. String comparison and dictionary order are covered further in Chapter 8.

Alternatively, Program 3-11 below can be used to check the Python ordering of a pair of strings.

Program 3-11

```
# Check Python ordering of strings
string1 = input('Enter string 1: ')
string2 = input('Enter string 2: ')
if string1 == string2:
    print('The strings are equal')
elif string1 < string2:
    print(string1, 'is before', string2)
else:
    print(string1, 'is after', string2)
```

Program 3-11 Example Run

```
Enter string 1: hello
Enter string 2: there
hello is before there
```

The table below summarizes the common string comparison operators, assuming *s* and *t* are string constants, variables, or expressions:

Expression	Meaning
s < t	The string *s* is before the string *t*
s <= t	The string *s* matches or is before the string *t*
s > t	The string *s* is after the string *t*
s >= t	The string *s* matches or is after the string *t*
s == t	The characters in the string *s* match the characters in string *t*
s != t	The characters in string *s* do not match the characters in string *t*

Table 3-3: String comparison operators

Terse `if/else`

Since simple `if/else` cases are common, Python provides a compact one-line statement that can test a conditional expression and return either a True or False result. The general format of this is:

```
<true-expression> if <condition> else <false-expression>
```

For example, referring back to Program 3-6, lines 7–10 can be replaced by:

```
interestRate = 3.25 if invest >= 1000 else 3.0
```

Though more concise, this can be harder to read than the multi-line statements in Program 3-6. To improve readability, the logical expression can be enclosed in parentheses. For example:

```
interestRate = 3.25 if (invest >= 1000) else 3.0
```

Section Summary

In this section, we looked at a few more aspects of decision structures in Python. The keywords `and`, `or`, and `not` allow you to write more involved conditional expressions. In addition to numbers, strings can be compared for equality, inequality, and dictionary order. A terse `if/else` statement is also provided by Python to handle simple two-way logic.

Chapter Review

Being able to choose between alternatives is a very common requirement in programs. In this chapter, we've seen how choices between alternatives can be handled in Python programs. There are three keywords used in Python decision structures: `if`, `elif`, and `else`. The `if` statement can be used to conditionally control whether certain code is executed. Two-way choices can be handled using `if` in combination with `else`. Three-way choices can be modeled using `if`, `elif`, and `else`. Four or more choices can be covered with multiple `elif` statements in combination with `if` and possibly `else`. A sequence of decisions can be modeled using nested statements, which allow you to indent `if`, `elif`, and `else` within other `if`, `elif`, and `else` statements. Logical statements use operators to determine at program runtime whether a condition is True in order to guide the program to execute certain statements. Composite logical statements can be formed using the keywords `and`, `or`, and `not`. Strings can be compared in logical expressions for equality, inequality, or dictionary order. For simple `if/else` cases, the terse `if/else` statement can be used.

Key Terms

conditional statements	Statements indented immediately below an if, `elif`, or else statement are conditionally executed, depending on the result of the logical expression(s).
elif	When paired with an `if` statement, the `elif` statement can evaluate a conditional statement to control whether a set of statements will be executed.
else	When paired with an `if` statement, the else statement defines one or more statements that may execute if the if (and any `elif`) statement is False.
if	The primary keyword for controlling whether certain statements are executed.
logical expression	A kind of expression that evaluates to True or False.

Programming Problems

1) Dollar Store

A store sells all items for $1. Customers buying 10 or more items receive a 5% discount. Additionally, there is 7.5% sales tax.

a) Write a program that prompts the user for the number of items purchased and calculates the total cost before and after tax. Here is an example run:

```
Enter number of items: 11
Gross cost: 11.0
Discount:   0.55
Net cost:   10.45
Tax:        0.78
After tax:  11.23
```

2) BMI—Target Weight

Write a program that prompts the user for their height (in inches) and weight (in pounds). The program should calculate BMI and then display one of three messages:

- the user is within the healthy BMI range (if their BMI is between 18.5 and 24.9)
- the number of pounds needed to lose in order to be at the top of the healthy BMI range (if their BMI is above 24.9)
- the number of pounds needed to gain in order to be at the bottom of the healthy range (if their BMI is below 18.5)

Here is an example run:

```
Enter your height: 68
Enter your weight: 165
Weight to lose for healthy BMI: 7 pounds
```

3) Vehicle Routing

An intercampus shuttle can take one of two routes. For each run, a dispatcher checks what the current traffic conditions are to find out the average speed on each route and determine which route the shuttle driver should take.

Write a program that:

- prompts for four inputs: route 1's distance in miles and speed in miles per hour as well as route 2's distance and speed.
- calculates the time each route would take
- determines and displays which route is faster and how long it will take in minutes

4) Quantity Discounts

A store sells items priced according to the following table:

# of items	Cost per item
1–19	$4.75
20–49	$4.50
50–99	$4.25
100 or more	$4.00

For example, an order for 25 items would cost (25 × $4.50) = $112.50

Write a program that takes as input the number of items, determines the cost per item, and displays the cost per item and total order cost.

5) Income Taxes

US federal income tax is based on a schedule of tax rates depending on income. For example, for those with a filing status of single, the following table applies:

Income	Income Tax
Up to $9,875	10% of income
$9,876–$40,125	$986 + 12% of income over $9,875
$40,126–$85,525	$4,618 + 22% of income over $40,125
$85,526–$163,300	$14,606 + 24% of income over $85,525
$163,301–$207,350	$33,272 + 32% of income over $163,300
$207,351–$518,400	$47,368 + 35% of income over $207,350
$518,401 or more	$156,235 + 37% of income over $518,400

The marginal tax rates are the percentages in the table above. The effective tax rate is the income tax divided by the income. Write a program that prompts for income and calculates and displays the marginal tax rate, effective tax rate, and income tax.

6) Hotel Charges

A hotel charges a standard rate of $100 per night. For platinum members (customers who have previously stayed 20 nights or more), the hotel offers a discounted price of $90 per night. Also, for platinum members staying 5 or more nights, 1 night is free. For example, a platinum member staying 5 nights would be charged $90 for 4 nights ($360).

Write a program that prompts the user for the number of nights for their stay and whether they are a platinum status member and then determines the cost for their stay. Also, if they are a platinum member and their stay is 5 or more nights, indicate this in a note. Additionally, if the customer is not a platinum member and their stay was 3 or more nights, display a note asking them to consider becoming a platinum member.

Here is an example run:

```
# of nights: 5
Platinum member (y/n)? y
Charge = $ 360
Note: charges discounted for one free night
```

Chapter 4

Repetition

Chapter Objectives

In this chapter, you'll learn:

- about repetition, which can process different scenarios with a common set of code
- how the `while` statement can be used to achieve repetition
- how the `for` statement, in combination with the `range()` function, can concisely provide repetition
- what infinite loops are and how to repair them
- how `break` and `continue` statements interrupt repetition flow, and how to target their use

Chapter Outline

4.1 About Repetition

In Chapter 2, each statement in the example programs is executed once. In Chapter 3, if the statement is part of a decision structure (`if/elif/else`), the statement may or may not execute. In this chapter, we'll look at ***repetition***, where statements may execute multiple times. Repetition, also referred to as ***looping***, is an important and common aspect of programs. For example, repetition can be used to process different scenarios or read and process each line from a file.

Python provides two keywords for repetition: `while` and `for`. In general, `for` is more concise than `while`. However, `while` can handle some cases that `for` can't—for example, if the number of repetitions depends on user input. Further, there are some cases that `while` and `for` can both handle, and you may choose one or the other based on which makes the code clearer or for another reason.

A key characteristic of repetition is that the program flow can move from one statement to a prior statement. This is different from previous chapters, where programs proceed from the top down. This reverse flow can take some getting used to, but is key to writing more complex programs.

In the next sections of the chapter, we'll look at the `while` and `for` statements and how they can be used in a variety of ways, such as to generate tables and compare scenarios. We'll then cover a potential bug in repetition, the infinite loop problem. The decision structures of the previous chapter can be combined with looping, and this is shown in the next section. Lastly, two keywords that can interrupt the looping flow, `continue` and `break`, are discussed in the last section.

Spreadsheet Compare/Contrast
Repetition vs. Copying Cells

Repetition in programming languages is very different from how spreadsheets handle repeating calculations. In a spreadsheet, you might define a formula to calculate a result for one row and then copy this down to other rows. In this way, a spreadsheet calculation can be repeated and applied to different cells. In contrast, in Python and other programming languages, the repeatedly used logic is indicated once, and the flow of the program controls the reuse of the logic for different scenarios.

4.2 The `while` Statement

Repeating with `while`

The `while` keyword is typically used in the following pattern:

```
initializing statement(s)
while <logical condition>:
   loop body—i.e., repeating statement(s)
   change statement(s)
```

The main concept when using `while` is that the statements indented below the `while` are repeated as long as the `while` condition is True. Note that the indenting pattern is similar to decision structures covered in the prior chapter.

To demonstrate this, let's consider the problem of generating a table of numbers with each number's square and cube. We'd like our squares and cubes program to generate this table for the sequence of integers from 1 to 5. Program 4-1 uses the `while` statement to generate this table.

Program 4-1 Video animation available.*

```
# Using while repetition, generate a table of squares and cubes

#initialization
i = 1 # starting value for the table
n = 5 # ending value for the table

# using repetition, generate the table
while i <= n: # condition
    # calculation and display
    print(i, i**2, i**3)

    # change value before restarting loop
    i = i + 1

print('Program ends')
```

Program 4-1 Output

```
1 1 1
2 4 8
3 9 27
4 16 64
5 25 125
Program ends
```

* Go to the Videos page at www.prospectpressvt.com/textbooks/groner-python.

In Program 4-1, line 8 tests the condition whether `i` is less than `n`. If this condition is True, Python executes the first statement indented under the `while` statement. If the condition is False, Python will jump over all the statements indented under the `while` statement. Further, after the last indented statement below the `while` executes, the program flow goes back up to the `while` statement, which again checks the logical condition. This jump back is what distinguishes repetition statements, and it is also shown in the flowchart diagram in Figure 4-1.

For this program, on line 8, the first time it is executed, the condition is True, since `i` is 1 and `n` is 5, so `i` is less than or equal to `n`, and lines 10 and 13 will execute. Then flow moves *back* to the `while` statement in line 8. As `i` is now 2, the condition is still True (2 <= 5), and the statements indented under `while` are executed again. The repetition and jump back happen a few more times, until `i` is 6. At that point, the `while` condition is checked, and since `i` is no longer less than or equal to 5, the condition is False, and the program no longer executes the statements indented under the `while`. Instead, Python moves the execution to the next non-indented line: in this case, line 15.

In Program 4-1, the variable `i` is used as the first column of the table, and it also controls whether the `while` statement will repeat the indented statements. Because it is involved in controlling the repetition, in this example, `i` is referred to as the ***control variable***.

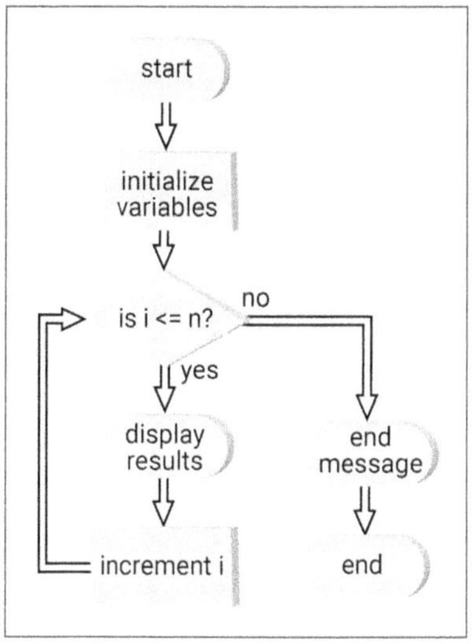

Figure 4-1: Flowchart for Program 4-1

Cumulating

Repetition can be used to total values. In this section, we'll look at a basic example that prompts a user for numbers, keeps track of the total, and displays the final count and total of values.

In Program 4-2, lines 4 and 5 initialize our tracking variables count and total. The key cumulation statement is line 10, which adds the user's last entered value to the variable `total`. In this way, `total` keeps track of the total of all user-entered values so far.

Program 4-2 Video animation available.

```
1  # Cumulate a set of user-entered numbers
2
3  # Initialize values before repetition
4  count = 0 # number of values entered by user
5  total = 0 # keep track of the running total of user-entered values
6
7  # repetition to prompt for and cumulate total
8  value = float(input('Enter a value: ')) # prompt for the first value
9  while value != 0:
10     total = total + value # keep a running total
11     count = count + 1 # keep track of number of values entered
12     value = float(input('Enter a value (or 0 to end): ')) # prompt for values
13
14 # Output results
15 print()
16 print('Number of values:', count)
17 print('Total of values:', total)
```

Program 4-2 Example Run

```
Enter a value: 10
Enter a value (or 0 to end): 30
Enter a value (or 0 to end): 25
Enter a value (or 0 to end): 0

Number of values: 3
Total of values: 65.0
```

In order to add a value to a variable, Python provides a shorthand syntax. Lines 4 and 5 could be replaced with:

```
total += value
count += 1
```

Program 4-2 also shows how you can continue repeating until a variable takes on a special value. In this case, the variable `value` functions as the control variable, and the special value is 0.

Maximum

We can also use repetition to find the largest (or smallest) of a set of values. Again we'll prompt the user for values, but rather than summing them, we'll use an `if` statement to detect a new largest value.

Program 4-3 Video animation available.

```
1  # Find the maximum of a set of user-entered numbers
2
3  # repetition to prompt for values and track largest
4  value = float(input('Enter an initial value: ')) # prompt for the first value
5  largest = value
6  while value != 0:
7      if value > largest: # is latest entry larger than the largest so far?
8          largest = value # if so, track this as the new largest value
9          print('New largest value:', largest)
10     # prompt for subsequent values
11     value = float(input('Enter another value (or 0 to end): '))
12
13 # Output result
14 print('\nLargest value:', largest)
```

Program 4-3 Example Run

```
Enter an initial value: 10
Enter another value (or 0 to end): 30
New largest value: 30.0
Enter another value (or 0 to end): 25
Enter another value (or 0 to end): 0

Largest value: 30.0
```

In Program 4-3, line 7 checks the largest value known so far and then checks if the latest user entry is larger. If so, the indented statements below the `if` statement will keep track of the new largest value and report that.

Using a specific control variable

In the prior examples, we used the variable `value` to get user input, as well as used this variable as the control variable by comparing it with the special value 0. Instead of using a special value to decide when to end repeating, we can define a specific variable that controls if the repeating steps will happen. Let's consider this in the next example. In this program, a BMI is calculated for different possible weights, as entered by the user. In this example, the control variable stores the answer to a yes or no question for whether the user wishes to calculate for another scenario.

Program 4-4 Video animation available.

```
1  # Using while repetition and a control variable,
2  # calculate BMIs for a height and different weights
3
4  height = float(input('Enter height (inches): '))
5  anotherWeight = 'y' # this is the control variable, for looping
6
7  while anotherWeight == 'y':
8
9     weight = float(input('\nEnter weight (pounds): '))
10
11    bmi = weight / height**2 * 703
12    print('BMI = ', round(bmi, 1))
13
14    anotherWeight = input('Enter another weight (y/n)? ')
```

Program 4-4 Example Run

```
Enter height (inches): 68

Enter weight (pounds): 150
BMI =  22.8
Enter another weight (y/n)? y

Enter weight (pounds): 145
BMI =  22.0
Enter another weight (y/n)? n
```

In Program 4-4, the control variable is `anotherWeight`, and its value is initially set by the program on line 5. This allows the loop to proceed for a first iteration. This variable is also set by the user on line 14 in order for the `while` statement to determine if the loop will repeat.

For a control variable, you can also use a True/False variable, which has the ***bool*** type. In Program 4-5, the variable `anotherWeight` is set to True on line 5, and reset on line 15 based on the user input. In this way, the repetition is controlled by a specific bool variable.

Program 4-5

```
1  # Using while repetition and a bool control variable,
2  # calculate BMIs for a height and different weights
3
4  height = float(input('Enter height (inches): '))
5  anotherWeight = True # this is the control variable, for looping
6
7  while anotherWeight:
8
9      weight = float(input('\nEnter weight (pounds): '))
10
11     bmi = weight / height**2 * 703
12     print('BMI = ', round(bmi, 1))
13
14     reply = input('Enter another weight (y/n)? ')
15     anotherWeight = (reply == 'y')
```

Program 4-5 Example Run

```
Enter height (inches): 68

Enter weight (pounds): 150
BMI =  22.8
Enter another weight (y/n)? y

Enter weight (pounds): 145
BMI =  22.0
Enter another weight (y/n)? n
```

Note that in all the examples so far, the control variable is:

- initialized before the `while` statement
- tested as part of the `while` condition
- changed within the indented statements under the `while`

> **Tip**
> **About Case Sensitivity**
>
> In Program 4-4, the while test for `anotherWeight` is case sensitive. For example, if the user replies with an uppercase `Y` to the prompt on line 14, the condition on line 7 will evaluate to False, since Python string compares are case sensitive, and the loop will not continue. In Chapter 8, we'll look more in depth at strings and see how the `lower()` method could be used to make your program case insensitive. For example, in Program 4-4, we could use the following statement on line 7 to allow both `y` and `Y` responses to allow the loop to repeat:
>
> ```
> while anotherWeight.lower() == 'y'
> ```

Using a special value

In the previous example, a particular variable, `anotherWeight`, is used to control the iteration of the looping. Having a specific (well-named) control variable is a good approach for use in looping. As we've seen, another approach is to designate a special value for a variable that will indicate when to stop looping. To help contrast these approaches, Program 4-6 shows the use of a special value. In particular, when `weight` is set to 0, this is an indication for the `while` statement to stop looping.

Program 4-6 Video animation available.

```
# Using while repetition and special end value,
# calculate BMIs for a height and different weights

# get input height (common for all calculations)
height = float(input('Enter height (inches): '))

# get initial weight to calculate with
weight = float(input('Enter weight (pounds): '))

# repeat calculating as long as entered weight is non-zero
while weight != 0:
    bmi = weight / height**2 * 703
    print('BMI = ', round(bmi, 1))

    weight = float(input('\nEnter weight goal (or 0 to end): '))
```

Program 4-6 Example Run

```
Enter height (inches): 68
Enter weight (pounds): 150
BMI =  22.8

Enter weight goal (or 0 to end): 145
BMI =  22.0

Enter weight goal (or 0 to end): 0
```

Repeating with `start`, `increment`, end

In the previous few examples, user input is provided within the loop as part of the repeated logic. In the next program, the looping scenarios are set up by prompting for minimum, maximum, and incrementing values. These inputs are gathered before the loop, and the program then generates the looping without further user input.

Program 4-7 generates a table of alternative interest rates and the corresponding interest earned for those rates.

Program 4-7

```
1  # Using while repetition, generate a table of
2  # CD interest rates and interest earned
3
4  invest = float(input('Enter initial investment ($): '))
5  years = int(input('Enter years: '))
6
7  rateMin  = float(input('Minimum rate: '))
8  rateMax  = float(input('Maximum rate: '))
9  rateIncr = float(input('Increment rate: '))
10
11 # print the header rows for the table
12 print()
13 print('Interest', 'Interest')
14 print('    Rate', '  Earned')
15
16 # generate a table of interest rates and corresponding interest earned
17 rate = rateMin # the initial rate for the table
18 while rate <= rateMax:
19     endBalance = invest * (1 + rate/100) ** years
20     interest = endBalance - invest
21
22     print(format(rate,'8.3f'), format(interest, '8,.2f'))
23
24     rate += rateIncr # the next rate
```

Program 4-7 Example Run

```
Enter initial investment ($): 1000
Enter years: 3
Minimum rate: 2
Maximum rate: 3
Increment rate: .25

Interest Interest
    Rate   Earned
   2.000    61.21
   2.250    69.03
   2.500    76.89
   2.750    84.79
   3.000    92.73
```

There are a few points to note in Program 4-7. The initial condition is set just before the `while` statement by initializing the variable `rate`. This variable is used in the `while` condition, and is adjusted at the end of the loop. In this way, the user's inputs, such as `rateMin`, are left as is (static), but the control variable may change over the execution of the program. That the user's inputs are static can be a good approach in case they may be needed later in the program.

Also of note is how the table headings are displayed in lines 12–14. These statements are ahead of (outside) the loop, as the headings are displayed once instead of repeatedly. Note the `print()` statement on line 12; this displays a blank line on the screen to better separate the user inputs from the program's outputs.

Last, note that line 22 uses the `format()` function, which was described briefly in Chapter 2. In prior examples, we have used the `round()` function to control the displayed number of decimal places. The `format()` function can also control the number of decimal places, as well as include commas in the displayed value. Further, `format()` can indicate how many positions the displayed item should have (in this case, 8), which is useful for lining up a column of numbers. The `format()` function is covered in more detail in Appendix C.

Conditions recap

In the examples in the chapter, we've seen a few comparison operators: ==, !=, <=, and <. As looping conditions are logical expressions, any of the comparison operators used in decision conditions, as discussed in Chapter 3, can be used in looping conditions. Additionally, composite conditions can be created using logical operators. Parentheses can also be used. Below are tables of comparison and logical operators.

Operator	Meaning
<	Less than
<=	Less than or equal to
>	Greater than
>=	Greater than or equal to
==	Equal to
!=	Not equal to

Table 4-1: Comparison operators

Operator	Meaning
and	True if both statements are True
or	True if either statement is True
not	Reverses the True/False result

Table 4-2: Logical operators

Section Summary

The `while` statement can be used to repeat execution of statements for different scenarios. One or more variables are initialized ahead of the `while` statement. Then the `while` statement checks an expression to see if the indented statements below it will execute. If they do, flow will return back to the `while` statement after the last statement in the indented repetition block executes.

Test Your Understanding

1) The `while` statement includes a condition whose result is a number.
   a) True
   b) False
2) When used with a `while` structure, a control variable is typically:
   a) initialized before the `while` statement
   b) tested as part of the `while` condition
   c) has its value changed in the body of the `while` block
   d) all of the above
3) Logical operators that are used in `if` statements can also be used in `while` conditions.
   a) True
   b) False

4.3 The `for` Statement, with `range()` Function

In the previous section, we looked at how the `while` statement can be used to achieve repetition in programs. The other keyword for Python repetition is `for`. The `for` statement can work with a sequence of things (e.g., integers, lines in a file, records in a database). In this section, we'll look at how `for` can be used with the `range()` function to work with a sequence of integers.

`for` with `range(start, end)`

The `range()` function provides a way to generate a list of integers. For example, `range(1,6)` generates the integers 1, 2, 3, 4, 5. Note that Python's convention is to start the integer list at the first value (1 in this case), and generate integers up to *but not including* the last value. Returning back to generating squares and cubes for a set of numbers, the `for` statement can be used in combination with the `range()` function as shown in Program 4-8.

Program 4-8 Video animation available.

```
1  # Using for repetition, generate a table of squares and cubes
2  n = 5 # number of rows in the table
3  # using for with range(), generate the table
4  for i in range(1, n+1):
5      print(i, i**2, i**3)
```

Program 4-8 Output

```
1 1 1
2 4 8
3 9 27
4 16 64
5 25 125
```

`range()` can be used with either one, two, or three arguments. We'll initially look at the two-argument case. For this, `range()` is called:

```
range(start, end)
```

where `start` is a starting integer value, and `end` controls the largest value generated by `range()`. Note that `range()` will generate the integers from `start` to `end-1`. This is why line 4 in the above example uses `n+1` as the second parameter.

In line 4, `for` and `range()` work together to control the repetition of the indented following statement. Essentially `for` and `range()` are in dialogue to set the target variable (in this case called `i`) and control the repetition of the body of the loop.

Tip
`for` and `range()` Dialogue

In Program 4-8, the `for` statement can be thought of as having a dialogue with `range()` in order to manage the repetition. `for`'s first question to `range()` is "Do you have a value?" In Program 4-8, `range()`'s answer is "Yes, the value is 1." `for` then sets `i` to this value, and then Python executes the indented statement. Then the flow returns back to `for`, which again asks `range()`, "Do you have another value?" `range()`'s answer is "Yes, the next value is 2." `for` then sets `i` to 2 and another repetition is performed. This continues until the sixth time, when `range()`'s response to `for` is "No, there are no more values." At this point, Python ends the program since there are no more statements.

Program 4-9 also uses `for` and `range()`. In this example, the annual balances and interest earned for a CD are generated and displayed. In this case, on line 16, the target variable `year` is initially set to 1. Each iteration of the loop sets `year` to an increasing value, up to the value of `years`.

Program 4-9

```
# Using for repetition, generate table of
# annual CD balances and interest

# get input parameters for the CD
initial = int(input('Initial investment ($): '))
years = int(input('Number of years: '))
interestRate = float(input('Interest rate (%): '))

# set up variables used in the loop
balance = initial
interestRateDecimal = interestRate / 100

print('\nyear', '   start', 'interest', '     end') # header for table

# for each year, calculate and display balances and interest earned
for year in range(1, years+1):
    interestEarned = int(balance * interestRateDecimal) # round down
    end = balance + interestEarned

    print(format(year,'4d'), format(balance, '8,d'),
          format(interestEarned, '8,d'), format(end, '8,d'))

    balance = end # roll over balance from year end to next year start
```

Program 4-9 Example Run

```
Initial investment ($): 1000
Number of years: 3
Interest rate (%): 5

year    start interest      end
   1    1,000       50    1,050
   2    1,050       52    1,102
   3    1,102       55    1,157
```

`range(count)`

The `range()` function can also be used with one argument. In this case, the first value of the generated integers is not specified but defaults to 0. For `range(n)`, the sequence of integers generated is from `0` to `n-1`. Program 4-10 demonstrates the use of `range(n)` in order to calculate BMI values for a set of `n` weights.

Program 4-10

```
# Using for repetition, prompt for weights and calculate several BMIs

# height is an input common to all calculations, so outside the loop
height = float(input('Enter height (inches): '))

# n will be the number of times the loop executes
n = int(input('Number of BMIs to calculate: '))

for i in range(n): # control the number of times the loop executes

    weight = float(input('\nEnter weight (pounds): '))

    bmi = weight / height**2 * 703
    print('BMI = ', round(bmi, 1))
```

Program 4-10 Example Run

```
Enter height (inches): 68
Number of BMIs to calculate: 3

Enter weight (pounds): 150
BMI =  22.8

Enter weight (pounds): 145
BMI =  22.0

Enter weight (pounds): 140
BMI =  21.3
```

`range(start, end, skip)`

The last variation of `range()` involves using three arguments. In this case, the first two arguments control the first and last values of the sequence. The third argument is the skip space between the generated set of integers. For example, `range(1, 6, 2)` would generate the numbers 1, 3, 5. Program 4-11 demonstrates the three-argument variation of `range()`. It generates a table of weights and corresponding BMIs, with the space between weights determined by user input and used in `range()`'s third argument.

Program 4-11

```
# Using for repetition, generate a table of weights and calculated BMIS

# height is an input common to all calculations, so get outside the loop
height = float(input('Enter height (inches): '))

# These variables control the beginning, steps, and end of the table
weightMin = int(input('Minimum weight: '))
weightMax = int(input('Maximum weight: '))
weightIncr = int(input('Increment weight: '))

print() # print a blank line
print('Weight', 'BMI') # heading for table

# use for statement to generate a table, using 3-part range()
for weight in range(weightMin, weightMax+1, weightIncr):
    bmi = weight / height**2 * 703
    print(weight, ' ', round(bmi, 1))
```

Program 4-11 Example Run

```
Enter height (inches): 68
Minimum weight: 140
Maximum weight: 150
Increment weight: 2

Weight  BMI
140     21.3
142     21.6
144     21.9
146     22.2
148     22.5
150     22.8
```

Tip
for Fully Handles the Target Variable

In all the programs in this section, the `for` statement sets its target variable, and there is no need to separately initialize the target variable or update it within the indented statements. As an example:

```
i = 1 # this statement isn't needed, i is set in for statement below
for i in range(1, 6):
    print(i, i**2)
    i = i +1 # this statement isn't needed, i is updated in for statement
```

Here the first and last statements aren't needed, since `for` and `range()` work together to set *i*'s value as needed.

Section Summary

The `for` statement can be used to repeat execution of statements for different scenarios. The `for` statement can be used together with the `range()` function to repeat an indented repetition body with a control variable set to different integer values. As with the `while` structure, flow will return back to the `for` statement after the last statement in the indented repetition block executes. The `range()` function can be used with one, two, or three arguments, as `range(count)`, `range(start, end)`, or `range(start, end, increment)` in order to generate an integer list for use in looping.

Test Your Understanding

1) Which function can be used to generate a list of integers?
   a) `int()`
   b) `range()`
   c) `sequence()`
   d) all of the above
2) Similar to `while`, the `for` statement uses a logical expression to test if its body will execute.
   a) True
   b) False
3) How many arguments can the `range()` function take?
   a) one
   b) two
   c) three
   d) all of the above

4.4 Using `while` vs. `for`

In some cases, repetition programs can be written with either `while` or `for`. In general, `for` will be more concise. For example, Programs 4-1 and 4-5 both generate a table of squares and cubes. Program 4-1 uses `while` and involves more statements, since the initial condition, repeating test, and variable change are each in separate statements. In contrast, Program 4-5 combines initialization, repeating test, and variable changing all into one statement.

Conversely, some programs should be written using `while`. In particular, if the number of repetitions depends on user input during looping, then `while` is a better fit than `for`. As an example, Program 4-4 asks the user if they would like to try another scenario, and this works well with `while`. In order to use `for`, we would need to ask the user how many scenarios they would like to try, which is a valid but different approach to the problem.

4.5 Infinite Loops

Programs that have loops will generally stop looping at some point—for example, when a `while`'s repetition condition evaluates to False. However, you may write a program that accidentally keeps the repetition condition always set to True. In this case, your loop will repeat over and over. This error in logic is referred to as an ***infinite loop***. Usually the correction is to add a statement that will update the variable checked in the repetition condition.

For example, Program 4-12 has a bug that causes it to have an infinite loop. In particular, the variable `i` is set to 1, and its value is never changed. As such, the while test (is `i <= n` ?) will always be True, and the program will keep looping until it is forced to stop. Your development tool will have a way to stop a running program. If your program is run from a command line, Control+C should stop the program.

Program 4-12

```
# This program demonstrates an infinite loop error

#initialization
i = 1 # starting value for the table
n = 5 # ending value for the table

# using repetition, generate the table
while i <= n: # condition will always be true
    # calculation and display
    print(i, i**2, i**3)

    # missing statement below here should increment variable i
    # instead, i is stuck at 1
```

The fix to correct Program 4-12 is to include a statement that will change the value of `i`. In this case, adding a line as the last indented statement, `i = i + 1`, will change `i` during each iteration of the loop, and the infinite loop bug will be repaired. In general, if your program has an infinite loop, review the variable(s) in the repetition condition to see how they are (or aren't) changed in the body of the repetition loop.

4.6 Adding Validation

In Program 4-4, different weight scenarios are evaluated in a `while` loop to calculate corresponding BMIs. As an enhancement, we would like to add validation for the entered weight. In particular, the weight entered by the user should be positive. Program 4-13 adds an `if/else` structure to check if the entered weight is positive. If it is, BMI is calculated and displayed. However, if the weight is not positive, the else block causes an error message to be displayed. In this way, a decision structure is nested within repetition to evaluate scenarios and validate input.

Program 4-13

```
# Using while repetition and control variable,
# generate a table of weights and BMIs
# Check that weight is > 0

height = float(input('Enter height (inches): '))
keepGoing = 'y' # this is the control variable, for looping

while keepGoing == 'y':

   weight = float(input('\nEnter weight (pounds): '))

   if weight > 0:
      bmi = weight / height**2 * 703
      print('BMI = ', round(bmi, 1))
   else:
      print('Error, weight must be positive')

   keepGoing = input('Enter another weight (y/n)? ')
```

Program 4-13 Example Run

```
Enter height (inches): 68

Enter weight (pounds): 0
Error, weight must be positive
Enter another weight (y/n)? y

Enter weight (pounds): 145
BMI =  22.0
Enter another weight (y/n)? n
```

Another approach to validation is to use repetition. For example, after a user enters a value, it can be checked, and if not valid, repetition can be used to ask for a correct input. Program 4-14 shows this technique by checking the value entered by the user for height. The `while` statement in line 6 checks for a non-positive value, and if it is True, the program will display an error and ask again for a value for height. This loop will continue repeating until the user enters a positive value for height.

Tip
About Validation

In this section, we assume the user is entering numeric values. In these example programs, if the user enters non-numeric input, these programs will display an error message and end execution. A more comprehensive approach to validation involves using exceptions, which is covered in Appendix D.

Program 4-14

```
# Using while repetition and special end value,
# generate a table of weights and BMIs

# get input height (common for all calculations)
height = float(input('Enter height (inches): '))
while height <= 0: # validate height
    print('Error, height must be > 0.')
    height = float(input('Enter height (inches): '))

# get initial weight to calculate with
weight = float(input('Enter weight (pounds): '))

# repeat calculating as long as entered weight is non-zero
while weight != 0:
    bmi = weight / height**2 * 703
    print('BMI = ', round(bmi, 1))
    weight = float(input('\nEnter weight goal (or 0 to end): '))
```

Program 4-14 Example Run

```
Enter height (inches): 0
Error, height must be > 0.
Enter height (inches): 68
Enter weight (pounds): 145
BMI =  22.0

Enter weight goal (or 0 to end): 140
BMI =  21.3

Enter weight goal (or 0 to end): 0
```

The technique used in Program 4-14 can be used for all user input. This is demonstrated in Program 4-15.

Program 4-15

```
1 # Using while repetition and special end value,
2 # generate a table of weights and BMIs
3
4 # get input height (common for all calculations)
5 height = float(input('Enter height (inches): '))
6 while height <= 0: # validate height
7     print('Error, height must be > 0.')
8     height = float(input('Enter height (inches): '))
9
10 # get initial weight to calculate with
11 weight = float(input('Enter weight (pounds): '))
12 while weight <= 0: # validate weight
13     print('Error, weight must be > 0.')
14     weight = float(input('Enter weight (pounds): '))
15
16 # repeat calculating as long as entered weight is non-zero
17 while weight != 0:
18     bmi = weight / height**2 * 703
19     print('BMI = ', round(bmi, 1))
20
21     weight = float(input('\nEnter weight goal (or 0 to end): '))
22     while weight < 0: # validate weight goal
23         print('Error, weight must be > 0.')
24         weight = float(input('Enter weight goal (or 0 to end): '))
```

Program 4-15 Example Run

```
Enter height (inches): 68
Enter weight (pounds): 0
Error, weight must be > 0.
Enter weight (pounds): 145
BMI =  22.0

Enter weight goal (or 0 to end): 140
BMI =  21.3

Enter weight goal (or 0 to end): 0
```

Note that the `while` on line 17 is controlling the scenarios for different weights, but the `while` on line 22 is checking for a valid weight input. In this way, the `while` on line 22 is nested under the `while` on line 17.

4.7 About `break` and `continue`

We conclude the chapter with two special statements related to repetition: `break` and `continue`. These statements interrupt the usual flow of a repetition loop. Because they cause a jump in flow, programs that use them can be harder to follow. In general, you should try to avoid using them in your programs—in fact, in some courses, assignments that use them might lose points! However, as they may be used by programs written by others that you may read, and they may be suitable to use in some limited cases, we'll look at these two statements.

The `break` statement causes the body of a repetition loop to stop executing, and the program's execution will continue at the first statement after the `while` or `for` indented block. The `break` statement is typically used with a `while` condition—for example, when something unusual happens. Program 4-16 demonstrates this.

Program 4-16

```
1  # Using while repetition and special end value, generate BMI scenarios
2  # includes use of break statement to end looping
3
4  # get input height (common for all calculations)
5  height = float(input('Enter height (inches): '))
6
7  # get initial weight to calculate with
8  weight = float(input('Enter weight (pounds): '))
9
10 # repeat calculating as long as entered weight is non-zero
11 while True:
12     if weight == 0:
13         break
14
15     bmi = weight / height**2 * 703
16     print('BMI = ', round(bmi, 1))
17
18     weight = float(input('\nEnter weight goal (or 0 to end): '))
19
20 print('Program ends')
```

Program 4-16 Example Run

```
Enter height (inches): 68
Enter weight (pounds): 150
BMI =  22.8

Enter weight goal (or 0 to end): 140
BMI =  21.3

Enter weight goal (or 0 to end): 0
Program ends
```

In the above program, line 13 checks if the user has entered a zero weight, and if so, the `break` statement will execute and cause the program to jump to line 20, which is the first statement in parallel with the `while`.

Program 4-16 is equivalent to Program 4-5, which checks if the weight is zero in the `while` condition. Program 4-5 is more concise, and since it doesn't use a `break` statement, it is preferred to Program 4-16.

Another statement that interrupts the flow of repetition is `continue`. Like `break`, `continue` is typically used with a `while` statement. The `continue` causes the program flow to jump back up to the `while` (or `for`) that it is indented within. In other words, the `continue` indicates to stop processing the remainder of the body of the loop, but instead proceed to the next iteration of the loop (if there is one). Program 4-17 demonstrates the use of `continue`, in order to skip calculating and displaying a BMI when the input weight is not positive.

Program 4-17

```
# Using while repetition, generate scenarios for weights and BMIs
# includes use of continue statement for invalid weight

# get input height
height = float(input('Enter height (inches): '))

# repeat calculating as long as entered weight is non-zero
weight = -1 # initialize weight so at least 1 repetition executes
while weight != 0:

    weight = float(input('Enter weight (pounds), or 0 to end: '))

    if weight < 0:
        print('Error, weight must be positive')
        continue

    elif weight > 0:
       bmi = weight / height**2 * 703
       print('BMI = ', round(bmi, 1))
```

Program 4-17 Example Run

```
Enter height (inches): 68
Enter weight (pounds), or 0 to end: -150
Error, weight must be positive
Enter weight (pounds), or 0 to end: 150
BMI =  22.8
Enter weight (pounds), or 0 to end: 0
```

Program 4-17 is similar to Program 4-13, which uses `if/else` to validate the weight. In general, Program 4-13 would be preferred, since it avoids use of the `continue` statement. However, if your program required validation of several variables, use of `continue` might be easier to follow instead of alternatives such as nested `if` statements.

Section Summary

Repetition flow typically proceeds from the top down to the last statement of the indented repetition block. However, two special statements, `break` and `continue`, can be used to interrupt this flow. The `break` statement will jump execute out of the repetition block. The `continue` statement will jump back up to the repetition control statement (`while` or `for`). Usually there is a preferred alternative approach—for example, using `if`—instead of using `break` or `continue`.

Test Your Understanding

1) The `continue` statement causes the program flow to stop repetition.
   a) True
   b) False

2) The `break` statement causes the program to generate an error.
   a) True
   b) False

Chapter Review

In this chapter, we've seen how repetition can be used to process different scenarios with a common set of code. Two types of repetition statements, `while` and `for`, were introduced. The `while` statement uses a condition to determine if the indented statements will be executed and possibly repeated. The `for` statement can be combined with the `range()` function in order to iterate a set of statements for a list of integers. Infinite loops may arise, usually due to a bug where the looping condition is always True. Last, the `break` and `continue` statements were covered to show how they interrupt the regular flow of a repetition of statements.

Key Terms

break	The `break` statement causes the program flow to skip the remaining statements in the loop body and pass execution to the next statement after the corresponding loop control statement (`while` or `for`).
continue	The `continue` statement causes the program flow to skip the remaining statements in the loop body, and the program flow moves back up to the corresponding `while` or `for` statement.
for	The `for` statement allows repetition of the indented statements below it by setting its target variable from items in a list of values, one at a time.
infinite loop	A set of repetition statements that continue executing, often in error, since the looping condition remains True.
logical expression	A kind of expression that evaluates to True or False.
loop body	Statements indented immediately below a `for` or `while` statement that may be repeatedly executed.
range()	A function that generates a list of integers and can be used as part of a `for` statement.
repetition	A programming structure for repeating execution of statements.
while	The `while` statement allows repetition of the indented statements below it by checking if its corresponding expression is True.

Programming Problems

1) Purchases

Write a program to determine the total cost for a set of purchases. The program should:

- use a while loop to prompt the user for each purchase's information, including for each item the number of items and the unit cost
- at the end of the while loop, get user input on whether there are more items
- after the while loop, calculate tax at 7.5%
- display the pre-tax cost, the amount of tax, and the total cost including tax

Here is an example session:

```
Item 1
Enter count of item 1: 2
Enter unit cost for item 1: 5.25
More items (y/n)? y
```

```
Item 2
Enter count of item 2: 1
Enter unit cost for item 2: 10.50
More items (y/n)? n

Enter sales tax rate (%): 7.5

Total cost (pre-tax):   21.00
Sales tax:              1.57
Total cost (with tax): 22.57
```

2) Purchases with Some Items Taxable

Looking back again at problem 1, assume that an item may or may not be taxed. In addition to the requirements for problem 1, for each item, also prompt for whether the item is taxed.

Here is an example session:

```
Item 1
Enter count of item 1: 2
Enter unit cost for item 1: 5.25
Is this item taxable (y/n)? y
More items (y/n)? y

Item 2
Enter count of item 2: 1
Enter unit cost for item 2: 10.50
Is this item taxable (y/n)? n
More items (y/n)? n

Enter sales tax rate (%): 7.5

Total cost (pre-tax):    21.00
Sales tax:               0.79
Total cost (with tax):  21.79
```

3) Vehicle Routing

An intercampus shuttle can take one of several routes. For each run, a dispatcher checks what the current traffic conditions are to find out the average speed on each route and determine which route the shuttle driver should take.

a) Write a program that, in a while loop:
   - prompts for two inputs: distance in miles and speed in miles per hour
   - calculates the time the route should take (in minutes) and displays this result
   - keeps track of which alternative route is the fastest and displays this after all routes are entered

Here is an example session:

```
Enter route 1 distance (miles): 30
Enter route 1 speed (miles/hour): 60
More routes (y/n)?: y

Enter route 2 distance (miles): 25
Enter route 2 speed (miles/hour): 55
More routes (y/n)?: n

Route 2 is fastest; 27 minutes
```

b) Assume current traffic conditions are as follows. Run your program using the data below to determine the fastest route.

Route Type	Distance (miles)	Current speed (miles/hour)
local	30	30
parkway	35	40
highway	48	55

c) Enhance your program to detect invalid inputs of 0 or negative values for distance or speed.

4) Route Time

Assume a route between two points has several segments. Each segment has a distance and estimated speed. Write a program to determine the total time for the route. For this program, prompt the user for the number of segments, and then use a for statement to loop over each segment, prompting the user for the distance and estimated speed for each segment, and determining and displaying the total time.

Here is an example session:

```
Number of legs: 2

Enter leg 1 distance (miles): 30
Enter leg 1 speed (miles/hour): 60

Enter leg 2 distance (miles): 10
Enter leg 2 speed (miles/hour): 30

Total time (minutes): 50
```

5) Investment Alternatives

Assume that you have several investment choices. For each investment, there are two equally likely end-of-year forecasts: the optimistic value (up_value), and the pessimistic value (down_value). Assume the expected return and risk of each investment is given by:

```
expected_end_value = (up_value + down_value) / 2
expected_return = (expected_end_value-initial_investment) / initial_investment
risk = (up_value-down_value) / 2
```

Write a program that evaluates several investment choices and then determines the choice with the highest expected return. Your program should initially prompt the user to enter how many choices there are as well as the initial investment amount. Then for each investment choice, prompt for 2 inputs:

- optimistic end value
- pessimistic end value

and calculate each investment's expected return and risk.

Here is an example run:

```
Enter number of investment options: 2
Enter start value: 1000

Option 1
 Enter high value: 1010
 Enter low value: 1000

Option 2
 Enter high value: 1100
 Enter low value: 930

Best investment:  2
Best average return = 1.5 %
Risk = 85.0
```

Chapter 5

Defining Functions

Chapter Objectives

In this chapter, you'll learn:

- how to design, write, and use your own functions to create well-organized programs
- how to use named parameters and multiple return values in your functions
- how to write targeted functions for getting keyboard input or displaying results to the screen
- how to define and use a `main()` function in your program
- about global variables and why you should generally avoid using global variables in functions

Chapter Outline

5.1 Introduction to Functions

About functions

In the previous chapters, we focused on variables, statements, and built-in functions as key building blocks for creating programs. In this chapter, we look more closely at functions, in particular how you can define and use your own functions. The built-in functions provided with Python are ones you can use "off the shelf." But you can also define and use your own functions to organize your programs better.

In the same way that paragraphs help organize an essay or a chapter, functions help organize programs into coherent units. As we'll see, functions typically accept one or more inputs, and often provide one output (the returned result). Based on the specific values for inputs provided to the function, the function will process and then provide a result. In other words, functions can be thought of as following the IPO (input-processing-output) model described in Chapter 1.

Before writing your own functions, it's useful to look more closely at some built-in functions you've already used. For example, let's look at the built-in function `float()`. The `float()` function takes one input ***argument*** (the value, variable, or expression inside its parentheses), and tries to convert this input to a floating point number. If it can, `float()` will provide as output a `float` result. If `float()` can't make this conversion, it will generate an error (a Python ***exception***). For example, if `name = 'Albert'`, then `float(name)` can't determine a corresponding number. In Appendix D, we look at exceptions in more detail. In general, a Python function when executed can have one of two outcomes: either a returned result (successful execution) or an exception (an unsuccessful execution). To convert a string to a number, the input/processing/output (IPO) recipe for `float()` is:

input: a Python string
processing: convert the string to a floating point number
output: a floating point number

As another example, the `round()` function takes one or two arguments (a number and optionally an integer) as input. `round()`'s processing determines a number rounded to a certain number of decimal places. The IPO recipe for `round()` can be seen as:

input: a Python number to round, and optionally a number of decimal places (as an `int`)
processing: determine the number that is rounded to the requested decimal places
output: a rounded Python number

To see how `round()` works, we can use it interactively. For example:

```
>>> round(1.234,2)
1.23
>>> round(1.234)
1
```

We can also see the type of result `round()` returns:

```
>>> type(round(1.234, 2))
<class 'float'>
```

Functions can be shown in an IPO diagram to highlight their input/processing/output pattern. For example, Figure 5-1 summarizes the `round()` function.

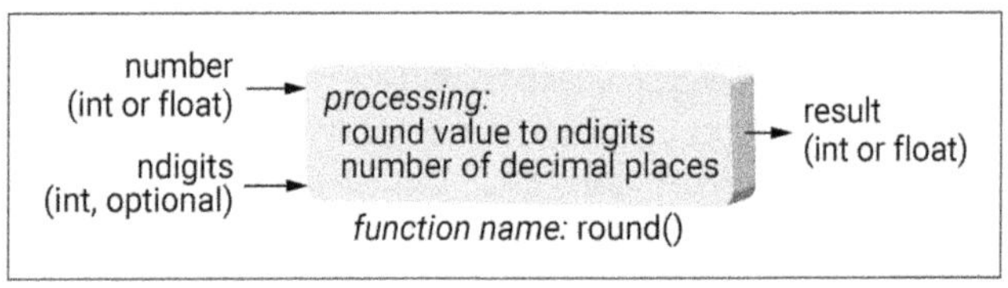

Figure 5-1: The `round()` function: Input/processing/output

Defining your own functions

Having reviewed some built-in functions, let's look at how you can define your own function. In Python, this involves deciding on the following:

1) the name of your function
2) your function's ***input parameters***—these are variables, so we'll need to decide their names and valid types
3) what processing your function will do
4) what error conditions your function may hit, and how to handle them
5) what result your function will return—that is, the answer your function gives

All these questions were considered and answered when the built-in functions in Python were designed and implemented.

Note that for step 2 above, we use the term *input parameter*. The function's input parameters allow it to receive values for variables that the function will use in its processing. Earlier we used the term *argument*. Parameter and argument are related terms. When we use a function, the values passed to it are referred to as arguments. The corresponding variables that receive the argument values are referred to as parameters.

Let's look at a first very simple function to address these five questions and see what the Python syntax is in order to code it. The function we would like to write will display a greeting to the user, including their name. For the five questions above:

1) We'll name the function `'hello'`. Like variables, all functions have a name that allows the program to find and use the function.
2) Our hello function will have one input parameter, a variable called `nameParameter`. This variable contains the person's name that will be included in the greeting displayed on the screen.
3) The function will display the message `'Hello'` and the name in the `nameParameter` variable followed by an exclamation point. For example, if `nameParameter` is `'Stefanie'`, the function should display to the screen the message:

```
Hello, Stefanie !
```

4) If `nameParameter` is not set to a string, the function should return an error.
5) Since our function displays a message to the screen, there is no result the function will return.

Having defined the above aspects of the function, we can look at the syntax we need to define this function in a Python program. The first aspect is the ***signature*** of the function. The signature is the first line of the function that includes:

- the keyword `def` (short for define)
- the name of the function (in our example, `'hello'`)
- in parentheses, the names of the function's parameter(s)—in this case `nameParameter`
- a colon

Combining all these elements, the signature (first line) for our function is:

```
def hello(nameParameter):
```

The remainder of the code for the function indicates the processing we want, which is referred to as the function's ***body***. These are Python statements indented under the signature. Note that the indentation pattern is like that used with conditional (if/`elif`/`else`) and repetition (while/for) statements. For our program, in order to display a line to the screen, we can use the statement:

```
    print('Hello,', nameParameter, '!')
```

In most function bodies there are multiple statements, but for our `hello()` function, we just need one statement. Combining the signature and the body, we have our complete `hello()` function:

```
def hello(nameParameter):
    print('Hello,', nameParameter, '!')
```

Note that having the above two lines in our program does not execute (run) the function. If you create and run a Python program with only the above two lines, nothing will appear to happen. Instead, the signature and body you write are read by Python and readied for use. But a function needs to be explicitly ***called*** in order to run. This explicit call is what you have done in order to use built-in functions. To call a function, you specify the name of the function, parentheses, and values/variables/expressions separated by commas inside the parenthesis as arguments. Further, if the function returns a value, that value can be assigned to a variable, passed to another function, or used in an expression.

In order to use our hello function, assuming our program has a variable called `name`, we need to execute it using a statement like:

```
hello(name)
```

Program 5-1 combines all these aspects to define `hello()`, prompt the user for their name, and use the `hello()` function to display a greeting on the screen.

Program 5-1 Video animation available.

```
# define and use a function to display a greeting

# the hello() function has 1 parameter: 'nameParameter'
def hello(nameParameter):
    print('Hello,', nameParameter, '!')

name = input('Please enter your name: ')
hello(name) # call the function with the argument 'name'
```

Program 5-1 Example Run

```
Please enter your name: Stefanie
Hello, Stefanie !
```

A diagram summarizing the `hello()` function in Program 5-1 is shown in Figure 5-2.

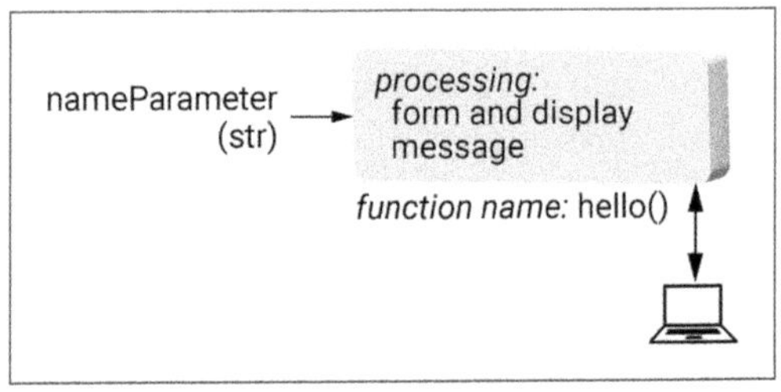

Figure 5-2: IPO function diagram for `hello()`

Spreadsheet Compare/Contrast
Spreadsheet Functions

Spreadsheets provide many built-in functions that, like Python functions, can take input arguments and provide a result. Python also provides built-in functions, some of which we've covered in Chapter 2. However, in contrast with spreadsheets, Python provides a straightforward way to define your own functions within your code.

A function that returns a result

In the prior section, we designed and coded the function `hello()`. This function displays a message to the screen, but it doesn't return a result* to the function's calling statement. In that sense, it is like the built-in `print()` function. In practice, most functions provide a result to the calling statement—for example, like `input()` and `float()`. In the case of `input()`, the returned result is a string that represents what the user typed. For `float()`, the returned result is based on the argument passed within parentheses and converted into a Python float. The results from functions can be assigned to variables, used in expressions, or passed as an argument to another function. An example of the latter is what we've seen in statements like:

```
height = float(input('Please enter your height: '))
```

In the above statement, the output result of `input()` is passed as the input argument to `float()`.

In this section, we look at how you can return a result to the calling statement of the function. The returned result is different from what a function may display on the screen. Many well-designed functions are "quiet" in that they don't display anything to the screen. Instead, a function's results are provided to the caller of the function by using the keyword `return`.

As an example, let's consider a bookstore that wants a program to calculate the cost of an order. We'll start simple and expand on this throughout the chapter. Let's initially assume all books cost $17.50. We'd like to calculate the cost of an order, given the number of books purchased. Program 5-2 prompts the user to input the number of books, and using a function, `calcCost()`, calculates the total cost of the order.

* Technically, the function returns the special Python value `None`.

Program 5-2 Video animation available.

```
 1 # one input, one output (assumes all books are $17.50)
 2
 3 # define the function
 4 def calcCost(numberItems):
 5     totalCost = 17.50 * numberItems
 6     return totalCost
 7
 8 # input
 9 count = int(input('Enter # of books: '))
10
11 # processing - call the function
12 total = calcCost(count)
13
14 # output
15 print('Total: $', total)
```

Program 5-2 Example Run

```
Enter # of books: 5
Total: $ 87.5
```

The new type of statement introduced in Program 5-2 is the `return` statement on line 6. It has the form:

```
return <result>
```

where `<result>` is a value, variable, or expression that provides the result of the function. In this program, the variable `totalCost` is set according to the prior statements. It is this variable's value that is returned to the calling statement. The calling statement is on line 12:

```
total = calcCost(count)
```

Here the `calcCost()` function determines a float result and returns it, and then this result is assigned to the variable `total`. `total` is then used in the last statement of the program on line 15 to display the result to the screen.

To determine the result, the `calcCost()` function creates the variable `totalCost` on line 5. After the `return` statement on line 6 executes, the `totalCost` variable is no longer defined. In other words, we can't refer to `totalCost` in statements that are not indented within the `calcCost()` function. The variable `totalCost` is referred to as a ***local variable***, since it is only known locally within `calcCost()`, as opposed to being available generally (globally) within our program.

It is worth noting that the `calcCost()` function has one return statement. While it is possible for a function to have multiple return statements, functions having a single return statement are often easier to follow.

Figure 5-3 shows the IPO diagram for the `calcCost()` function in Program 5-2.

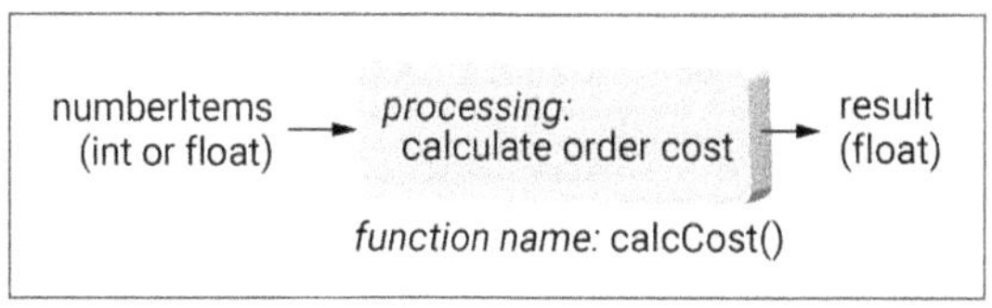

Figure 5-3: IPO function diagram for `calcCost()` in Program 5-2

Section Summary

Functions are key building blocks in programs. Built-in functions are provided with Python, and you can also write your own functions. By writing your own functions, you organize your programs into well-defined sections of code that are easier to understand and enhance. The keyword to introduce a function is `def`, short for define.

A function can have input parameters, which along with the function name, are declared in the signature of the function. The body of the function is made up of statements that perform the processing for the function. A `return` statement provides the function's result to the calling statement. However, a function doesn't require a return statement. For example, a `return` statement isn't needed if a function simply displays output to the screen.

Test Your Understanding

1) The first line of the function is called the signature, and begins with the keyword:
   a) function
   b) begin
   c) def
   d) define
2) A function you write is executed only if a statement calls that function.
   a) True
   b) False
3) The keyword for ending a function and returning a result is:
   a) result
   b) return
   c) end
   d) none of the above

5.2 Multiple Parameters

In the earlier sections of this chapter, we looked at some basic programs that defined and used functions. In this section, we'll expand on this by introducing multiple input parameters.

Programs 5-1 and 5-2 both have a single input parameter. However, functions can have multiple input parameters. In this case, the variable names are separated by commas in the signature of the function. Importantly, the order of the arguments in the statement calling the function must match the order of the parameter's in the function's signature.

Returning to the bookstore calculation, in the next program, we will decide to use a function to calculate the order cost based on the number of books and the cost per book. The signature of the function can be defined as:

```
def calcTotal(numberItems, unitCost):
```

where `numberItems` is an integer, and `unitCost` can be a float or integer.

Program 5-3 Video animation available.

```
# two inputs, one output

# define the function
def calcTotal(numberItems, unitCost):
    totalCost =  numberItems * unitCost
    return totalCost

# get inputs
count = int(input('Enter # of books: '))
cost = float(input('Enter cost per book ($): '))

# processing - call the function
total = calcTotal(count, cost)

# display the result
print('Total: $', total)
```

Program 5-3 Example Run

```
Enter # of books: 10
Enter cost per book ($): 20.50
Total: $ 205.0
```

When using a function with multiple parameters, the calling statement needs to have the right order of arguments within the parentheses. For Program 5-3, the `count` variable is provided first, and this will be assigned to the `numberItems` variable when the function is called. The second argument, `cost`, will be assigned to the `unitCost` parameter.

We've seen the concept of argument order previously with built-in functions. For example, when using the `round()` function, we first provide a value to `round()`, followed by the number of decimal places desired. The argument order in this case matters—if we reverse it we get an error result. For example, if you try:

```
x = 1.23
y = round(1, x)
```

you will get:

```
TypeError: 'float' object cannot be interpreted as an integer
```

This is because the `round()` function expects the second argument to be an integer corresponding to the number of decimal places desired.

In other cases, if the argument order is incorrect, you may not get a Python error message but instead get an erroneous result.

Tip
About Function Design

Among the design choices for your programs are what functions you want to have and what each function's inputs, processing, and output are. There is usually not a single correct answer for the set of functions to have. Outlining and grouping all the functionality needed in your program may be helpful to find the blocks of related logic that would lend themselves to being implemented as functions. This design is not unlike the outlining process you might do to organize an essay into paragraphs.

About argument types

In Program 5-3, the input arguments provided were expected to be numbers. In fact, either `int` or `float` values could be provided when calling the function. This works since `calcTotal()` is multiplying the provided values—although a fractional number of books doesn't make sense. Further, in Program 5-3, line 9 uses the `int()` function to convert user input from a string to an integer, so a fractional reply for number of books, like 0.5, will cause an error.

Similarly, the `round()` function takes two numeric values. In this case, the second argument must be an integer. In general, when functions have multiple parameters, each provided argument should be of an appropriate type.

Constants as arguments

In the previous programs, the arguments passed to the function were variables. This is common but not required. In fact, arguments can be constants, variables, or expressions. For example, we have used statements like:

```
height = float(input('Please enter your height: '))
```

In this case, as the result of `input()`, a string is passed as the argument to the `float()` function.

To show an example in which a constant value is passed to a function, Program 5-4 extends the bookstore calculation to include a tax rate. The function `calcTotal()` here has three parameters, and the calling code uses the constant .075 to set the tax rate for the call to `calcTotal()`.

Program 5-4

```
# 3 inputs (1 passed as constant)
def calcTotal(numberItems, unitCost, taxRate):
    pretaxCost =  numberItems * unitCost
    aftertaxCost = pretaxCost * (1 + taxRate)
    return aftertaxCost

# get inputs
count = int(input('Enter # of books: '))
cost = float(input('Enter cost per book ($): '))

# processing - call the function
total = calcTotal(count, cost, .075)

# display the result
print('Total: $', total)
```

Program 5-4 Example Run

```
Enter # of books: 10
Enter cost per book ($): 17.50
Total: $ 188.125
```

The ability to use constants as arguments to functions makes it easy to interactively test your functions, as well as experiment with built-in functions. For example, if you wanted to see if the built-in `max()` function will work with words, you can interactively enter:

```
max('apple', 'banana')
```

to see what it does. Similarly, you can test your own functions with different combinations of constants for arguments to make sure they are working as you expect.

Section Summary

A function can have multiple input parameters. In this case, the caller of the function must use the correct order of the input arguments. The input parameters may be of different types, and for each input parameter, one or more specific types may be valid. Further, the caller of the function can provide each argument as a constant, variable, or expression.

Test Your Understanding

1) The statement that calls a function must use variables as the arguments for the function.
   a) True
   b) False

2) When calling a particular function, the calling statement should provide arguments of appropriate Python types to pass to the input parameters of the function.
   a) True
   b) False

5.3 More Function Features

In this section, we look at some optional features that can be used with functions. ***Named parameters*** are another way to provide an input to a function when there is a good choice for a default value. Another feature, returning multiple results, provides a way of returning more than one result in a single return statement.

Named optional parameters

For some functions, you may have a good default value for an input, but still want to allow for some flexibility for this input. Named parameters are a way to provide this. The syntax is simple. In the signature of a function, you can add an equals sign and value following the last parameters of the function:

```
paramname=defaultvalue
```

Program 5-5 shows this by refining the `calcTotal()` function for our bookstore example. In the signature:

```
decimals=2
```

indicates that `decimals` is an optional input argument for the function's caller, and if not provided by the caller, the function will use the value 2. The caller can set another value by indicating the name of the parameter and a value. For example:

```
calcTotal(count, cost, .075, decimals=0)
```

This is shown in Program 5-5. Note that when the caller provides values for named parameters, this is done after (to the right) of the set of unnamed arguments.

Program 5-5

```
# named parameter

# define the function, including a named parameter
def calcTotal(numberItems, unitCost, taxRate, decimals=2):
    pretaxCost = round(numberItems * unitCost, decimals)
    aftertaxCost = round(pretaxCost * (1 + taxRate), 2)
    return aftertaxCost

# inputs
count = int(input('Enter # of books: '))
cost = float(input('Enter cost per book ($): '))

# processing - call the functions
total = calcTotal(count, cost, .075, decimals=0)

# output
print('Total: $', total)
```

Program 5-5 Example Run

```
Enter # of books: 10
Enter cost per book ($): 17.50
Total: $ 188.12
```

Multiple returned results

A well-designed function has a clear output. In many cases, this is a single value, such as a number or string. In other cases, it may be a list of values (we'll look at lists in the next chapter) or other composite results.

Python offers a shorthand for returning multiple results. In the return statement, values, variables, or expressions can be combined by separating them with commas, and these are returned together to the calling code.

Program 5-6 shows an example of this. Here we would like the `calcTotal()` function to return the pre-tax cost as well as the sales tax. As these are two related values, we can calculate and return them from one function.

Program 5-6

```
1  # return two values from a function
2
3  # define the function
4  def calcTotal(numberItems, unitCost, taxRate):
5      pretaxCost = round(numberItems * unitCost, 2)
6      salesTax = round(pretaxCost * taxRate, 2)
7      return pretaxCost, salesTax # return 2 results to the caller
8
9  # inputs
10 count = int(input('Enter # of books: '))
11 cost = float(input('Enter cost per book ($): '))
12
13 # call the function, receiving two results on the left-hand side
14 pretax, tax = calcTotal(count, cost, .075)
15
16 # output
17 print('Pre-tax: $', pretax)
18 print('Tax:     $', tax)
19 print('Total:   $', pretax+tax)
```

Program 5-6 Example Run

```
Enter # of books: 10
Enter cost per book ($): 17.50
Pre-tax: $ 175.0
Tax:     $ 13.12
Total:   $ 188.12
```

The calling statement of the function can use comma-separated variables to receive the multiple results of the function, as shown in line 14 above. Technically Python is returning a ***tuple*** (an unchangeable list), which we are unpacking into individual variables. Tuples will be covered in the next chapter.

Section Summary

Functions can have named parameters that are optional for the calling statement to provide. Named parameters are used to set default values for a function's input variables, but also provide a way to allow for an override of the default. Another function feature is the ability to provide multiple returned results. In this case, the caller can use corresponding multiple variables to receive the results.

5.4 Functions and the Console: Input and Output

In this section, we look at functions designed to interact with the console (keyboard and screen). This is different from "quiet" functions that get their inputs exclusively from their input parameters (between parentheses in the signature) and provide their results via a return statement.

Functions for displaying output

For programs that need to display results in a particular format, it may be useful to write one or more functions that focus on output. In Program 5-7, there are now two functions: `calcTotal()` still handles the calculation, while the new function `displaySales()` is introduced to display the results. Note that the `return` statement on line 14 doesn't indicate a value to return. In this case, Python returns the special value `None`. We could also have omitted this line, since functions without a `return` statement will implicitly return `None` when they complete successfully. The `displaySales()` function is called on line 24 with `pretax` and `tax` as input arguments. Figure 5-4 shows the IPO diagram for this program.

Program 5-7

```
# 2 functions (including a display function)

# calculate the pre-tax cost and tax
def calcTotal(numberItems, unitCost, taxRate):
    pretaxCost = round(numberItems * unitCost, 2)
    salesTax = round(pretaxCost * taxRate, 2)
    return pretaxCost, salesTax

# display the results
def displaySales(pretax, tax):
    print('Pre-tax: $', pretax)
    print('Tax:     $', tax)
    print('Total:   $',  pretax+tax)
    return

# inputs
count = int(input('Enter # of books: '))
cost = float(input('Enter cost per book ($): '))

# processing
pretax, tax = calcTotal(count, cost, .075)

# output
displaySales(pretax, tax)
```

Program 5-7 Example Run

```
Enter # of books: 10
Enter cost per book ($): 17.50
Pre-tax: $ 175.0
Tax:     $ 13.12
Total:   $ 188.12
```

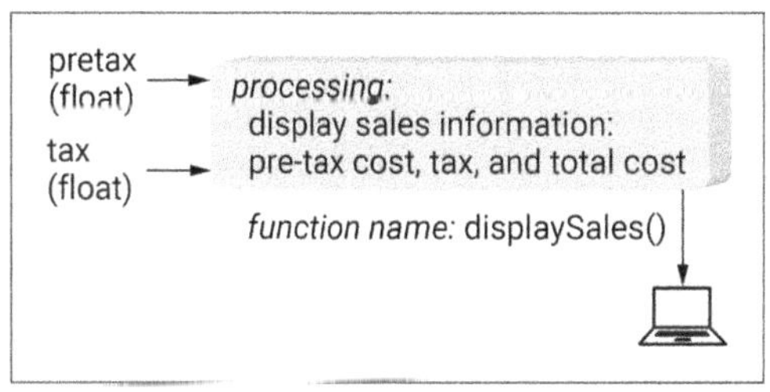

Figure 5-4: IPO diagram for `displaySales()` function

Functions for getting input

In some cases, a function may focus on input. For our bookstore example, the inputs are simple, but in more involved cases, the input may come from a file, database, or the internet. To show an input-oriented function, we introduce `getSalesInfo()` in Program 5-8. Note that `getSalesInfo()` doesn't have any variable names within the signature's parentheses—in other words, there are no parameters in this function's signature. This is because the information for this function will be collected using the `input()` function. Since two data items (number of books, cost per book) are collected, this function returns two results as opposed to one.

Program 5-8

```
# 3 functions (including an input function w/ no parameters)

# prompt user for inputs
def getSalesInfo():
    count = int(input('Enter # of books: '))
    costPerItem = float(input('Enter cost per book ($): '))
    return count, costPerItem

# calculate the pre-tax cost and tax
def calcTotal(numberItems, unitCost, taxRate):
    pretaxCost = round(numberItems * unitCost, 2)
    salesTax = round(pretaxCost * taxRate, 2)
    return pretaxCost, salesTax

# display the results
def displaySales(pretax, tax):
   print('Pre-tax: $', pretax)
   print('Tax:     $', tax)
   print('Total:   $',  pretax+tax)

# input, processing, output
count, costPerBook = getSalesInfo()
pretax, tax = calcTotal(count, costPerBook, .075)
displaySales(pretax, tax)
```

Program 5-8 Example Run

```
Enter # of books: 10
Enter cost per book ($): 17.50
Pre-tax: $ 175.0
Tax:     $ 13.12
Total:   $ 188.12
```

Figure 5-5 shows the IPO diagram for Program 5-8.

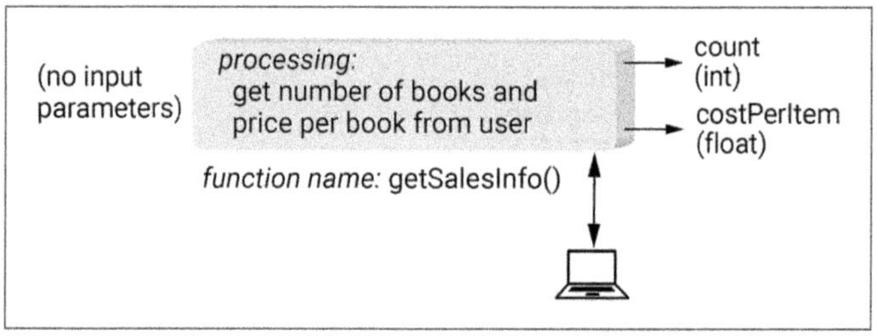

Figure 5-5: IPO diagram for `getSalesInfo()` function

Section Summary

As you develop more complex programs, consider separating your code into functions that handle different tasks. One approach is to write separate functions for handling input, processing, and output.

As functions do not require input parameters, a function that collects user input using `input()` doesn't need to have any input parameters in its function signature. In other words, functions may get their inputs from the keyboard using `input()` and/or get input through variables in the function signature.

As well, functions that focus on displaying results to the screen don't require a `return` statement. Functions may display results to the screen and/or return results using the `return` statement—these are two different ways a function can provide its results.

Test Your Understanding

1) A function that includes calls to `input()` in its body can also have input parameters in its signature.
   a) True
   b) False

2) A function that displays results to the screen cannot have a return statement.
   a) True
   b) False

5.5 A Function Calling Another Function

A function can call another function. We have seen this in functions you write that call built-in functions. For example, in the previous example, Program 5-8's `displaySales()` function calls the built-in `print()` function. Similarly, functions you define can call other functions you define.

Program 5-9 introduces a user-defined function, `main()`, that calls three other user-defined functions.

Program 5-9

```
# 3 functions plus main() (function calling other functions)

# prompt user for inputs
def getSalesInfo():
    count = int(input('Enter # of books: '))
    cost = float(input('Enter cost per book ($): '))
    return count, cost

# calculate the pre-tax cost and tax
def calcTotal(numberItems, unitCost, taxRate):
    pretaxCost = round(numberItems * unitCost, 2)
    salesTax = round(pretaxCost * taxRate, 2)
    return pretaxCost, salesTax

def displaySales(pretax, tax):
   print('Pre-tax: $', pretax)
   print('Tax:     $', tax)
   print('Total:   $', pretax+tax)

# call each function for input, processing, and output
def main():
   count, cost = getSalesInfo()
   pretax, tax = calcTotal(count, cost, .075)
   displaySales(pretax, tax)

main()
```

Program 5-9 Example Run

```
Enter # of books: 10
Enter cost per book ($): 17.50
Pre-tax: $ 175.0
Tax:     $ 13.12
Total:   $ 188.12
```

In some programming languages, a main function is required to indicate where the program should start. Python does not have this requirement, but a `main()` function can be written to group the initial and/or highest level statements together. This can clarify the main steps and overall flow of the program. In this arrangement, `main()` can be the first executed statement, simply as:

```
main()
```

Lastly note there is a sequence of function calls in Program 5-9. The `main()` function calls `getSalesInfo()`, which in turn calls `input()`. We can think of this as a chain: `main()` → `getSalesInfo()` → `input()`. For function call chains, Python keeps track of the full sequence of calls, so that as each function completes, the chain is unwound a step at a time. In general, we don't need to be concerned with this since Python is keeping track of the call sequence details as needed.

Section Summary

A function's body can call one or more other functions. The called functions could be built-in functions or functions you've written. A complex program may have a chain of functions calling functions. As the program executes, these chains are followed and unwound according to the logic of the program.

5.6 Some Further Design Points for Functions

Function design

The choice of what functions you decide to write is a design question. Ideally each function does a well-defined task and has a well-defined set of inputs and outputs. A function with only one statement is valid and can be useful. Conversely, a function that is more lines than a screenful can be harder to understand and might be considered "too large." If you decide a function is too large, you can rewrite it to break it up into two or more functions. As with an essay, the size of a paragraph has no hard rule, but good paragraphs can convey an idea or related set of ideas.

Functions and local vs. global variables

In this chapter, all the examples have functions that are self-contained in terms of variables. In other words, the statements in the function body use the function's input parameters and/or variables the function defines. This is a good practice, preferable to a function referring to a ***global variable***. A global variable is one that is not defined within any function and available to all functions in the .py file. In general, you should avoid the use of global variables by functions so that functions are self-contained and more generally useful.

5.7 Combining Concepts: Functions and Decisions

For most of this chapter, we've kept our functions simple. In practice, functions may have more involved logic. For example, in Chapter 3 we covered decisions and showed how functions can have `if`, `else`, and `elif` statements to handle alternatives. For repetition, as covered in Chapter 4, functions can have `while` and `for` statements.

In this last section of the chapter, we look at a program that uses a function with the `if` statement. For example, if our bookstore wants to offer a discount for purchases of 10 or more books, we can include an `if` statement within a function to cover this logic. Program 5-10 is a variation of Program 5-6 that includes discounting logic.

Program 5-10

```
1  # use if within a function
2
3  # define the function
4  def calcTotal(numberItems, unitCost, taxRate):
5      if numberItems < 10:
6          discountFactor = 1.0
7      else: # discount of 10% off for 10 or more items
8          discountFactor = .90
9      pretaxCost = round(numberItems * unitCost * discountFactor, 2)
10     salesTax = round(pretaxCost * taxRate, 2)
11     return pretaxCost, salesTax # return 2 results to the caller
12
13 # inputs
14 count = int(input('Enter # of books: '))
15 cost = float(input('Enter cost per book ($): '))
16
17 # call the function, receiving two results on the left hand side
18 pretax, tax = calcTotal(count, cost, .075)
19
20 # output
21 print('Pre-tax: $', pretax)
22 print('Tax:     $', tax)
23 print('Total:   $', pretax+tax)
```

Program 5-10 Example Run

```
Enter # of books: 10
Enter cost per book ($): 17.50
Pre-tax: $ 157.5
Tax:     $ 11.81
Total:   $ 169.31
```

The `calcTotal()` function in Program 5-10 is refined to calculate using the discount for larger purchases. Lines 5–8 set a discount factor depending on the number of books being purchased. In this way, the `calcTotal()` function is handling more involved logic. We could further enhance this function with more input parameters for the discount rate and/or the order size required in order to receive the discount.

Section Summary

A function's body can contain more involved logic. Most types of Python statements that can be written outside of a function can be written in a function's body. This includes decision statements and repetition statements that we've covered previously as well as other types of statements we'll see in upcoming chapters.

Chapter Review

Functions are key building blocks in programs. In Chapter 2, we used several built-in functions, such as `print()` and `input()`. In Python, you can also write your own functions, and this is a key way to organize programs. A function can have none, one, or several input parameters, which along with the function name, are declared in the signature of the function. The body of the function are statements that perform the processing for the function. A return statement provides the function's result to the calling statement. A function may not have a `return` statement—for example, if the function provides its results to the screen; in this case the function's return value is the special Python value `None`.

For functions that have multiple input parameters, the calling code of the function must correctly order its arguments to match the function parameters' order. The input argument types must be valid for the function being called.

Functions can have named parameters that are optional for the caller to specify. These are useful if there is a default value for a function's input variable but you'd like to provide a way to override the default. Another function feature is the ability to provide multiple returned results. In this case, the caller can use corresponding multiple variables to receive the results.

Functions may have input parameters specified in the function signature; these are different from inputs acquired from the keyboard. Similarly, functions return results with the return statement; this is different from a function displaying results to the screen.

A function can call another function, which can be a built-in function or a function you've written. A complex program may have a chain of functions calling functions. As the program executes, these chains are followed and unwound according to the logic of the program.

Key Terms

argument	A value, variable, or expression that is provided to a function as part of the caller's use of the function.
calling statement	A statement that uses (calls) a function.
def	The `def` keyword is used to introduce a function signature.
function	A signature and body that defines a block of code, potentially with one or more inputs and a returned result.
function body	The set of statements that comprise the processing performed by a function.
global variable	A variable that is not defined in a particular function but can be used by functions.
local variable	A variable that is defined and used within a function.
parameter	An input to a function as a variable defined in the function signature.
return	The keyword that ends the function's execution and returns a result to the calling statement.
signature	The opening line of a function that defines the name of the function as well as the name of the parameters (variables) that can be used in the function's body.

Programming Problems

1) Tuition

A college charges tuition according to the following:

12 or more credits: $20,000 for semester
1–11 credits: $1,200 + $1,700 per credit

Write a program that prompts the user for a number of credits and determines and displays the tuition for the semester. As part of your program, define and use a function with the signature:

```
def calcTuition(credits)
```

which returns the tuition for a given number of credits.

2) Parking Garage

Create a program to compute the fee for parking in a garage for a number of hours. The program should:

- Prompt the user for how many hours parked
- Calculate the fee based on the following rules:
  $2.50/hour
  minimum fee is $6.00
  maximum fee is $20.00
- Print the result

For your program, define and use the following two functions:

```
# use input() to prompt for the # of hours parked, and return this value
def getParkingHours():
# takes as input hours (a number), and returns the parking fee
def calcParkingFee(hours):
```

3) Hotel Charges

Write a program to solve Chapter 3's problem 7. For your program define and use two functions, with the following signatures:

```
calcCharges(nights, isGoldMember, regularRate, goldRate)
```

inputs:

nights: number of night for the stay
isGoldMember: the value `True` if the guest is a gold member, otherwise `False`
regularRate: the cost per night for a regular guest
goldRate: the cost per night for a gold member guest

outputs:

the total charges for the stay
a Boolean value (True or False) indicating if there was a free night for the guest

```
printCheckout(charge, isFreeNight, isGoldMember)
```

inputs:

charge: the total charge for the stay
isFreeNight: True if there was a free night for the guest's stay
isGoldMember: True if the guest is a gold member

outputs:

this function displays messages to the screen (see sample sessions below) as opposed to returning any results to the calling statement

Example session 1:

```
# of nights: 5
Gold member (y/n)? y
Charges = $ 360
Note: charges discounted for one free night
```

Example session 2:

```
# of nights: 2
Gold member (y/n)? n
Charges = $ 200
Consider becoming a Gold Member!
```

4) Economic Order Quantity

Program 2-9 calculates the economic order size, adjusted for a minimum order size. Revise that program by defining and using a function with the following signature:

```
def eoq(demand, reorderCost, holdingCost, minOrder=0)
```

which returns the economic order quantity (or the minimum order size, if it is higher)

5) Investments—Return and Risk

Chapter 2's problem 6 calculates expected return and risk for an investment given two scenarios. Write a program that solves this problem, including a function with the signature:

```
def calcReturnAndRisk(startValue, highValue, lowValue)
```

which returns a pair of values: the investment's expected return and its risk.

6) ATM Waiting Time

Refer to Chapter 2's problem 8, related to calculating the average waiting time for customers, assuming an average arrival rate and service rate. Write a program that prompts the user for the arrival rate and service rate, and calculates average wait time (in minutes). As part of your program, define and use functions with the following signatures:

function	description	Input arguments	Returned result
`getInputs()`	Use `input()` to prompt user for average arrival rate and service rates		Return arrival rate and service rate as a pair of results
`calcWaitTime(lamda, mu)`	Use arrival rate and service rate to calculate the average customer wait time	lamda: arrival rate mu: service rate	Average customer wait time in hours
`displayResult(expWaitTime)`	Convert waiting time from hours to minutes, display result to screen	expWaitTime: average customer wait time in hours	
`main()`	Calls each function above		

Example session:

```
Enter arrival rate (customers/hour): 30
Enter service rate (customer/hour): 45

Average wait time = 2.7 minutes
```

7) Dollar Store

A store sells all items for $1. Customers buying 10 or more items receive a 5% discount. Additionally, there is 7.5% sales tax. Write a program that prompts the user for the number of items purchased, and calculates the total cost before and after tax. Here is an example session:

```
Enter number of items: 10
Net Cost:     9.50
Tax:          0.71
After tax:   10.21
```

As part of your program, define and use these functions:

function	description	Input arguments	Returned result
`calcCost(count)`	Calculate the pre-tax cost	Count: an integer for the number of items purchased	Pretax total cost of items purchased, accounting for discount if any
`displayLine(label, amount)`	Display one line of the output, formatted to a given width and number of decimal places	Label: the label for the displayed line Amount: the amount for the displayed line	None (function displays to screen)
`display(cost, tax)`	Display net cost, tax, and after-tax amounts using calls to `displayLine()` function	Cost: the pre-tax cost, calculated by `calcCost()` Tax: the tax	None (function displays to screen)

Chapter 6

Lists

Chapter Objectives

In this chapter, you'll learn:

- how you can use lists to represent an ordered set of items
- how to create lists with a predetermined set of items, or by adding items one at a time
- how your program can visit each item in a list, and how your code can refer to an individual item in a list
- how to use slicing, to get a particular part of a list
- how to find a list's count of items, as well as a list's minimum and maximum values
- variations for maintaining lists
- how to order a list
- how to represent a table as a list of lists
- about tuples, which are similar to lists but whose items are unchangeable

Chapter Outline

6.1 About Lists

In the previous chapters, each value and variable represented a single data item—for example, a single number or a single string. In most programs, there is a need to keep track of related items together. In this chapter, we'll look at Python lists, which is a way your program can keep track of an ordered set of items.

Lists have many features, and many will be covered in this chapter. This includes how to create lists and access individual items in a list. Lists allow you to change them, including adding, changing, and removing items. You can ask certain questions of a list—for example, how many items a list has or whether a certain value is contained in a list. Another powerful feature of Python lists is the ability to get subsets using ***slicing***.

A list is one of several composite data structures that Python provides. As you write more involved programs, lists will be a key ingredient for solving problems.

Spreadsheet Compare/Contrast
Lists vs. Cell Ranges

A list in Python is similar to a range of cells in a spreadsheet. For example, in a spreadsheet, you may sum the 10 values in A1:A10. This set of cells and their values is similar to a Python list. However, as with single valued variables, in Python, you provide a name for this list.

6.2 Creating Lists

There are two primary ways to create a list. You can define the items in the list directly in code. We'll cover this approach first. Another way is to define an empty list and then add items to it—we'll look at that second.

Lists as a fixed set of items in code

Square brackets are used in several ways to work with lists. For example, to define a list of the primary colors of light, you can use the following statement:

```
colors = ['red', 'green', 'blue']
```

This defines the variable `colors`, which points to the ordered set of values `'red'`, `'green'`, and `'blue'`.

Tip
Lists Interactively

Many of the list features can be performed in a single statement. This lends itself to trying them interactively in the Python shell. In many Python tools, the interactive prompt is indicated by >>>. You can enter expressions and see the result, or assign the result to a variable, then use the variable subsequently. For example, you can enter the above statement and then enter the variable `colors` to see the list with its values.

```
>>> colors = ['red', 'green', 'blue']
>>> colors
['red', 'green', 'blue']
```

Python list items can be strings, numbers, or in fact, any Python type. For example, the following statement defines a list of numbers.

squares = [1, 4, 9, 16]

Tip
List Element Types

Python is very flexible with types. For example, a list could contain different types of items:

```
mixedList = ['hello', 123, 45.6, True]
```

The above list is a heterogeneous list. For this chapter, we'll use homogenous lists that contain a common type—this is perhaps the most common use of lists.

In order to refer to one of the items of a list, you also use square brackets. Specifically, the combination of the name of the list, square brackets, and an integer value, variable, or expression inside the square brackets refers to a particular item of the list. Further, the first item in the list is identified as item 0.

```
>>> colors = ['red', 'green', 'blue']
>>> colors[0]
'red'
```

The numbered location of a list item is called its ***index***. In the above example, colors[0] refers to the item in the `colors` list at index 0. The first index of all Python lists is 0. Figure 6-1 shows a diagram view of the `colors` list. Python also allows for negative indexes, which counts from the end of the list. For example, `colors[-1]` refers to `'blue'`.

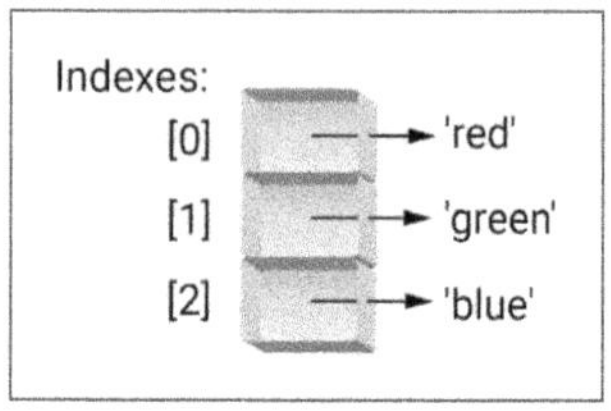

Figure 6-1: List of colors

Lists are often used with the `for` statement. This combination allows your program to visit each item in the list one at a time. What the program does with each item depends on the problem being solved.

Our first program in this chapter uses lists and `for` in order to solve the following problem related to air quality. According to the EPA, there are six ranges of the Air Quality Index (AQI) related to six warning levels. The lowest level (AQI <= 50) corresponds to the "good" level and is labeled green. Table 6-1 shows the different AQI ranges and their corresponding colors and levels of concern. We would like to write a program that, for a given input AQI, can look up and tell us the corresponding color and concern.

AQI range	Color	Concern
0–50	green	Good
51–100	yellow	Moderate
101–150	orange	Unhealthy for Sensitive Groups
151–200	red	Unhealthy
201–300	purple	Very Unhealthy
Above 300	maroon	Hazardous

Table 6-1: Air Quality Index (AQI) ranges and corresponding colors and concerns

Program 6-1 Video animation available.

```
1 # Translate Air Quality Index (AQI) from number to color and concern
2 # https://www.airnow.gov/aqi/aqi-basics/
3
4 # lists of levels and corresponding colors and concerns
5 levels = [301, 201, 151, 101, 51, 0]
6 colors = ['Maroon', 'Purple', 'Red', 'Orange', 'Yellow', 'Green']
7 concerns = ['Hazardous', 'Very Unhealthy', 'Unhealthy',
8             'Unhealthy for Sensitive Groups', 'Moderate', 'Good']
9
10 # input
11 level = int(input('Enter an AQI level (integer): '))
12
13 # processing
14 for i in range(len(levels)):
15     if level >= levels[i]:
16         color = colors[i]
17         concern = concerns[i]
18         break
19
20 # output
21 print('AQI color:', color)
22 print('AQI concern level:', concern)
```

Program 6-1 Example Run

```
Enter an AQI level (integer): 75
AQI color: Yellow
AQI concern level: Moderate
```

In Program 6-1, three parallel lists are created in lines 5, 6, and 7. The user is prompted for an AQI level on line 11. The use of the lists on lines 14–18 finds the color and concern for the user-entered AQI level. Line 14 uses the `len()` function, which returns the number of items in the list variable levels. Note the syntax on lines 15–17, which access particular items of the lists using brackets and the integer variable `i`—for example, `levels[i]`.

Tip

Plural Variable Names for Lists

As a list usually represents a set of items, using a plural variable name can be a good convention for list variable names. For example, in Program 6-1, the variable names for lists are `levels`, `colors`, and `concerns`. Other variables that store singular values are named in the singular: `color` and `concern`.

Creating an empty list and adding items with `append()`

In the prior section we saw how to create a list and initialize its items in one statement. If you have a fixed set of items, this is a suitable approach. However, often the items in a list aren't known ahead of time. For example, your program may prompt a user for values, or read values from a file or database. In these cases, the program can define an empty list, and add values to the list as the program executes.

In order to define an empty list, the syntax is:

```
listName = []
```

where listName is a Python variable name you choose. After this statement executes, the variable points to an empty list. An empty list has no items in it yet, but is ready for items to be added to it.

In order to add an item to the end of a list, the `append()` method can be used. As described in Section 2.7, methods are like functions but are used following a variable (or constant or expression) and period. The syntax for `append()` is:

```
listName.append(item)
```

where item can be a constant, variable name, or expression. In this chapter, we will use a few methods that are compatible with lists.

Tip

Recap: Methods vs. Functions

We encountered functions early on, in Chapter 2, such as `print()` and `float()`. Methods are similar to functions in that they have a name and may have input arguments and a result. However, methods, unlike functions, are connected to a variable (or constant or expression) with a period. Each type in Python has a set of methods that are suitable for that type. We'll see more examples of methods related to strings in Chapter 8, and other types in other chapters.

Now that we have the syntax for creating an empty list and adding items to it, let's look at an example. For this and some subsequent examples, we'll introduce Dave Sigma, a quality control manager at a food company. Dave is responsible for making sure the processing machines in the company are filling boxes with the right amount of product. Scales in the company's plant provide data as to the weight of each package. The longer-term goal is to have a system that automatically collects weight data to ensure each package has the right amount of product.

The system will be built in stages. The initial goal is straightforward—to allow a user to enter a series of weights, and determine their average. Program 6-2 shows an example program that solves this problem with lists.

Program 6-2

```
# show how to create an empty list and add elements to it

# prompt for a series of values and build and return a list
def getValues():
    result = [] # define an empty list
    entry = input('Enter a value, or <return> when done: ')
    while entry != '':
        result.append(float(entry)) # add a value to end of list
        entry = input('Enter a value, or <return> when done: ')
    return result

# calculate the mean of a list
def calcMean(alist):
   n = len(alist) # determine the number of elements in the list
   if n == 0: return None # handle an empty list
   sum = 0 # sum will cumulate the list's values
   for i in range(n):
       sum += alist[i]
   return sum / n # calculate and return the average

measures = getValues()
mean = calcMean(measures)
print('Mean=', mean)
```

Program 6-2 Example Run

```
Enter a value, or <return> when done: 9.8
Enter a value, or <return> when done: 10.1
Enter a value, or <return> when done: 10.7
Enter a value, or <return> when done:
Mean= 10.2
```

In Program 6-2, the key list statements are on lines 5 and 8. Line 5 creates an empty list named result. In line 8, values are added one at a time to the end of this list, as the user enters numbers.

Program 6-2 also shows how lists can work with functions. Line 10 returns a list as the result of the `getValues()` function. Also line 22 shows that lists can be passed as an argument to a function. In this case, we pass the list of values to the `calcMean()` method, in order for it to access each item in the list and ultimately calculate the average of the lists' values.

Section Summary

Lists are an ordered set of items. Variable names point to lists, and individual list items can be accessed using square brackets—for example, `a[i]`. The integer location of an item in a list is referred to as the index.

Items can be added to a list all at once, or added one at a time using the `append()` method. The `for` statement with the `range()` method is one way to visit all items in a list.

Test Your Understanding

1) The items in a list have a particular order.
   a) True
   b) False
2) The index location of an item must be an integer value.
   a) True
   b) False
3) Items can be placed in a list:
   a) when the list is created
   b) one at a time, after an empty list is created
   c) (a) and (b)

6.3 Iterating over a List to Get a List's Items

In earlier programs, we used `for` and the `range()` function to iterate over each of the items in a list. Program 6-3 does this as well in order to calculate the mean of a list of numbers.

Program 6-3

```
# demonstrate use of for and range(), to iterate over the list's elements

# calculate the mean of a list of numbers
def calcMean(alist):
    n = len(alist)
    if n == 0: return None # handle an empty list
    sum = 0
    for i in range(n):
        sum += alist[i] # access item i from the list
    return sum / n

measures = [10, 10.1, 9.9, 9.6, 10.1, 9.9, 9.8, 9.9, 10.3, 10]
mean = calcMean(measures)
print('Mean=', mean)
```

Program 6-3 Output

```
Mean= 9.96
```

Iterating over all items in a list is very common, so Python provides a simplified way to do it. The form is:

```
for <variable> in <list variable>:
```

Program 6-4 is a variation of Program 6-3 using the simplified iteration described above.

Program 6-4

```
1  # demonstrate the use of for, to iterate over a list's elements
2
3  # calculate the mean of a list of numbers
4  def calcMean(alist):
5      sum = 0
6      for value in alist: # iterate over the elements of the list
7          sum += value
8      return sum / len(alist)
9
10 measures = [10, 10.1, 9.9, 9.6, 10.1, 9.9, 9.8, 9.9, 10.3, 10]
11 mean = calcMean(measures)
12 print('Mean=', mean)
```

Program 6-4 Output

```
Mean= 9.96
```

In Program 6-4, line 6 is the new syntax. The `for` statement uses a variable named `value`, which is set in turn to each value in the list for each iteration of the loop. (Here `value` is the name chosen for the variable, as opposed to a keyword, but another variable name could be used—for example, `number`.) Note that the `value` variable is used in line 7 in order to create the sum of the list. This line also uses the += operator, as a shorthand to the equivalent statement:

sum = sum + value

6.4 Getting Part of a List with Slicing

In this section, we look at how to generate a list with some items from a source list. Python provides a very concise and powerful way to do this called ***slicing***.

Slicing involves using a list with square brackets [ ] as well as one or two colons. In one variation, you specify the index location of the initial item of interest, a colon, and a second location, as in:

```
alist[start:end]
```

For example:

```
>>> alist = [1, 4, 9, 16, 25]
>>> alist[0:3]
[1, 4, 9]
```

Note that slicing returns a list starting at the first item, and going up to *but not including* the index of the last item. In the above statements, `alist[0:3]` returns a list with the items at index locations 0, 1, and 2.

Similar to the `range()` function, slicing allows for an optional third argument, indicating how to skip items. Continuing the statements above:

```
>>> alist[0:4:2]
[1, 9]
```

You can omit the start argument for the slice, in which case the slice starts at item 0.

```
>>> alist[:2]
[1, 4]
```

Similarly, you can omit the end argument for the slice, in which case the slice's end is the last item in the list.

```
>>> alist[3:]
[16, 25]
```

To further demonstrate list slicing, let's return to the quality control example. Dave Sigma has a new request—he would like to see how the average weight of products evolves over time. This calculation is called a "growing average." For example, as the first weight is known, the average is just that value. Once the second weight is known, the next result is the average of the first and second weights.

Program 6-5 calculates and displays the list of growing averages for a list of weights.

Program 6-5

```
 1 # demonstrate slicing a list
 2
 3 # calculate the mean of a list of numbers
 4 def calcMean(alist):
 5     n = len(alist)
 6     sum = 0
 7     for i in range(n):
 8         sum += alist[i]
 9     return sum / n
10
11 # calculate a list of averages for a growing window of a list
12 def calcGrowingAverage(alist, r=2):
13     result = []
14     for i in range(len(alist)):
15         sublist = alist[0:i+1] # take a slice of elements from 0 to i
16         print(sublist)
17         mean = calcMean(sublist) # calculate the mean of the slice
18         result.append(round(mean, r)) # add the slice mean to the result
19     return result
20
21 measures = [10, 10.1, 9.9, 9.6, 10.1, 9.9, 9.8, 9.9, 10.3, 10]
22 growingAverage = calcGrowingAverage(measures)
23 print()
24 print('Growing average:', growingAverage)
```

Program 6-5 Output

```
[10]
[10, 10.1]
[10, 10.1, 9.9]
[10, 10.1, 9.9, 9.6]
[10, 10.1, 9.9, 9.6, 10.1]
[10, 10.1, 9.9, 9.6, 10.1, 9.9]
[10, 10.1, 9.9, 9.6, 10.1, 9.9, 9.8]
[10, 10.1, 9.9, 9.6, 10.1, 9.9, 9.8, 9.9]
[10, 10.1, 9.9, 9.6, 10.1, 9.9, 9.8, 9.9, 10.3]
[10, 10.1, 9.9, 9.6, 10.1, 9.9, 9.8, 9.9, 10.3, 10]

Growing average: [10.0, 10.05, 10.0, 9.9, 9.94, 9.93, 9.91, 9.91, 9.96, 9.96]
```

In Program 6-5, line 15 takes a slice of the list, which is displayed by line 16, and used to calculate the average on line 17. In this way, through repetition, the growing average is built by the `calcGrowingAverage()` function.

6.5 Asking Questions of a List

Earlier in the chapter, in Program 6-1, we introduced the `len()` function, which can determine the number of items in a list. Other functions useful for lists include `min()`, which returns the smallest value in the list, and `max()`, which returns the largest value. Program 6-6 shows how these two functions can be used to determine the range of a list.

Program 6-6

```
# use min() and max() functions to calculate the range of a list
measures = [10, 10.1, 9.9, 9.6, 10.1, 9.9, 9.8, 9.9, 10.3, 10]
range = max(measures) - min(measures)
print("Range=", round(range, 1))
```

Program 6-6 Output

```
Range= 0.7
```

Note that `min()` and `max()` will give an error if the argument is an empty list.

Python provides the `in` keyword in order to determine if a value is in a list. Program 6-7 shows how this can be used to see if a user entered value is in the list.

Program 6-7

```
# Check if a value is in a list using: in
measures = [10, 10.1, 9.9, 9.6, 10.1, 9.9, 9.8, 9.9, 10.3, 10]
value = float(input('Enter a value to check if in list: '))
print(value in measures)
```

Program 6-7 Example Run

```
Enter a value to check if in list: 10
True
```

Note that `in` can be used together with the `not` keyword to check if a value isn't in a list.

6.6 Maintaining a List

In this section, we will look at a few ways to maintain a list. In particular, a value can be added to a list, removed from a list, or changed in a list.

Changing a list's element

A list item's value can be revised if square brackets are used on the left side of the equals sign. Program 6-8 shows an example of this. In this program, a function determines how many list values are above and below a given value.

Program 6-8

```
# demonstrate list indexing on the left-hand side

# determine counts of items above or below a certain value
def aboveBelow(alist, splitValue):
    result = [0, 0] # this list will track counts below/above splitValue
    for i in range(len(alist)):
       if alist[i] <= splitValue: result[0] += 1
       else: result[1] += 1
    return result

measures = [10, 10.1, 9.9, 9.6, 10.1, 9.9, 9.8, 9.9, 10.3, 10]

result = aboveBelow(measures, 10)

print(result)
```

Program 6-8 Output

```
[7, 3]
```

On line 7 in Program 6-8, the count is increased by one to track how many values are at or below a given value `splitValue`. Note the use of `result[0]` on the left side of the equal sign, which allows the first item in the result array to be updated. Similarly, on line 8, the second item in the list, `result[1]`, is updated to track the count of values above `splitValue`.

Methods for adding elements: `append()`, `insert()`, `extend()`

There are several methods for adding items to a list. The `append()` method will add an item to the end of a list, as we have seen earlier in the chapter. For placing an item at a particular location in a list, the `insert()` method can be used. And to place a list at the end of a list, the `extend()` method can be used. Program 6-9 shows the use of `extend()`. In this program, the individual means of two lists are calculated. Then the combined mean is calculated by using `extend()` to create a combined list from the two individual lists.

Program 6-9

```
1  # Compare means of two lists, and their combinations using extend()
2
3  # calculate the mean of a numeric list, rounded to r decimal places
4  def calcMean(alist, r=2):
5     n = len(alist)
6     if n == 0: return None # handle an empty list
7     sum = 0
8     for i in range(n):
9         sum += alist[i]
10    return round(sum / n, r)
11
12 # 2 lists of measures
13 measures1 = [10, 10.1, 9.9, 9.6, 10.1, 9.9, 9.8, 9.9, 10.3, 10]
14 measures2 = [9.9, 10.0, 9.9, 9.6, 10.1, 9.5, 9.7, 9.8, 10.1, 9.9]
15
16 # calculate means of each list
17 mean1 = calcMean(measures1)
18 mean2 = calcMean(measures2)
19
20 # Use extend to consolidate the list and determine an overall mean
21 measures1.extend(measures2)
22 meanAll = calcMean(measures1)
23
24 print(mean1, mean2, meanAll)
```

Program 6-9 Output

```
9.96 9.85 9.9
```

In the above program, line 21 uses the `extend()` method to combine the two lists in order to get an overall mean of the combined set of numbers.

Removing elements: `del`, `remove()`, `pop()`, `clear()`

There are several ways that an item can be removed from a list. The `del` keyword allows you to delete an item from a list. Program 6-10 shows this by trimming a list of values that lie outside a certain range. Note that this program works backward through the list, since removing items will shift what values are at each index.

Program 6-10

```
# demonstrate the use of del to remove elements from a list

# based on a provided list, create a new list
# that excludes values outside of a certain range
def trimOutliers(alist, lo, up):
    list2 = alist.copy() # make a distinct copy of the list argument
    for i in range(len(alist)-1, -1, -1): # work backwards through the list
       # if the list element value is outside the range, remove it
       if (alist[i] < lo) or (alist[i] > up):
           del list2[i] # del removes this item from the list
    return list2

# an original list
measures = [10, 10.1, 9.9, 9.6, 10.1, 9.9, 9.8, 9.9, 10.3, 10]

# create a new list with values remove that are outside of [9.9, 10.1]
measures2 = trimOutliers(measures, 9.9, 10.1)

# display original and trimmed list
print(measures)
print(measures2)
```

Program 6-10 Output

```
[10, 10.1, 9.9, 9.6, 10.1, 9.9, 9.8, 9.9, 10.3, 10]
[10, 10.1, 9.9, 10.1, 9.9, 9.9, 10]
```

Here are some other ways to take items out of a list:

- The `clear()` method will set the list to be empty.
- The `pop()` method, without any arguments, takes the first item out of the list, and returns it.
- The `remove()` method removes a value from a list (or generates an error if that value is not found).

Copying a list

There is a fine point about lists we will now cover in this section. Note that if you assign a list variable to another list variable, these two variables are actually pointing to the same set of items, as shown in Figure 6-2. As a result, changes to the items through one variable will impact the list as seen through the other variable. You can see this interactively:

```
>>> a = [1, 2, 3]
>>> b = a
>>> a[1] = 10
>>> b
[1, 10, 3]
```

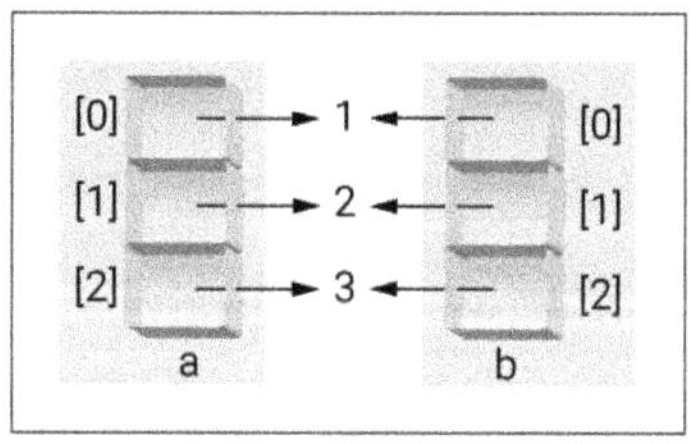

Figure 6-2: Lists sharing items: The result of b = a

If the sharing of items between variables is not what you want, you can instead make a copy of one list and assign the copy to another variable. For example:

```
b = a.copy()
```

If a list is copied using the `copy()` method, then the original and copied list are separate, and changes to the original list's items do not show up as changes in the copied list's items.

6.7 Ordering a List: `sorted()`, `sort()`, `reverse()`

Python provides the `sorted()` function, which can be used to generate a new list that is ordered. Note that when using `sorted()`, the original list is unchanged. Lists also have a `sort()` method to order a list "in place"—in other words, without generating a new list. Another method, `reverse()`, generates a new list in reverse sorted order.

Program 6-11 shows the use of both `sorted()` and `sort()`.

Program 6-11

```
# demonstrate sorting of a list
measures = [10, 10.1, 9.9, 9.6, 10.1, 9.9, 9.8, 9.9, 10.3, 10]

# create a new sorted list, separate from a provide list
measuresSorted = sorted(measures)
print(measuresSorted)
print(measures)

# sort a list in place
measures.sort()
print(measures)
```

Program 6-11 Output

```
[9.6, 9.8, 9.9, 9.9, 9.9, 10, 10, 10.1, 10.1, 10.3]
[10, 10.1, 9.9, 9.6, 10.1, 9.9, 9.8, 9.9, 10.3, 10]
[9.6, 9.8, 9.9, 9.9, 9.9, 10, 10, 10.1, 10.1, 10.3]
```

6.8 Representing a Table in a List

As tables are a very common way to represent data, in this section, we look briefly at one way to represent tables. This can be done by using a list of lists. In this arrangement, the main list points to child lists; each child can be thought of as a row of the table. If the child lists are of equal length, this can represent a rectangular table.

To demonstrate, we can interactively create a table of first and last names using:

```
matrix = [['John', 'Jones'], ['Jane', 'Doe'], ['Jim', 'Johnson']]
```

Here the matrix is a list of lists; the child lists are two-item lists each containing a first and last name. We can see it visually as a matrix using this code:

```
for row in matrix:
    for cell in row:
        print(cell, end='|')
    print()
```

which will display the result:

```
John|Jones
Jane|Doe
Jim|Johnson
```

The child lists can also be of unequal length. This is demonstrated in Program 6-12, which creates two child lists of the parent list `result`.

Program 6-12

```
# demonstrate a list of lists

# determine which indexes have values below/above a certain value
def aboveBelow(alist, splitValue):
    result = [[], []] # initialize a list of lists
    for i in range(len(alist)):
       if alist[i] <= splitValue: result[0].append(i)
       else: result[1].append(i)
    return result

measures = [10, 10.1, 9.9, 9.6, 10.1, 9.9, 9.8, 9.9, 10.3, 10]

result = aboveBelow(measures, 10)

print('below/at:', result[0])
print('above:    ', result[1])
```

Program 6-12 Output

```
below/at: [0, 2, 3, 5, 6, 7, 9]
above:    [1, 4, 8]
```

Tip

Looking Ahead—Matrices and the pandas Package

Although lists of lists allow you to represent a table in a Python program, there is another approach that we'll look at in Chapter 10. This involves using the pandas package, a very powerful library for representing tables using its DataFrame structure.

Spreadsheet Compare/Contrast

Cell Ranges and List of Lists

In spreadsheets, each sheet (tab) is a table. You may also arrange data in a table—for example, with a header row and subsequent rows with the data. Although tables are a natural arrangement in spreadsheets, you may initially feel tables are a little less natural in Python. On the other hand, Python can readily represent cubes of data (a list of lists of lists) or higher order dimensions. In future chapters, we'll also see other data structures, such as dictionaries and pandas DataFrames, which are highly flexible and powerful data structures. That you can choose the right data structures when solving a problem is a strength of Python relative to spreadsheets.

6.9 About Tuples—Unchangeable Lists

We close out this chapter with this brief section about tuples. Tuples are like lists except their items can't be changed once the tuple is created. This can help make a program more efficient in case you have an unchangeable ordered set of items. Tuples can be created using parentheses (as opposed to square brackets for lists). For example, the following statement:

```
levels = (301, 201, 151, 101, 51, 0)
```

could be used in Program 6-1 for line 5 in order to create the unchangeable tuple. Similarly, the following statement could be used instead of line 5:

```
colors = ('Maroon', 'Purple', 'Red', 'Orange', 'Yellow', 'Green')
```

You can verify the unchangeable nature of tuples. If you interactively enter the above line, then try to change an item using:

```
colors[0] = 'Brown'
```

you will get:

```
TypeError: 'tuple' object does not support item assignment
```

Chapter Review

In this chapter, we introduced lists, an ordered collection of items. A list can be created by indicating the items all at once, or a list can start as empty then items can be added. The `for` statement is one way to iterate over the items in a list. Slicing is a concise and powerful way to get a subset of a list. Some questions can be asked of a list—for example, its length, minimum, maximum, as well as whether a particular value is (or isn't) in the list. Several approaches are available to maintain a list, including changing and deleting items. Lists can be ordered using the `sorted()` function or the `sort()` method. Tables can be represented as a list of lists. Tuples are similar to lists, though the individual items in a tuple can't be changed.

Key Terms

element	A particular item within a list, accessible through a variable name as well as the index location of the element.
index	The integer numeric location of an item in a list.
item	Lists are composed of zero, one, or more items.
iterate	Logic to visit each item—for example, in a list.
list	An ordered collection of items, where each item can be accessed via an integer index.
method	A method is similar to a function, but is specified as a trailing name after a period and after a variable (or constant or expression).
slice	A subset of list, created using square brackets and a colon.

Programming Problems

1) Min and Max

Write a function, `minmax()`, that takes a list argument, and determines and returns the minimum and maximum values in the list. For this function, do not use the built-in Python functions `min()` and `max()`. The signature of the function should be:

```
def minmax(alist)
```

Note that you can return a pair of values from a function with the syntax:

```
return a, b
```

Test your function with the following statements (after your function):

```
# test the function on a list of numbers
lo, hi = minmax([1, 2, 3])
print(lo, hi)

# test the function on a list of words
lo, hi = minmax(["what", "are", "the", "min", "and", "max"])
print(lo, hi)

# test the function on an empty list
print(minmax([]))
```

2) Net Present Value

The net present value of a set of cashflows is the total discounted value of all flows, where each flow is discounted by the interest rate compounded by time.

Assuming each cashflow occurs in year *t* (*Ft*), and a single discount rate in decimal (*r*), the present value *P* is given by:

$P = \Sigma_t (F_t / (1 + r)^t)$

For example, the net present value the cashflows [20, 50, 90] discounted at a rate of 5% is:

$P = 20 / (1.05) + 50 / (1.05)^2 + 90 / (1.05)^3 = 142.14$ (rounded to two decimal places)

Write a function `NetPresentValue(flowlist, rate)` that takes a list of values, and an interest rate (in decimal), and returns the net present value. The list of values is assumed to be annual flows, so for example, [10, 20] represents a flow of 10 at end of year 1, and 20 at end of year 2. Zero is a valid value, so for example, [0, 20] represents no flow at the end of year 1, and a flow of 20 at the end of year 2.

Test your function with the following statements (after your function):

```
print(NetPresentValue([0, 100], .05)) # 2-year 0-coupon bond
print(NetPresentValue([3, 3, 3, 103], .05)) # 4-year bond paying 3% a year
print(NetPresentValue([], .05)) # no cashflows, should return 0
```

3) Mean and Standard Deviation Excluding Outliers

A quality control manager wants to determine the mean and standard deviation of a list of product weights. The mean and standard deviation should be based on values in the list that are inside a certain range (low and high values).

Write a program that, given a list of numbers, determines the mean and standard deviation of a list of numbers that fall within a certain range.

To solve this problem, write and use two functions:

`filterOutliers(alist, arange)`: Given a list of numeric values in argument alist, return a list of values that fall within the values arange[0] and arange[1]. Note that the result returned from this function should be separate from alist, in other words, after calling this function, the list alist should not be modified.

`calcStats(alist)`: Given a list of numeric values in alist, return the mean and standard deviation as a result pair.

Test your functions with the following statements (these statements should be placed after, not inside, your functions):

```
inliers = filterOutliers([9.9, 10.0, 9.7, 10.1], [9.85, 10.15])
print(inliers)
print(calcStats(inliers))
```

For standard deviation, use:

sd = sqrt(Σ_i (x_i - mean)^2 / n)

where:

x_i: the ith value of the list
mean: the average of the values in the list
n: the number of values in the list

4) Virus Spread

A state is planning for a vaccination program for a virus. It wants to compare the spread of the virus with and without the vaccine program. The virus is expected to spread each week at a rate r_t, such that the expected number of people infected in a week is given by the number of people infected in the prior week times r_t. In other words:

$n_t = n_{t-1} \times r_t$

a) Write a function to calculate the expected number of people infected each week, given a list of weekly assumptions for r_t. As part of your program, write and use a function, `calcInfected()`:

```
def calcInfect(start, weeklyRates)
```

where start is the initial number of people infected, and weeklyRates is a list of the estimated infected rates for each week. The function should return a list of expected people infected each week.

Test your function assuming that r is 3 for the next eight weeks.

b) Write a program that compares two scenarios. In the first scenario, the infection rate is r = 3 for the eight weeks ahead. In the second scenario, the infection rate is 3 for the next four weeks, and after vaccinations, the infection rate is 0.8 for the subsequent four weeks.

Sample output:

```
week     r=3  vaccine
  1        3        3
  2        9        9
  3       27       27
  4       81       81
  5      243       65
  6      729       52
  7    2,187       41
  8    6,561       33
```

5) Quality Control

A quality control manager would like to find the average weight of a list of weights. Prior experience indicates that the first two measures are atypical, and the practice is to exclude these from the average calculation. Write a program that calculates the average of a list of numbers that are input by the user, excluding the first three values in the average. To solve this problem, use Python list slicing. For your program, define and use the following functions:

`getValues(n)`: Prompt the user for n values, and return these in a list.
`calcAverage(alist)`: Calculate the average of a list.

Chapter 7

Reading and Writing Files

Chapter Objectives

In this chapter, you'll learn:

- how you can use Python functions and methods to read a data file
- how repetition with `while` and `for` can be used to read lines of a file
- how to read file lines with multiple fields into your program
- how to write results to a file
- how to handle errors with Python exceptions

Chapter Outline

7.1 About Files

In the prior chapters, our programs have used the keyboard to get input and the screen to display prompts and output. For small programs with a limited amount of data, this is fine, but in practice, you may work with more data. There are a few alternatives for working with larger datasets, and in this chapter, we'll look at how Python can work with files. In particular, Python can read a file of data as well as write results to a file.

There are different ways that files may represent data. This chapter focuses on ***text files***. A text file contains data that can be viewed in a text editor. Text files can be read or written by Python with a few built-in functions and methods beyond those we've already seen. In contrast with text files, ***binary files*** may contain data that do not correspond with specific characters. Although Python can read and write binary files, they are beyond the scope of this book.

The next section shows how your program can read a single line from a file into your program, introducing a function and methods for this. In practice, you will want to read multiple lines from a file, and the following section shows how you can use repetition to do this. Section 7.4 shows how multiple fields in a file can be read and processed.

Writing results to a file can easily be done using the built-in `print()` function, and this is covered in Section 7.5. The last section of this chapter introduces exceptions, which can handle errors such as missing files. Exceptions are covered in more detail in Appendix D.

7.2 Reading a File's Line

The steps you can use in Python for reading a text file are:

1) Associate the file with a variable, using the built-in `open()` function. The variable is used subsequently to work with the file. We'll refer to this as the ***file variable***.
2) Use the file variable with the `readline()` method, to read a line of the file into a string variable.
3) Use the string `rstrip()` method to take off the special ***newline*** character from the string. The newline character marks the end of each line of a text file.
4) When finished reading the file, use the `close()` method to indicate the file reading is complete.

Program 7-1 shows these steps in order to read the first line of the measures.txt file.

Program 7-1 Video animation available.

```
1  # read one line from a file and display it to the screen
2
3  # use the open() function to locate the file and connect it to variable f
4  f = open('measures.txt', 'r')
5
6  # read the first line of the file into the string variable called aline
7  aline = f.readline()
8
9  # remove the trailing newline that Python returns
10 aline = aline.rstrip('\n')
11
12 print(aline)
13
14 # close the file
15 f.close()
```

Program 7-1 Output

```
10.1
```

To run Program 7-1, you need to place the file measures.txt in the same directory as your program (.py or .ipynb) file. If you don't, you will get a FileNotFoundError.

The first statement on line 4 associates the file named measures.txt with the Python variable `f`. (`f` is a commonly used variable name when working in Python with files, but you could use another variable name such as `file` or `fileIn`.) The `open()` function makes this connection, and its second argument set to `'r'` indicates that the file will be read. (In a subsequent section we'll use `'w'` to indicate the file is being written to.) The statement on line 4 does not read data from the file; instead, this is done on line 7 using the `readline()` method. `readline()` returns a string with data read from the file. By convention, `readline()` includes the special newline character `\n` at the end of strings it returns, corresponding to the line end marker in the data file. Line 10 uses the `rstrip()` method to remove that special character from the end. On line 12, we use the familiar `print()` function to display the string as read from the file. Finally, line 15 indicates that we are finished working with the file by calling the `close()` method.

In practice, it is most common to read all lines of a file, and we'll look at this in the next section.

Tip

What open() Returns

The `open()` function returns a Python type of `TextIOWrapper`. This built-in Python type allows Python to work easily with text files. In practice, you don't need to know about this type, but it provides the available methods such as `readline()` and `close()` that are used for working with a text file.

Section Summary

A file's data can be read into Python variables using certain built-in functions and methods in combination. The `open()` function can be set to a variable that references a file. That variable can subsequently be used with the `readline()` method to read a line of data from a file. The `readline()` method returns string data that include the newline character; `rstrip()` can be used to remove this character. The `close()` method can be used to indicate that the program has completed reading the file.

Test Your Understanding

1) In order to read a file in Python, the `open()` function can be used to connect a variable with the file.
   a) True
   b) False
2) The `open()` function will read the contents of a file into a variable.
   a) True
   b) False
3) Which of the following methods can be used to read a line of a file into a string variable?
   a) `getline()`
   b) `readline()`
   c) `open()`
   d) all of the above

7.3 Reading a File's Lines with Repetition

In the previous section, we looked at how to read a line from a file. In this section, we'll use repetition to read all the lines of a file. Either `while` or `for` can be used to read lines from a file.

Using `while` to read a file

The `while` statement can be used in combination with `readline()` to read all lines of a file. The convention that Python uses is for `readline()` to return an empty string when all lines in the file have been read. Program 7-2 shows how to use `while` repetition with this condition to read and display each line of a file.

Program 7-2 Video animation available.

```
# with while, read a file and echo its lines to the screen

# use the open() function to locate the file and connect it to the variable f
f = open('measures.txt', 'r')

# read the first line of the file into the string variable called line
line = f.readline()

# while there are more lines in the file
while line != '':
    # remove the trailing newline that Python returns
    line = line.rstrip('\n')

    print(line)

    # read the next line of the file, if any
    line = f.readline()

# close the file
f.close()
```

Program 7-2 Output

```
10.1
9.9
9.8
```

As with Program 7-1, Program 7-2 uses the `open()` function to associate a variable with the file. As well, on line 7, `readline()` is used to read the first line of the file. Line 10 is the key repetition statement, to check if there was a nonempty string returned from the `readline()` method. As noted earlier, `readline()` will return an empty string after it has read all the lines in the file. Line 17 reads all subsequent lines in the file.

Using `for` to read a file

As we saw in Chapter 4, the `for` statement can provide a more concise way for repetition as compared with `while`. This is also the case for reading files. Given a file variable, you can use the syntax:

```
for <string variable> in <file variable>:
```

In this approach, the `for` statement will populate the `<string variable>` with each line of the file. You can then use indented statements below `for` to process the data.

Program 7-3 uses the `for` statement with a file variable, to read each line of a file and display them on the screen. It is very similar to (and more concise than) Program 7-2.

Program 7-3 Video animation available.

```
# with for, read a file and echo its lines to the screen

# use the open() function to locate the file and connect it to the variable f
f = open('measures.txt', 'r')

# for each line in the file
for line in f:
    # remove the trailing newline that Python returns
    line = line.rstrip('\n')

    print(line)

# close the file
f.close()
```

Program 7-3 Output

```
10.1
9.9
9.8
```

Tip

Reading a Header Line

Text files may include a first line with labels for the file's subsequent data. If your program needs to skip this line, you can use the `readline()` method before the repetition loop. This can be done for either `while` or `for` repetition. For example:

```
f = fopen('measures.txt', 'r')
f.readline() # read the file's header line
for line in f:
# program continues ...
```

In the second statement above, `f.readline()` will read a line from the file without doing anything further with it, in order to skip over the first line of the file.

Program 7-3 is minimal, as it just displays the file's lines to the screen. In practice, you will want to process the data read from the file. Program 7-4 adds some statements in order to calculate the average of all values read from the file.

Program 7-4

```
# with for, read a file of numbers
# and calculate and display their average

total = 0 # the total of all values read
i = 0 # the number of values read

f = open('measures.txt')
for line in f:
    line = line.rstrip('\n').strip()
    if line != '': # ignore blank lines
        value = float(line) # the variable line is of type str, convert to number
        print('value ', i+1, ': ', line, sep='')
        total += value
        i += 1

f.close()
average = total / i
print("average =", round(average, 2))
```

Program 7-4 Output

```
value 1: 10.1
value 2: 9.9
value 3: 9.8
average = 9.93
```

Section Summary

The Python repetition statements can be used to read each line of a file into a string variable. In particular, the `while` statement can be used in combination with the `readline()` method to read a file's data. Alternatively, the `for` statement can be used to more concisely read file data into a string variable.

Test Your Understanding

1) The `readline()` method will return an empty string when there is no more data to read from a file.
   a) True
   b) False

2) The result of the open function can be used as part of a `for` statement to loop over and read each line of a file.
   a) True
   b) False

7.4 Parsing a File's Fields

In the prior sections, our file contained one value on each line. Another common layout is to have multiple values on each line. In this section, we look at this multi-field case.

Assume we have a file in which each line has values related to a product: a product identifier, a demand forecast (in sales per year), an order cost (in $ per order), and a holding cost (in $ per item per year). We would like to write a program that reads each line of the file, and computes and displays the economic order quantity (EOQ) for each product. The formula for EOQ is covered in Section 2.6.

A convenient way to handle multi-field lines is the string `split()` method. This method takes a string value, and based on a separator, returns a list of string values. For example, given the string:

```
data = 'p1,100,5.50,2.75'
```

the expression:

```
data.split(',')
```

will return a list:

```
['p1', '100', '5.50', '2.75']
```

Used together with `readline()` and `int()` or `float()`, the `split()` method provides a concise way to parse the values of a file. Program 7-5 shows this by reading a file of product data, and calculating and displaying the economic order quantity (EOQ) for each product.

Program 7-5

```
1  # read a file with product information and calculate product EOQ
2
3  # calculate an economic order quantity for given parameters
4  def calcEOQ(demand, orderCost, holdingCost):
5      result = (2 * demand * orderCost / holdingCost) ** .5
6      return result
7
8  f = open('eoq.txt', 'r')
9  f.readline() # skip over the file's header line
10 for line in f:
11    line = line.rstrip('\n').strip()
12    if line != '': # skip over lines with no data
13       fields = line.split(',')
14       id = fields[0]
15       demand = int(fields[1])
16       orderCost = float(fields[2])
17       holdingCost = float(fields[3])
18       print("product fields:", id, demand, orderCost, holdingCost)
19       eoq = calcEOQ(demand, orderCost, holdingCost)
20       print("EOQ for", id+":", round(eoq,0))
21 f.close()
```

Program 7-5 Output

```
product fields: P1 100 10.0 2.0
EOQ for P1: 32.0
product fields: P2 200 15.0 2.0
EOQ for P2: 55.0
```

In Program 7-5, the `split()` method takes the data and provides a list of strings, based on where commas occurred in the file. In this case, our file has commas separating the data. The split method can also be used without any arguments, in which case it will assume the separator is any ***whitespace*** character. Whitespace characters are certain invisible characters such as space, tab, newline, and others.

Since the `split()` method returns a list of strings, you may need to convert these list items to numbers using `int()` or `float()`, as is done in lines 15, 16, and 17. The combination of splitting and converting to numbers is a simple example of ***parsing*** data. To highlight parsing as a set of related actions, Program 7-6 introduces a method `parseLine()` in order to reorganize Program 7-5.

Program 7-6

```
# read a file with product information and calculate their EOQ

# calculate an economic order quantity for given parameters
def calcEOQ(demand, orderCost, holdingCost):
    result = (2 * demand * orderCost / holdingCost) ** .5
    return result

# parse an input line into separate fields and convert to numeric type
def parseLine(line):
    fields = line.split(',')
    return fields[0], int(fields[1]), float(fields[2]), float(fields[3])

f = open('eoq.txt', 'r')
f.readline() # skip over the header line
for line in f:
   line = line.rstrip('\n').strip()
   if line != '':
      fields = line.split(',')
      id, demand, orderCost, holdingCost = parseLine(line)
      print("product fields:", id, demand, orderCost, holdingCost)
      eoq = calcEOQ(demand, orderCost, holdingCost)
      print("EOQ for", id+":", round(eoq,0))
f.close()
```

Program 7-6 Output

```
product fields: P1 100 10.0 2.0
EOQ for P1: 32.0
product fields: P2 200 15.0 2.0
EOQ for P2: 55.0
```

Spreadsheet Compare/Contrast

Spreadsheet Import vs. File Read

Spreadsheets allow you to import data from a file into cells. You can also import comma-separated data from a data file, or split imported file data in other ways. In Python, a combination of statements can be used to read data from a file and parse its fields into variables. In this section, we looked at the functions and methods that can be used together to read and parse a data file.

Section Summary

The `split()` method can be used to process a string, and based on a delimiter, return a list of strings. This method is useful when reading fields from a text file.

If fields provided by `split()` contain numeric data, these can be further processed with the `int()` and `float()` functions to create numeric values.

Test Your Understanding

1) The string `split()` method returns a string result.
   a) True
   b) False

2) The string `split()` method only recognizes the comma as a delimiter.
   a) True
   b) False

7.5 Writing a File

The previous sections have focused on reading a file. In this section, we look at how to create a file and write results to it. In order to write to a file, we use the `open()` function introduced earlier, but with a second argument of `'w'` to indicate writing, as opposed to reading. We can then use the familiar built-in `print()` function, but direct it to place the items in a file, as opposed to displaying them on the screen. This file variation for `print()` is done using its named parameter called `file`. In general, this is:

```
print(item1, item2, … , file=fout)
```

where `fout` is a file variable from a previous call to the `open()` function.

Program 7-7 shows a simple example of writing to a file. It prompts for a name, and then creates a file with a greeting message.

Program 7-7 Video animation available.

```
1  # Prompt for input and write a single line to a file
2
3  name = input('Enter your name: ')
4
5  # open a file called greeting.txt for writing, and associate variable fout
6  fout = open('greeting.txt', 'w')
7
8  # use print() function to write a line to the file
9  print('Hello,', name, file=fout)
10
11 # close the file
12 fout.close()
```

Program 7-7 Example Run

```
Enter your name: Adam
```

Program 7-7 Output File (greeting.txt)

```
Hello, Adam
```

In Program 7-7, line 6 uses the `open()` function, but instead of using the second argument of `'r'` (for reading), it indicates `'w'` for writing. Line 9 is the key statement that uses the `print()` function to write a line to the file. Note that `open()` with the `'w'` argument will erase the file with that indicated filename if it already exists.

With these basic mechanics for writing a file, we can revise our EOQ example. Assume we want to read a file of product demand and cost data as before, but save the calculated economic order quantities to a file. The changes from prior programs are minor: we can still use `print()` but in this case direct it to write the results to a file instead of displaying the results to the screen. Program 7-8 uses this approach.

Program 7-8

```
1  # read a file with product information and calculate their EOQ
2  # write the results to a file
3
4  # calculate an economic order quantity for given parameters
5  def calcEOQ(demand, orderCost, holdingCost):
6      result = (2 * demand * orderCost / holdingCost) ** .5
7      return result
8
9  # parse an input line into separate fields and convert to numeric type
10 def parseLine(line):
11     fields = line.split(',')
12     return fields[0], int(fields[1]), float(fields[2]), float(fields[3])
13
14 fin = open('eoq.txt', 'r')
15 fin.readline() # skip over the input file header line
16 i = 1 # counter for lines read
17
18 fout = open('eoqout.txt', 'w')
19 print('id', "eoq", file=fout, sep=',') # header line for output file
20 j = 1 # counter for lines written
21
22 for line in fin:
23    line = line.rstrip('\n').strip()
24    if line != '':
25       # parse, calculate, and save
26       id, demand, orderCost, holdingCost = parseLine(line)
27       eoq = calcEOQ(demand, orderCost, holdingCost)
28       print(id, round(eoq,0), sep=',', file=fout)
29       j += 1
30    i += 1
31 print(i, 'lines read;', j, 'lines written')
32 fin.close()
33 fout.close()
```

Program 7-8 Output to Screen

```
3 lines read; 2 lines written
```

Program 7-8 Output File

```
id,eoq
P1,32.0
P2,55.0
```

In Program 7-8, the key statements for writing results to a file are on line 18 (to create the file) and on lines 19 and 28 (to write data to the file). Note that line 19 writes a header record to the file; as this is only done once, this statement is ahead of, instead of within, the `for` repetition statements.

Section Summary

The `print()` function can be used to write data to a text file by setting its `file` parameter. The value of the file parameter can be set to a variable set from the `open()` function. `open()` is used for writing to files by setting its second parameter to `'w'`. As with reading files, the `close()` method is called to indicate the program is done writing data to a file.

Test Your Understanding

1) A function that can be used to write data to a file is:
   a) `write()`
   b) `print()`
   c) `save()`

2) The `open()` function can be used in order to read as well as write to files.
   a) True
   b) False

7.6 Handling Errors with Exceptions

In this last section of the chapter, we'll look at Python's approach to handling errors, also called exceptions. Exceptions occur when something unusual happens during the program's processing. File-related actions can come across a few errors—for example, if your program tries to read a file that doesn't exist. In this section, we'll look at a simple example related to exceptions and then apply it to reading files. Exceptions are covered in more detail in Appendix D.

In prior chapters, our code has had little or no consideration for unexpected situations that might cause the program to crash. For example, if the user is prompted to enter a number, but instead enters a word, our program will crash when the program tries to convert input using the `int()` or `float()` functions. Program 7-9 demonstrates this.

Program 7-9

```
# Number entry with no checking
entry = input('Enter a number: ')
value = float(entry)
print(value)
```

Program 7-9 Example Run

```
Enter a number: hello
Traceback (most recent call last):
  File "program7-9.py", line 3, in <module>
    value = float(entry)
ValueError: could not convert string to float: 'hello'
```

Python provides a few keywords that allow our programs to anticipate and react to problems such as invalid entries and other cases. The primary keywords are `try` and `except`. The `try` keyword introduces a block of code that is to be executed. If a statement in this block hits an error, the flow of code will jump to the code block indented under the corresponding `except` keyword. Program 7-10 shows this arrangement for our simple numeric entry example.

Program 7-10

```
# Number entry with checking
entry = input('Enter a number: ')
try:
    value = float(entry)
    print(value)
except:
    print('Error converting entry to number')
print("Program ends")
```

Program 7-10 Example Run

```
Enter a number: hello
Error converting entry to number
Program ends
```

In effect, the `except` block acts as a backup plan if the `try` block above it runs into an error. This pattern of `try` and `except` can be used in many other situations. For example, when trying to open a file for reading, if the file does not exist, the `open()` function will generate an exception. Similar to the numeric entry case above, we can use `try` and `except` to anticipate a missing file, and handle the error rather than having the program crash. Program 7-11 demonstrates this.

Program 7-11

```
# with for, read a file and echo its lines to the screen
# Add try/except in case file is not found

def echo(f):
    for line in f:
        line = line.rstrip('\n')
        print(line)

try:
    f = open('measures0.txt', 'r')
    echo(f)
    f.close()
except:
    print("File not found")
```

Program 7-11 Example Run

```
File not found
```

In Program 7-11, the `open()` statement as well as subsequent file-related activity is indented with a `try` block. The corresponding except block will run in case the file is not found (or if there is any other exception). In this way, the program will not crash if the measures.txt file does not exist when the program is run.

Section Summary

Python exception handling can be used to anticipate errors and provide code if they happen. The primary keywords for this are `try` and `except`. The code indented with the `try` section is executed but if an exception occurs, the program flow will jump to the corresponding code with the `except` section. Exception handling can cover a number of potential errors, including those related to files, such as when a file is not found.

Test Your Understanding

1) The statements in an except block will always get executed.
   a) True
   b) False

2) A try block can be used to detect and handle a file not found error, when trying to open a file for reading.
   a) True
   b) False

Chapter Review

Text files can be read and written in Python using a few functions and methods. The `open()` function can be used in a statement to associate a file with a file variable. The first argument for `open()` is the filename, the second argument can be `'r'` to indicate reading or `'w'` to indicate writing. The file variable is subsequently used to read from or write to a file. For reading a file, repetition can be used with the `readline()` method to read a text file's lines into a string variable. If the string contains multiple fields, the `split()` method is one way to generate a list of values corresponding to the fields.

You can write data to a text file using the `print()` method, by specifying its file parameter. The parameter value can be set to a variable assigned from a prior call to `open()`.

The `try` and `except` keywords can be used to anticipate and handle exceptions during the program's execution. This approach can cover a variety of error conditions, including those that may occur during file-related processing.

Key Terms

delimiter	A character used to separate data fields—for example, a comma, space, or tab.
except	The `except` keyword introduces code that will be executed if there is an exception detected in the corresponding try block.
exception	An error during the flow of the program.
file variable	A variable that corresponds to a file—for example, that is set by calling the `open()` function.
open() function	The built-in function that can be used to associate a file with a variable.
parsing	Dividing data into parts—for example, to split a line from a file into multiple fields.
readline() method	A method that can read a line from a text file and return the data in a string.
split() method	A string method that returns a list of strings based on a delimiter.
try	The `try` keyword introduces a block of code that may generate an error when run; if so the code in the corresponding except block is executed.
whitespace	Invisible characters such as space, tab, and newline.

Programming Problems

1) Average and Range—Three Values

For the file measures.txt, write a program that reads the first three lines, then computes and displays the average and range of these values. *Range* is defined as the maximum value minus the minimum value.

2) Average and Range—All Values

Assume a file contains one numeric value on each line. Write a program that reads each line, then computes and displays the average and range of these values. *Range* is defined as the maximum value minus the minimum value.

3) Running Average

Assume a file contains one value on each line. Write a program that displays for each line in the file: the line number, the value read, and the running average up to and including that value. The displayed values should be in comma-separated format. For example, if the file contains the values:

```
10.1
9.9
9.7
```

the program would display:

```
1,10.1,10.1
2,9.9,10.0
3,9.8,9.9
```

4) File Copy

Write a program that reads a file one line at a time and writes each of these lines to a new file called echo.txt

5) Filter Valid Lines

Assume an input file has one or more values separated by commas. We would like to create a file that contains those lines that have two values and report to the screen those lines that have other than two values.

For example, if the input file has:

```
10.1,10.4
9.9,10.0,10.1
9.8,9.9
```

the output file should have:

```
10.1,10.4
9.8,9.9
```

and the screen should display:

```
Line 2: 9.9,10.0,10.1
```

6) Filter Valid Lines with Numeric Values

Assume an input file has one or more numeric values separated by commas. We would like to create a file that contains those lines that have two numeric values and report to the screen those lines that have other than two values as well as lines with non-numeric values.

For example, if the input file has:

```
10.1,10.4
9.9,10.0,10.1
9.8,9.9x
```

the output file should have:

```
10.1,10.4
```

and the screen should display:

```
Line 2: not 2 values: 9.9,10.0,10.1
Line 3: non-numeric value: 9.8,9.9x
```

7) Detect Invalid User Input

Write an interactive BMI calculator that prompts for height (in inches) and weight (in pounds), and calculates BMI as:

weight / height**2 * 703

If the user enters a height or weight that is not numeric, display an error message.
Example run:

```
Enter height (inches): 66
Enter weight (pounds): 150x
Error: inputs must be numeric
```

8) Product Weight Batches

For quality control purposes, a company keeps track of product weight measurements in a file. Each line of the file represents a batch of weights measured. The company would like to determine which batch has the highest average weight. Here is an example file:

```
batch1:10,9.9,10.1
batch2:10.1,10.2,10.3
```

a) Write a program that reads a weight batch file, calculates the average weight for each batch, and determines and displays the batch name with the highest average weight.
b) Enhance your program to detect and handle the possibility that the weight data in the file may be non-numeric.

9) Patient File and BMI

A text file contains patient ID, height (in inches), and weight (in pounds). For example:

```
patient-id,height,weight
P0123,66,150
P4354,68,164
```

a) Write a program that reads each line of the file, calculates the BMI for each patient, and displays each patient's ID and their BMI to the screen. The equation for BMI is:
   BMI = weight / height^2 × 703
b) Enhance your program so that the patient ID and BMI are saved to a file called bmi.txt.
c) Enhance your program so that if the height or weight data in the file are not numeric, the program will display an error message to the screen including the patient's ID.

Chapter 8 'str'

Strings in Detail

Chapter Objectives

In this chapter, we'll cover:

- a recap of the basics for using strings in your programs
- how slicing can be used to concisely get part of a string
- how to use Python's built-in methods to search for characters within strings
- how to generate variations of strings—for example, making a string all lowercase or uppercase
- how to check if a string meets certain criteria—for example, that it is all digits, which is useful for validation

Chapter Outline

8.1 String Basics Recap

In this chapter, we'll recap some aspects of strings in Python and introduce other useful features of strings. Strings are a key element of programs, and as we'll see in this chapter, Python has many features to let you concisely work with strings.

Let's go back to Program 2-4:

```
name = input('Please enter your name: ')
print('Hi', name)
```

In the `print()` statement above, there are two string-related items. `'Hi'` is a ***string literal***, which is a string constant in the program. The variable `name` is a string variable. Both of these are of type `str`. The `str` type provides many features, including more than 40 methods related to strings. In this chapter, we'll look at many of the string-related methods.

A string contains a sequence of characters. Square brackets can be used to access a character of the string. However, square brackets can only be used to get—but not set—a character in a string. This is because strings are ***immutable***—once created, you can't directly change any of the characters in a string. For example, if `s = 'Hello'`, if you try:

```
s[0] = 'h'
```

you will get a `TypeError`. However, as we'll see in this chapter, we can reassign a string variable to a different string.

For strings, `len()` returns the number of characters in the string. For example, `len('Hello')` returns 5.

There is one special case of a string called the ***empty string***, which is a string with no characters. We can use two adjacent apostrophes to define an empty string. For example:

```
s = ''
```

sets `s` to an empty string. For an empty string, `len()` returns 0.

The plus sign can be used to join (concatenate) strings. For example:

```
>>> s = 'Hello'
>>> t = 'there'
>>> r = s + ' ' + t
>>> r
'Hello there'
```

In addition to using + to join two strings, += can also be used to append a string to another string, as in:

```
s += t
```

Strings containing numbers can be used in the `int()` and `float()` functions in order to create numeric values. Conversely, the `str()` function can be used to convert numbers (and other types) to strings.

There are a few special characters in Python strings that are indicated using the backward slash. Commonly used special characters are tab (`\t`) and newline (`\n`).

Strings can be compared for equality using ==, and for inequality using `!=`. Strings have an order based on the ASCII character values shown in Table 8-1. Since characters have an order, strings can be compared for order using <, >, <=, and >=. Note that uppercase letters have a lower numeric value than lowercase letters.

<table>
<tr><th>Range</th><th>Characters</th></tr>
<tr><td>0–31</td><td>special characters: e.g., tab, backspace</td></tr>
<tr><td>32</td><td>space</td></tr>
<tr><td>33–47</td><td>symbols1: ! " # $ % & ' ( ) ^ + , -. /</td></tr>
<tr><td>48–57</td><td>digits: 0, 1, . . . , 9</td></tr>
<tr><td>58–64</td><td>symbols2: : ; < = > ? @</td></tr>
<tr><td>65–90</td><td>uppercase letters: A, B, . . . , Z</td></tr>
<tr><td>91–96</td><td>symbols3: [\ ] ^ _ `</td></tr>
<tr><td>97–122</td><td>lowercase letters: a, b, . . . , z</td></tr>
<tr><td>123–126</td><td>symbols4: { | } ~</td></tr>
<tr><td>127</td><td>DEL</td></tr>
</table>

Table 8-1: ASCII character values

In this chapter, we'll write and evolve code for a simple text processor that allows you to enter and modify some text. Though its features are targeted, it will show a few string techniques and how concise they are. At the end of the chapter, we'll look at a more involved example, showing how to read and search a file of product price information.

Program 8-1 uses strings in a variety of ways. The string variable `text` is initially set to an empty string. This variable will keep track of the current value of our text. The `input()` function displays a string prompt, and returns the string that the user enters. Depending on the first character of the user's string, as accessed by `[0]` on line 6, the program will set the text, augment the text using concatenation, or reset the text to an empty string.

Program 8-1 Video animation available.

```
1  # Simple text processor with assignment and concatenation
2  text = '' # initialize as empty string
3
4  entry = input('> ') # input() returns a string
5  while entry != '': # string compare
6      command = entry[0] # character access
7      if   command == '=': text = input('>> ')   # assignment
8      elif command == '+': text += input('>> ')  # concatenation
9      elif command == '-': text = ''             # reset to empty string
10
11     print(text)
12     entry = input('> ')
```

Program 8-1 Example Run

```
> =
>> Python is great
Python is great
> +
>>  for working with strings.
Python is great for working with strings.
>
```

In the above program, strings are compared for inequality using != on line 5, and compared for equality using == on lines 7 through 9. String concatenation is done with += on line 8.

Spreadsheet Compare/Contrast
Spreadsheet Text

In spreadsheets, text can be entered in cells, and these are similar to string literals in programs. As well, spreadsheets have built-in functions for working with text, and these are similar to some of the string methods that will be covered in this chapter. As we'll see, Python can be more concise when working with strings.

8.2 Substrings (Slicing)

Python allows you to use slicing to generate substrings. The syntax is the same as lists (covered in Chapter 6), using square brackets and colons. For example:

```
s[i:j]
```

will create a string starting with the `i`th character of `s` and ending with the character just before the `j`th character. If `s='Hello'`, `s[0:2]` is `'He'`. Slicing doesn't alter the original string; it creates a new one. After we take the slice `s[0:2]`, `s` is still `'Hello'`. Figure 8-1 shows this example slice.

If the slice position before the colon is omitted, then 0 is implied. So `s[:j]` creates a string from the substring of `s` starting at its beginning and up to (but not including) position `j`. For example, if `s='Hello'`, then `s[:2]` is `'He'`. Similarly, `s[i:]` creates a string using the substring of `s` starting at position `i` and including the rest of the string. If `s='Goodbye'`, then `s[4:]` creates the string `'bye'`.

Program 8-2 enhances our text processor by using string slicing. This program now lets the user enter an action and some text in one step instead of two. This is done by checking the first character of the entered string. If it is `'+'`, the rest of the entered string is determined by taking a slice, and a concatenation will be done to

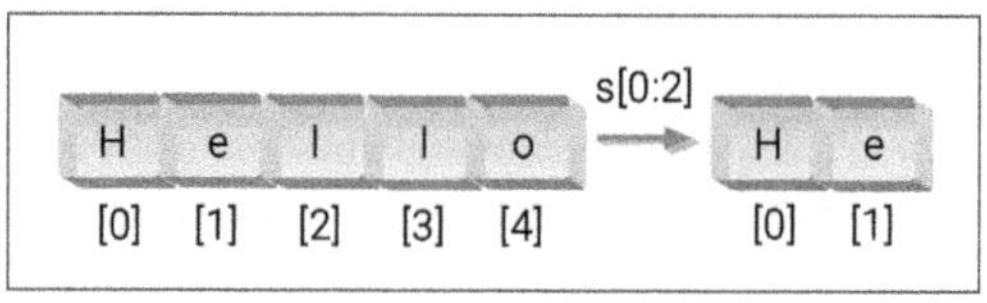

Figure 8-1: Example string slice

augment the stored text. If the first character is '-', the stored text will be cleared. If the user's entry does not begin with + or -, the entry is used to reset the stored text.

Program 8-2

```
# Simple text processor with slicing
text = ''

entry = input('> ')
while entry != '':
    command = entry[0]
    if   command == '+': text += entry[1:] # slice and concatenate
    elif command == '-': text = ''
    else:                text = entry
    print(text)
    entry = input('> ')
```

Program 8-2 Example Run

```
> Python is great
Python is great
> + for working with strings.
Python is great for working with strings.
>
```

Section Summary

Slicing is a feature in Python for generating a substring from a string. The syntax uses square brackets with integer values and colons within. The same syntax is used by lists for taking a subset of a list. For example, for a string variable `s`, `s[1:3]` would return a string with the second and third characters of `s`. Note that when you take a slice of a string, the original string is unchanged, and a new string is generated.

Test Your Understanding

1) Taking a slice of a string changes that string.
   a) True
   b) False
2) Taking a slice of a string generates a list of strings.
   a) True
   b) False
3) Given a string `s` with value `'Hello'`, what is the value of `s[1:3]`?
   a) `'He'`
   b) `'Hel'`
   c) `'el'`
   d) `'ell'`

8.3 Searching Parts of Strings

Several methods allow your program to check if one string is within another string. These are summarized in Table 8-2.

Expression	Description
`s.startswith(t)`	returns True if *s* starts with the string *t*
`s.endswith(t)`	returns True if *s* ends with the string *t*
`s.find(t)`	returns the position of the first occurrence of *t* in *s*, or -1 if there is none
`s.rfind(t)`	returns the position of the last occurrence of *t* in *s*, or -1 if there is none
`s.count(t)`	returns the number of occurrences of *t* in *s*
s in *t*	returns True if *s* occurs in *t*
s not in *t*	returns True if *s* is not in *t*

Table 8-2: Search-related expressions for strings

Program 8-3 enhances our text processor by adding a substring search command and a substring count command:

- If the user enters a ? followed by a string, the program uses the `in` keyword to see if the user-supplied string exists in the `text` variable.
- If the user enters a # followed by a string, the program uses the `count()` method to determine the number of occurrences of that string in the `text` variable.

Program 8-3

```
# Simple text processor to demonstrate some string features
text = ''

entry = input('> ')
while entry != '':
    display = False
    command = entry[0]
    if   command == '+':
        text += entry[1:] # slice and concatenate
        display = True
    elif command == '-':
        text = '' # reset to empty string
    elif command == '?':
        print(entry[1:] in text) # search
    elif command == '#':
        print(text.count(entry[1:])) # count
    else:
        text = entry

    if display:
       print(text)
    entry = input('> ')
```

Program 8-3 Example Run

```
> Python is great
> + for working with strings.
Python is great for working with strings.
> ?Python
True
> #in
2
>
```

Tip
String Method Return Types

The Python string methods covered in this section return either an `int` or `bool` (True or False) result. String methods covered in the next section return string results. When working with strings, you may be able to find a single method to get a desired result, or you may need to use more than one method in sequence. Considering what type is returned by the string methods may help you find a concise way to solve a particular task in your program.

Section Summary

The string type provides several methods that allow Python programs to check if a string contains another string. Some of these return `True` or `False`, others return the index of the match (or -1 if no match). Another method, `count()`, determines the number of times a string is contained within another string.

Test Your Understanding

1) For a string `s` with value `'Hello'`, what is the result of `s.find('ll')`?
   a) -1
   b) True
   c) 2
   d) 3

2) For a string `s` with the value `'Hello'`, what is the result of `s.count('l')`?
   a) 0
   b) 1
   c) 2

8.4 Methods That Return String Variations

Several string methods return a variation of a given string. This is useful to adjust input data to conform in certain ways. Note that these methods return a new string, as opposed to changing the given string. In order to update a string, you need to include that variable on both sides of the equals sign. For example:

```
s = s.upper()
```

will return the uppercase value of `s`, and then reassign that result to `s`.

If you want an uppercase representation of `s`, but also want to leave `s` as is, you can create a new variable like:

```
t = s.upper()
```

Category	Method	Description
Strip	`s.strip()`	Remove leading and trailing whitespace
	`s.lstrip()`	Remove leading whitespace
	`s.rstrip()`	Remove trailing whitespace
Case	`s.upper`	Convert letters to uppercase
	`s.lower()`	Convert letters to lowercase
First letter	`s.capitalize()`	Capitalize the first letter
	`s.title()`	Capitalize the first letter of each word
Justification	`s.center()`	Center the string
	`s.ljust()`	Left justify
	`s.rjust()`	Right justify
Replace	`s.replace(old, new)` `s.replace(old, new, count)`	Replace occurrences of old with new; if count is provided, limit to that many replacements

Table 8-3: Methods returning string variations

Program 8-4 further enhances our text processor by adding an uppercase command (using the `upper()` method), and a substring replace command (using the `replace()` method). Note that the program also uses the `split()` method, in order to split the user's entered string into parts. The `split()` method is described in Section 8.6.

Program 8-4

```
# Simple text processor to demonstrate some string features
text = ''

entry = input('> ')
while entry != '':
    display = False
    command = entry[0]
    if  command == '+':
        text += entry[1:] # slice and concatenate
        display = True
    elif command == '-':
        text = ''
    elif command == '?':
        print(entry[1:] in text) # search
    elif command == '#':
        print(text.count(entry[1:])) # count
    elif command == '^':
        text = text.upper() # uppercase and reset
        display = True
    elif command == '/':
        fields = entry.split('/') # parse entry from string to list of strings
        if len(fields) == 3:
            text = text.replace(fields[1], fields[2]) # replace
            display = True
    else:
        text = entry

    if display:
        print(text)
    entry = input('> ')
```

Program 8-4 Example Run

```
> Python is great for working with strings.
> /great/excellent
Python is excellent for working with strings.
> ^
PYTHON IS EXCELLENT FOR WORKING WITH STRINGS.
>
```

Tip

Methods Don't Directly Change Strings

The Python string methods covered in this section return a variation of the string. However, these methods don't directly change the string. This means if you use a statement like:

```
s.lower()
```

the value of `s` is unchanged, even though you intended for the string to be changed to lowercase. Instead, make sure you use a statement like:

```
s = s.lower()
```

in order to update the variable.

Section Summary

The Python string type provides several methods to generate variations of a string. These methods do not change the string but instead create a new string. Table 8-3 summarizes a number of these methods.

Test Your Understanding

1) For a string `s` with value `'Hello'`, what is the result of `s.lower()`?
   a) `'Hello'`
   b) `'hello'`
   c) `'HELLO'`
2) If a program contains the statements:
   ```
   s = 'Hello'
   s.upper()
   ```
   what is the value of `s` after these statements are executed?
   a) `'hello'`
   b) `'Hello'`
   c) `'HELLO'`
   d) none of the above

8.5 Checking Strings

The string type provides several methods that check if the string fits a certain pattern. This is useful for validating input data. For example, the `isdigit()` method lets your program check if a string contains all digits. Table 8-4 summarizes some of these methods. Each of these methods returns either True or False.

Method	Description
`isalnum()`	Are all characters letters or digits?
`isalpha()`	Are all characters letters?
`isdigit()`	Are all characters digits (0, 1, . . . , 9)?
`isidentifier()`	Is the string a valid Python identifier: e.g., could it be used as a variable name?
`islower()`	Are the characters lowercase letters (a, b, . . . , z)?
`isspace()`	Are the characters all whitespace characters?
`isupper()`	Are the characters uppercase letters (A, B, . . . , Z)?

Table 8-4: Selected string checking methods

Program 8-5 adds validation to the text processor. It uses the `isalpha()` method to check the user's entry before reassigning the entry to the `text` variable.

Program 8-5

```
# Simple text processor to demonstrate some string features
text = ''

entry = input('> ')
while entry != '':
    display = False
    command = entry[0]
    if   command == '+':
        text += entry[1:] # slice and concatenate
        display = True
    elif command == '-':
        text = '' # empty string
    elif command == '?':
        print(entry[1:] in text) # search
    elif command == '#':
        print(text.count(entry[1:])) # count
    elif command == '^':
        text = text.upper() # upper case with reassignment
        display = True
    elif command == '/':
        fields = entry.split('/') # split string into list of strings
        if len(fields) == 3:
           text = text.replace(fields[1], fields[2]) # replace
           display = True
    elif command.isalpha() or command == ' ' :
        text = entry
    else:
        print('Unknown command')

    if display:
       print(text)
    entry = input('> ')
```

Program 8-5 Example Run

```
> Python is great
> + for working with strings.
Python is great for working with strings.
> $
Unknown command
> -
>
```

8.6 An Example with Strings, Lists, and a File

In this last section, we'll look at how to read a comma-separated data file containing product prices, and allow for searches on its data. The program for this uses features from strings, lists, and files.

The file of interest, products.txt, has three fields: product ID, company division that sells the product, and product price. The file also has an initial header line identifying the fields. Here are the lines of the file:

```
product-id,division,price
P-001,diva,5.50
P-001x,diva,5.95
P-002,diva,6.50
P-002x,diva,6.95
P-010,divb,10.25
P-010x,divb,10.75
Q-011,divb,11.50
Q-011x,divb,11.95
```

Program 8-6 reads these data into a list of lists. Figure 8-2 shows the layout of this data structure.

Note that in order to run Program 8-6, the program and the data file need to be in the same directory.

The program uses `for` repetition to read the data in lines 8 through 13. Line 11 uses the string `split()` method to divide the line's data into a list of three items. This sublist is then appended to the overall list in line 13. The built list of lists is returned from the `loadProductData()` function on line 14.

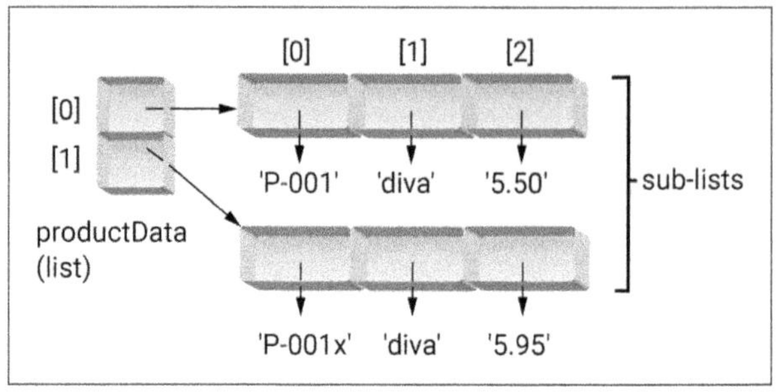

Figure 8-2: List of lists in Program 8-6

The user is prompted for which product(s) are of interest. The data are searched using a `for` repetition (line 28) with a filter condition using string `startswith()` and `lower()` methods.

Program 8-6

```
1  # search for matching products from a file of price data
2
3  # read a file of price data into a list of lists
4  def loadProductData():
5     fin = open("products.txt", 'r')
6     productData = [ ]
7     fin.readline() # read and skip header line
8     for line in fin: # read each subsequent line
9        line = line.rstrip('\n') # remove trailing newline
10       if line[0] != '#' and line.count(',') == 2:
11          productID, division, price = line.split(',')
12          # add sublist to list
13          productData.append((productID, division, price))
14    return productData
15
16 # get user interest for product of interest
17 def getProductFilter():
18    productFilter = input("Enter product ID (starts with): ")
19    productFilter = productFilter.strip().lower() # adjust entry for matching
20    return productFilter
21
22 # load product prices into a list of lists
23 productData = loadProductData()
24
25 # based on user's request, display matching product data
26 productFilter = getProductFilter()
27 while productFilter != 'quit':
28     for productRecord in productData:
29        if productRecord[0].lower().startswith(productFilter): # check for match
30           print(productRecord[0], productRecord[1], productRecord[2], sep='\t')
31     productFilter = getProductFilter()
```

Program 8-6 Example Run

```
Enter product ID (starts with): P-010
P-010   divb    10.25
P-010x  divb    10.75
Enter product ID (starts with): Q
Q-011   divb    11.50
Q-011x  divb    11.95
Enter product ID (starts with): quit
```

Chapter Review

Strings are an essential ingredient in programs, and Python provides a string type with many features. These include the ability to take substrings using slicing, as well as methods for searching within strings, generating string variations, and checking strings. The tables in this chapter provide several of the many methods and techniques for working with strings. Used together, they provide a concise and powerful set of tools for working with strings in your programs.

Key Terms

empty string	A string with no characters.
newline	The special character denoted as \n, which represents a new line.
slicing	Generating a substring from a string using brackets and colons—for example, `s[0:5]`.
str	The Python type for representing strings as a sequence of characters, which provides methods for working with strings.
string literal	A string constant, defined as a set of characters between apostrophes or quotes.
string method	A method that can work with a string to generate a result; depending on the method the result is of a particular type such as string, Boolean, integer, or list.
string variable	A variable that refers to a string.
whitespace	A character such as space, tab, or newline.

Programming Problems

1) Hello

Write a program that greets the user with "How are you?" If the response is "OK" (upper- or lowercase), then display "Good to hear!" Handle (ignore) leading and trailing spaces. Useful string methods are `lower()` or `upper()` and `strip()`.

Example run:

```
How are you?  OK
Good to hear!
```

2) Integer Check

Write a program that prompts the user to enter an integer and then checks if the entry is an integer. For this problem, do not use try/except; instead, use string methods. As part of your program, define a function `isInteger(s)` that has a string parameter `s` and returns True or False. Useful methods to consider are `strip()` and `isdigit()`.

Example run:

```
Please enter an integer: 123x
That isn't an integer
```

3) Distance Conversion

Write a program that prompts the user to enter a distance in feet or meters and displays the equivalent distance in meters or feet. The user should enter a number, one or more spaces, and then a word beginning with either M or F (upper- or lowercase). If an invalid word or number is entered, the program should display an error message.

Example run:

```
Enter distance in meters or feet: 1 meter
3.28084 feet
```

4) Word Cleaner

Text analysis may include a step to find words in a body of text. Write a program that reads a file of sentences and writes a file that converts words to lowercase and omits all punctuation. For your program, assume punctuation includes: . , ; : ? !

Your program should output words that have only letters. For example, if the file contains b2b, that should not be included in the output file. Useful string functions include `replace()` and `isalpha()`.

5) Date Conversion

Write a program that translates a date of form mm/dd/yyyy to the form: Month dd, yyyy. The program should prompt the user for a date formatted as mm/dd/yyyy and create and display the second form. For example, if the user enters 03/22/2018, the program should construct and display the string March 22, 2018.

To solve this problem, use two lists to associate the month numbers with month names and maximum number of days of each month. Also as part of your solution, define and use a function, `dateToLabel(indate)`, that takes a string of the form mm/dd/yyyy and returns the form Month dd, yyyy. If the string `indate` is not of a valid form, your function should detect it and raise an `Exception`; for example:

```
if len(fields) != 3:
    raise Exception('Incorrect format for date')
```

Detect and report certain error conditions:

- User input was not formatted as mm/dd/yyyy.
- User entered input like mm/dd/yyyy, but mm, dd, and/or yyyy were not numeric.

For example, 01/22/2019 is valid, but jan/22/2019 is invalid.

- Month (mm) was not a valid month.
- Day (dd) was not a valid day (you do not need to distinguish between leap and nonleap years).

Chapter 9

Dictionaries and Sets

Chapter Objectives

In this chapter, you'll learn:

- how to use dictionaries, an alternative to lists, to store collections of related key/value pairs
- how to create dictionaries
- how to find values in dictionaries
- how `for` can work with a dictionary to access its values
- how to maintain data in dictionaries
- about sets for collections of unique items

Chapter Outline

9.1 About Dictionaries

In Chapter 6, we looked at lists as a way to collect related data. Collections of data are an important aspect of programs, and lists are one way that data can be collected. Dictionaries are another important and flexible type of collection, which we'll look at in this chapter.

While lists number each item using an index, dictionaries allow you to name each item. (In this chapter, we'll use strings as dictionary item names, though other Python types can be used.) And like lists, you can use any Python type as the dictionary's values, so dictionaries are a very flexible type of collection.

Dictionaries store a collection of ***keys***, and each key is associated with a value. In other words, just as an index is used to access an item in a list, a key is the access to a dictionary's value. As keys are used to access items, each key needs to be unique in a given dictionary.

To make this more tangible, let's consider a program that will keep track of contacts, including email addresses and phone numbers. The program might allow you to enter an email address and find out the corresponding phone number for that person. The data might be stored in a file or a database.

In order to store email addresses and phone numbers, you could use two parallel lists. Then in order to look up a phone number for a given email address, the program could search the list of email addresses, determine the index if a match is found, and then use the index to locate the corresponding phone number. The data organization in this list arrangement would look like Figure 9-1.

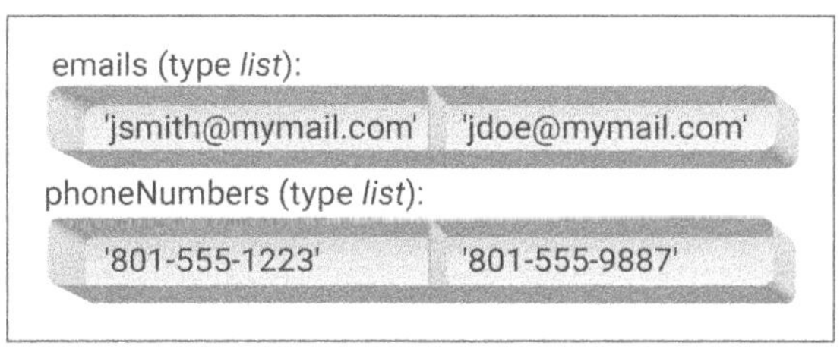

Figure 9-1: Contact information as parallel lists

Two parallel lists are OK but not ideal. First, the data are related but split into two variables. Also, searching for a particular email requires Python to potentially scan the entire email list. For a large set of data, this would be inefficient.

In contrast with parallel lists, a dictionary organization could keep pairs of data, with the email ID as the key and the phone number as the value. Figure 9-2 shows this arrangement.

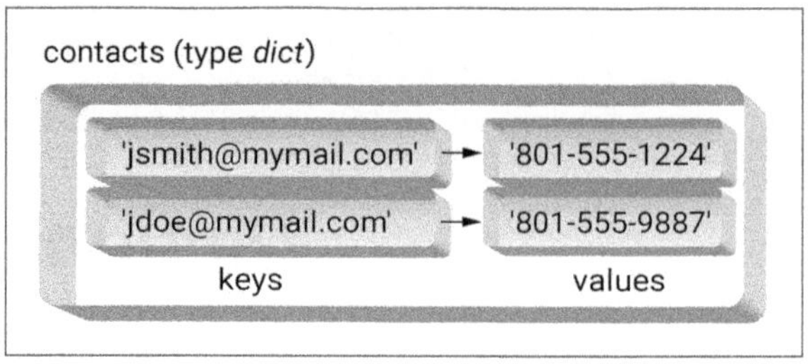

Figure 9-2: Contact information as key/value pairs

The dictionary arrangement has a few benefits over the list arrangement. First, lookup is faster in dictionaries, since they are structured so that a scan of potentially all entries is not needed, as is the case for lists. Second, the data in the dictionary are more tightly connected—in this example, one instead of two variables are needed. Lastly, if we want to capture additional information (for example, first and last name), we can use a dictionary of dictionaries, which would allow us to name each of the data fields. Figure 9-3 shows this arrangement.

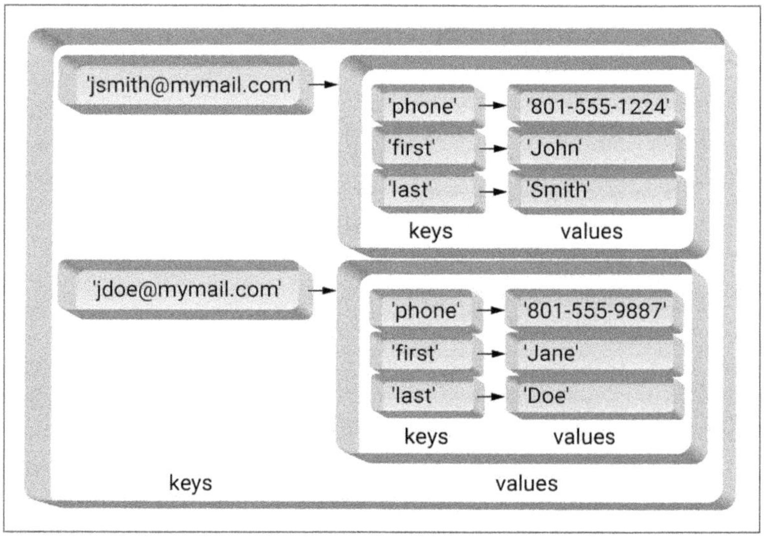

Figure 9-3: Dictionary of dictionaries for contacts

In the next section, we'll look at how to create dictionaries. Then in Section 9.3, we'll see how data can be searched in a dictionary. Section 9.4 describes how records of data can be stored in a dictionary by using named values. Section 9.5 looks at a few ways the `for` statement works with dictionaries to access their keys and values. Section 9.6 describes how data can be changed and deleted in dictionaries. Another type of collection, sets, which store a distinct collection of values, is covered in Section 9.7. The last section shows how to read a data file to create a dictionary dynamically.

Spreadsheet Compare/Contrast
Labeling and Naming Cells

In spreadsheets, you can use text cells to label values in adjacent cells. You can also name a cell. These layouts are roughly similar to the idea of a dictionary in Python. However, Python dictionary value types are not limited to simple types such as text, numbers, and dates. This expressiveness of dictionaries—that values can be lists, dictionaries, or any other Python type—allows Python to model complex data layouts in ways that can be hard to do in spreadsheets.

Section Summary

A dictionary is a collection type that stores values; each value can be accessed by its key. This can be a suitable alternative to lists where there is a unique key for the data, such as an ID. Dictionaries provide flexibility, as their values can be any Python type; this allows you to store a hierarchy of data. Further, dictionaries are efficient for finding data given how their keys are stored.

9.2 Creating Dictionaries

Like lists, dictionaries can have their items indicated when they are created, or can start empty and have their contents subsequently added. In this chapter, most examples use dictionaries with their items hard-coded when the dictionary is created, so the code is self-contained. In practice, it is common to start with an empty dictionary and add items to it during the program's execution.

In order to define a dictionary in code, curly braces {} are used. For example, for the contacts data in the prior section, these could be defined in code as:

```
contacts = {'jsmith@mymail.com': '801-555-1223',
            'jdoe@mymail.com': '801-555-9887'}
```

In the above statement, the key and value are separated by a colon, and each key/value pair is separated by a comma.

Alternatively, an empty dictionary can be created, and items added to it. To create an empty dictionary, use curly braces like so:

```
contacts - { }
```

Key/value pairs can be added to a dictionary using the syntax:

```
dictionaryVariable[key] = value
```

For example:

```
contacts["jsmith@mymail.com"] = "801-555-1223"
```

The approach of adding to an initially empty dictionary is shown in Program 9-1. This program builds the contacts dictionary based on user input. The key for the data is email address, and the value associated with each email address is phone number.

Program 9-1 Video animation available.

```
1  # build a contact dictionary dynamically from user input
2
3  contacts = {} # create an empty dictionary
4
5  print('Enter email and phone # at > prompt (<return> when done)')
6  entry = input('> ')
7  while entry != '':
8      email, phone = entry.split()
9      contacts[email] = phone # add entry to dictionary
10     print(contacts)
11     entry = input('> ')
```

Program 9-1 Example Run

```
Enter email and phone # at > prompt (<return> when done)
> jsmith@mymail.com 801-555-1223
{'jsmith@mymail.com': '801-555-1223'}
> jdoe@mymail.com 801-555-9887
{'jsmith@mymail.com': '801-555-1223', 'jdoe@mymail.com': '801-555-9887'}
>
```

In Program 9-1, the first dictionary-related statement is line 3, which creates an empty dictionary. An email address and phone number are added to the dictionary on line 9, with email as the key and phone as the value. With each update, the dictionary is displayed by line 10.

By the end of the program's run, the dictionary's contents are as shown in Figure 9-4.

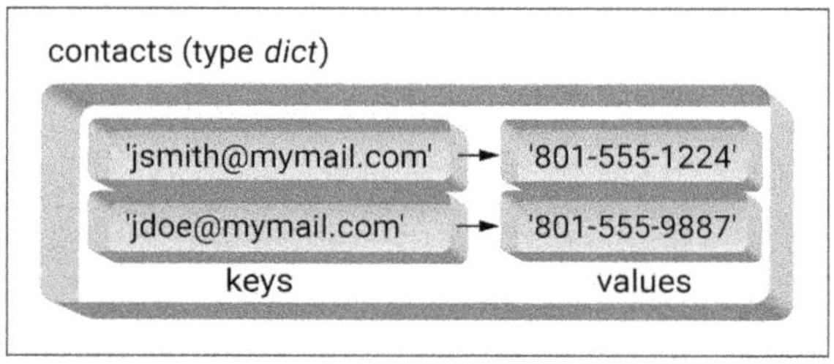

Figure 9-4: Example contacts dictionary created with Program 9-1

Section Summary

A dictionary can be created using curly braces. An initially empty dictionary can be defined, and items can be subsequently added. Alternatively, you can create a dictionary with initial values by indicating its keys and values within curly braces. Dictionary data can be added by statements like dictionary[key] = value. In this chapter, we'll use strings as keys. Certain other types can be used as keys—for example, integers; this will be discussed in the last section of the chapter.

Test Your Understanding

1) What will happen after these statements are executed?

```
colors = { }
colors['r'] = 'red'
```

a) an error will be generated
b) a dictionary will be created with the key `'r'` pointing to the value `'red'`
c) a dictionary will be created with the key `'red'` pointing to the value `'r'`
d) a list will be created with the values `'r'` and `'red'`

2) What will result after this statement is executed?

```
colors = {'r':'red', 'g':'green',
  'b':'blue'}
```

a) an error will be generated
b) a list will be created with six values
c) a dictionary will be created with the keys `'r'`, `'g'`, `'b'`
d) a dictionary will be created with the keys `'red'`, `'green'`, `'blue'`

9.3 Finding Data in a Dictionary

Once a dictionary has been populated, it is easy and efficient for your program to locate data using a key. The dictionary variable with the key inside square brackets can be used to locate the data value associated with that key. The general syntax is:

```
value = adict[key]
```

Note that this syntax is similar to how items in a list are found using an index. However, for dictionaries, the value within square brackets is not limited to an integer as it is when working with lists.

Program 9-2 provides code to look up a person's phone number given their email address.

Program 9-2

```
1  # Dictionary example to look up phone (value) from email (key)
2
3  # create the dictionary
4  contacts = {'jsmith@mymail.com': '801-555-1223',
5              'jdoe@mymail.com': '801-555-9887'}
6
7  print('Enter an email address to look up phone (or <return> when done)')
8  email = input('> ')
9  while email != '':
10     phone = contacts[email] # use email as key to look up phone
11     print(phone)
12     email = input('> ')
```

Program 9-2 Example Run

```
Enter an email address to look up phone (or <return> when done)
> jdoe@mymail.com
801-555-9887
> jsmith@mymail.com
801-555-1223
>
```

The key statement in Program 9-2 in line 10, which uses dictionary lookup syntax to search the dictionary for an email address, and return the corresponding phone number.

If you run Program 9-2 and enter an email address that is not in the dictionary, the program will generate a `KeyError` and end. This is because dictionary lookups with a key that is not in the dictionary will result in an error. This is similar to using an index of a list that is past the end of the list. To protect against this error, we can use the `in` phrase to check if a key is in a dictionary. In general,

```
<key> in <dictionary>
```

returns `True` if the <key> is contained in the `<dictionary>`. Similarly, `not in` can be used to see if a key isn't in a dictionary:

```
<key> not in <dictionary>
```

Program 9-3 uses this approach to enhance Program 9-2 so that keys not found will be detected and reported, as opposed to generating an error and ending the program.

Program 9-3

```
1  # Dictionary example to look up phone from email
2
3  contacts = {'jsmith@mymail.com': '801-555-1223',
4              'jdoe@mymail.com': '810-555-9887'}
5
6  print('Enter an email address to look up phone, or <return> when done')
7  email = input('> ')
8  while email != '':
9      if email in contacts: # check if the email is a key in dictionary
10         phone = contacts[email] # if so, look up value from key
11         print(phone)
12     else:
13         print('** not found **')
14     email = input('> ')
```

Program 9-3 Example Run

```
Enter an email address to look up phone, or <return> when done
> jsmith@mymail.com
801-555-1223
> jjones@mymail.com
** not found **
>
```

In Program 9-3, line 9 uses `in` to check if the user-entered email address is a key in the dictionary.

Section Summary

Dictionary data values can be found by using the syntax dictionary[key]. If the key exists, its associated value (of the key/value pair) will be returned. However, if the key does not exist, this expression will generate an exception. To check if a key is or isn't in a dictionary, the `in` and `not in` phrases can be used.

Test Your Understanding

1) What will happen after these statements are executed?

```
colors = {'r':'red', 'g':'green',
  'b':'blue'}
color = colors['b']
```

a) the value of color will be set to `'b'`
b) the value of color will be set to `'blue'`
c) an exception will be generated

2) What will happen after these statements are executed?

```
colors = {'r':'red', 'g':'green',
  'b':'blue'}
color = colors['y']
```

a) the key `'y'` will be added to the dictionary colors
b) an exception will be generated
c) the user will be prompted for the value to associate with the key `'y'`

9.4 Using `for` to Access Dictionary Items

As with lists, `for` can be used to access the items in a dictionary. However, as dictionaries contain keys and values, there are a few variations depending on whether you need to get keys, values, or both. Table 9-1 shows these alternatives.

`for key in dict.keys():`	Iterate through the keys of the dictionary.
`for value in dict.values():`	Iterate through the values of the dictionary.
`for key, value in dict.items():`	Iterate through the key/value pairs of the dictionary.

Table 9-1: Using `for` with a dictionary

For example, returning to the contacts example, Program 9-4 will display the email IDs and phone numbers.

Program 9-4

```
# Dictionary example to display all keys and values

contacts = {'jsmith@mymail.com': '801-555-1223',
            'jdoe12@mymail.com': '801-555-9887'}

for email, phone in contacts.items():
    print(email, phone)

print(len(contacts), 'contacts in dictionary')
```

Program 9-4 Output

```
jsmith@mymail.com 801-555-1223
jdoe12@mymail.com 801-555-9887
2 contacts in dictionary
```

Program 9-4 also shows that the `len()` function works with dictionaries. Note that the `len()` function returns the number of keys in the dictionary, as opposed to the number of keys and values.

Section Summary

In previous chapters we've seen how the `for` statement can be used to iterate over a set of values. Similarly, `for` can be used to access the data in a dictionary. Since dictionaries have keys and values, there are some variations for using `for` with dictionaries in order to get the keys, the values, or both. These involve using the `keys()`, `values()`, or `items()` method in combination with the dictionary and `for` statement.

Test Your Understanding

1) What will happen after these statements are executed?

```
colors = {'r':'red', 'g':'green',
  'b':'blue'}
for item in colors.keys():
    print(item, end=' ')
```

a) an exception will be generated
b) the values `r g b` will be displayed on the screen
c) the values red green blue will be displayed on the screen
d) a table of the dictionary's keys and values will be displayed

2) What will happen after these statements are executed?

```
colors = {'r':'red', 'g':'green',
  'b':'blue'}
for k, v in colors.items():
    print(k, v)
```

a) an exception will be generated
b) the values `r g b` will be displayed on the screen
c) the values red green blue will be displayed on the screen
d) a table of the dictionary's keys and values will be displayed

9.5 Storing Data Fields in Dictionaries

One use of dictionaries is to collect a set of data fields with their names. In this arrangement, the field names act as the keys of the data. Program 9-5 shows an example of this for our contacts data, where the phone, last, and first fields are labeled values using the key/value structure of dictionaries.

Program 9-5

```
1  # contacts as a dictionary of dictionary records
2
3  # establish a dictionary of dictionaries
4  contacts = {
5      'jsmith@mymail.com':
6        {'phone': '801-555-1223', 'last':'Smith', 'first':'John'},
7      'jdoe12@mymail.com':
8        {'phone': '801-555-1223', 'last':'Doe', 'first':'Jane'},
9  }
10
11 # display the data as a table
12 for email, data in contacts.items():
13     print(email, data['first'], data['last'], sep='\t')
```

Program 9-5 Output

```
jsmith@mymail.com       John    Smith
jdoe12@mymail.com       Jane    Doe
```

In Program 9-5, lines 4 through 9 create a dictionary of dictionaries. The primary dictionary is the variable `contacts`, whose keys are email addresses, and whose values are secondary dictionaries. The secondary dictionaries contain three key/value pairs, where the keys are the names of the fields (`"phone"`, `"last"`, `"first"`). The keys point to the specific values for that entry.

Section Summary

One use of dictionaries is to store a collection of named data fields. This approach differs from storing data in lists, which can be less clear as data are accessed by index number. In this section, we looked at an example illustrating how a dictionary of dictionaries can be used to store contact data, where the key is email ID and the sub-dictionaries contain phone number, first name, and last name.

9.6 Maintaining Dictionaries

Earlier in the chapter, we reviewed how dictionaries can be created, including how to add keys and their values. In this section, we'll look further at how to maintain the data in a dictionary.

We've seen that using `dictionaryVariable[key] = value`, we can add a key and its value to a dictionary. The same syntax is used to change a key's value. That is, if the dictionary already has the key indicated in brackets, then its value is updated for that key.

Deleting a key and its value from the dictionary can be done with the `del` keyword. The syntax is:

```
del dictionary[key]
```

A dictionary can be combined with another dictionary using the update method, as in:

```
dict1.update(dict2)
```

If a key exists in both dictionaries, the value from `dict2` will be taken by `dict1` as part of the update.

Returning to the contacts example, Program 9-6 uses a few dictionary maintenance statements to allow interactive revisions and display of a dictionary.

Program 9-6

```
# contacts as a dictionary, with maintenance

contacts = {
    'jsmith@mymail.com': '801-555-1223',
    'jdoe12@mymail.com': '801-555-9887'
}

def add(contacts):
    id = input('email of contact to add/update: ')
    phone = input('phone of contact to add/update: ')
    contacts[id] = phone

def delete(contacts):
    key = input('email of contact to delete: ')
    if key in contacts: del contacts[key]
    else: print('Error, email not found in contacts')

def display(contacts):
    for email, phone in contacts.items():
        print(email, phone, sep='\t')

def menu():
    print('+ add   -delete   * display   / quit   ? help')

menu()
command = input('action: ')
while command not in ('', '/'):
    if   command == '+': add(contacts)
    elif command == '-': delete(contacts)
    elif command == '*': display(contacts)
    elif command == '?': menu()
    else: print('Unknown command')
    command = input('action: ')
```

Program 9-6 Example Run

```
+ add   - delete   * display   / quit   ? help
action: +
email of contact to add/update: djones@mymail.com
phone of contact to add/update: 801-555-5687
action: -
email of contact to delete: jsmith@mymail.com
action: *
jdoe12@mymail.com 801-555-9887
djones@mymail.com 801-555-5687
action: /
```

Section Summary

Dictionary data can be maintained by adding or deleting a key/value pair, as well as by changing a key's value. Adding data was covered in an earlier section. If a key already exists in a dictionary, changing that key's value can be done with a statement like `dictionary[key] = value`. Deleting a key/value pair from a dictionary can be done with the `del` keyword.

Test Your Understanding

1) What will happen after these statements are executed?

```
colors = {'r':'red', 'g':'green',
  'b':'blue'}
colors['y'] = 'yellow'
```

a) an exception will be generated
b) the dictionary colors will have four key/value pairs
c) the variable `'y'` will be set to `'yellow'`
c) none of the above

2) What will happen after these statements are executed?

```
colors = {'r':'rd', 'g':'green',
  'b':'blue'}
colors['r'] = 'red'
```

a) an exception will be generated
b) the dictionary colors will have three key/value pairs
c) the dictionary colors will have four key/value pairs
d) the variable `'r'` will be set to `'red'`

9.7 Sets

A Python set is a collection whose items are distinct. A set is somewhat like a dictionary with keys but no values. Sets are useful if you need your program to have a collection of distinct items. You can also concisely work with pairs of sets to, for example, find the union or intersection of two sets. This section covers some of the features of sets.

An empty set can be created by using the `set()` function. Items can be added to a set using the `add()` method. Alternatively, if the items are known in the program, the curly brace { } syntax can be used, like in dictionaries. For example:

```
colors = {'red', 'green', 'blue'}
```

This is equivalent to the statements:

```
colors = set()
colors.add('red')
colors.add('green')
colors.add('blue')
```

Although in the above statements, we add items to the set in a certain order (`'red'`, `'green'`, then `'blue'`), Python disregards your order and keeps set items in its own order. This allows sets to be efficient when determining if a certain item is in the set. If you need to maintain a certain order to items, then you should use a list or tuple.

In order to remove an item, the set `remove()` method can be used.

The `for` statement can be used to iterate over all the items in a set.

There are some useful operations for two sets; these are summarized in Table 9-2.

Operation	Syntax	Description
Union	a | b	Set of items in *a* or *b*
Intersection	a & b	Set of items in both *a* and *b*
Difference	a - b	Set of items in *a* that are not in *b*
Symmetric difference	a.symmetric_difference(b)	Set of items in *a* and *b* but not in both

Table 9-2: Set operations

For example, below we find the union and intersection of two sets:

```
>>> set1 = {'green', 'blue', 'red'}
>>> set2 = {'green', 'blue', 'yellow'}
>>> set1 | set2
{'blue', 'red', 'yellow', 'green'}
>>> set1 & set2
{'blue', 'green'}
```

Program 9-7 shows an interactive program that allows you to add and remove entries from a set.

Program 9-7

```
# build a contact set dynamically from user input

aset = set() # create an empty set

print('Enter email at > prompt (or <return> when done)')
entry = input('> ')
while entry != '':
    if entry[0] == '-': # a leading - indicates remove
        item = entry[1:].strip()
        if item in aset:
            aset.remove(item)
        else:
            print('** error, entry not found in set')
    else:
        aset.add(entry) # add entry to set (if not already in)
    print(aset)
    entry = input('> ')
```

Program 9-7 Example Run

```
Enter email at > prompt (or <return> when done)
> jsmith@mymail.com
{'jsmith@mymail.com'}
> jdoe12@mymail.com
{'jsmith@mymail.com', 'jdoe12@mymail.com'}
> jsmith@mymail.com
{'jsmith@mymail.com', 'jdoe12@mymail.com'}
> -jdoe12@mymail.com
{'jsmith@mymail.com'}
>
```

In addition to strings, Python sets can keep track of certain other types. These types are discussed in the last section of this chapter.

The four primary Python built-in collections are lists and tuples (covered in Chapter 6) and dictionaries and sets. Other types of collections are covered in later chapters, notably the pandas DataFrame collection for representing matrices in the next chapter.

Section Summary

Sets provide another Python collection type, which can be used for storing distinct items together. A set can be created with the `set()` function or with specific values within curly braces. As with lists and dictionaries, sets provide ways to maintain their data, including concise methods and symbols.

9.8 An Example with Dictionaries and Files

In this section, we'll look at a program that uses a dictionary to combine the data from two files. The goal is to display the total sales for each product, based on product price data in one file and unit sales data in another file. Because both files use product IDs, we can use these IDs as the keys in our dictionary to correlate the data and display a report of total sales by product.

The price data are in the file products.txt presented in Section 8.6. Here are the first few lines of the file:

```
product-id,division,price
P-001,diva,5.50
P-001x,diva,5.95
```

The sales data are in the file sales.txt and contain unit sales data with these fields: date, product ID, region, and count. Here are the first few lines of this file:

```
date,product-id,region,count
2021-01-05,P-001,north,5
2021-01-05,P-001x,south,10
```

In order to total sales for each product, Program 9-8 first creates a dictionary based on the products.txt file. Then it reads the sales.txt file and, for each row, adds to the sales in the dictionary.

The first dictionary-related statement is line 15, which creates an empty dictionary. Then line 21 adds to this dictionary with the product ID as the key and a sub-dictionary as the value. The sub-dictionary stores the product price, as well as creating a key `'sales'` with an associated value initialized to 0.

The `readSalesFile()` function reads each line from the sales file, and cumulates the sales data in the dictionary. Specifically line 32 uses the product ID to look up the price in the dictionary. Then sales are cumulated on line 33. Because we are using a dictionary of dictionaries, lines 32 and 33 use two keys—first the product ID, and then the name of the specific field in the sub-dictionary (`'price'`, `'sales'`).

On line 42 the string `ljust()` method is used to ensure the product ID is displayed left justified with added blanks if needed to display six positions in order to provide a tabular output.

Program 9-8

```
# read a product file into a dictionary, keyed by product ID

# supporting function to parse product file line
def parseProductLine(line):
    fields = line.split(',')
    return fields[0], fields[1], float(fields[2])

# supporting function to parse sales file line
def parseSalesLine(line):
    fields = line.split(',')
    return fields[0], fields[1], fields[2], int(fields[3])

# read product file into a dictionary of dictionaries
def readProductFile(filename):
   adict = {}
   fin = open('products.txt')
   fin.readline() # skip header
   for line in fin:
      line = line.rstrip('\n')
      id, div, price = parseProductLine(line)
      adict[id] = {'price': price, 'sales': 0} # add sub-dictionary
   fin.close()
   return adict

# read sales file, cumulating sales by product
def readSalesFile(filename, products):
   fin = open(filename)
   fin.readline() # skip header
   for line in fin:
      line = line.rstrip('\n')
      date, id, region, count = parseSalesLine(line)
      price = products[id]['price']
      products[id]['sales'] += price * count # cumulate sales
   fin.close()
   return products

products = readProductFile('products.txt')
readSalesFile('sales.txt', products)

# display results
for id in products.keys():
   print(id.ljust(6), format(products[id]['sales'], '8.2f'))
```

Program 9-8 Output

```
P-001     27.50
P-001x    59.50
P-002    130.00
P-002x    34.75
P-010    102.50
P-010x   107.50
Q-011    138.00
Q-011x    23.90
```

Section Summary

In this section, we looked at how dictionaries can be used in combination with files to relate pricing and sales data. In the example presented, a dictionary of dictionaries is the data structure used to cumulate total sales for each product. Because dictionaries use key values to organize their data, they can be useful for joining data together.

9.9 About the Hashable Type

In this chapter, we've used strings as dictionary keys. While strings are a common choice for dictionary keys, a Python dictionary can have other types as keys. For example, an integer can be a dictionary key. However, dictionary key types are limited to Python types that are ***hashable***. A hashable type is one that contains a stable value and can be compared to another value.

For example, let's say you want to use a two-part key—for example, an email username and the domain. A list with two values might seem like an option. We can test this with:

```
>>> {['jsmith', 'mymail.com']: 0}
```

But in this case, we get an error:

```
Traceback (most recent call last):
  File "<pyshell>", line 1, in <module>
TypeError: unhashable type: 'list'
```

This error is indicating that lists are not hashable and so one can't be used as a Python dictionary key. This is because lists aren't stable—their items can be changed. If you want to use a multiple-part key, a tuple is an alternative choice. Tuples are like lists, but once formed their items can't be changed. We can check this:

```
>>> {('jsmith', 'mymail.com')}
```

In this case, we don't get an error, confirming that a tuple is a valid Dictionary key.

The items in Python sets are also limited to hashable types. For example, you can create a set of tuples, but not a set of lists.

Section Summary

Dictionary keys are limited to types that are hashable. These are stable, comparable values. Some examples are `str`, `int`, and `tuple`. As well, set items are limited to hashable types. You can test whether a type is hashable by interactively trying to create a small dictionary or set.

Chapter Review

The dictionary is a Python collection type for storing key/value pairs. This is a useful alternative to lists if the data have a unique key, if a named set of fields is desired, or if a tree structure of data is useful. Dictionaries provide ways to add, change, and delete data. Additionally, the `for` statement can be used to iterate over the keys and/or values of a dictionary. A Python set is another collection type; sets contain items without any duplicate values.

Key Terms

dictionary	A Python collection type for storing key/value pairs.
empty dictionary	An empty dictionary can be created using curly braces, as in d = { }.
Hashable	A Python type that is stable and comparable.
key	The key of a dictionary is associated with a value and can be used to look up the value.
key/value pair	A dictionary contains a set of keys; each key points to a value.
set	A Python collection type for storing distinct items.

Programming Problems

1) Country Statistics Lookup—One Measure

Assume a program begins with the statement:

```
popData = {
'US': 325.7, 'CA': 36.5, 'MX': 129.2}
```

Write subsequent statements that (in a loop) prompt the user for a country code (US, CA, or MX), then look up in the above dictionary for the corresponding population value, and display that value. Example session:

```
Please enter a country code: CA
CA population = 36.5
```

Your program should handle the cases where a user-entered country code is not found.

2) Country Statistics Lookup—Multiple Measures

Assume a program begins with the statement:

```
allData = {
'US': {'pop':325.7, 'gdp': 19.39, 'ccy': 'USD', 'fx': 1.0},
'CA': {'pop': 36.5, 'gdp': 1.65, 'ccy': 'CAD', 'fx': 1.35},
'MX': {'pop':129.2, 'gdp': 1.15, 'ccy': 'MXN', 'fx': 19.68}
}
```

Write subsequent statements that (in a loop) prompt the user for a country code (US, CA, or MX), then prompt for a measure name (pop, gdp, ccy, or fx), then look up in the above dictionary for the corresponding value, and display that value. Example run:

```
Please enter a country code: CA
Please enter a statistic: pop
CA pop = 36.5
```

Your program should handle the cases where a user-entered country code or measure name is not found.

3) Word Frequency from File

Write a program that reads words from a file and determines and displays the number of occurrences of each distinct word. Assume that each line of the file can have one or more words and that each word may be in lowercase, uppercase, or mixed case.

4) Total Sales from Transaction File

Write a program that reads a file of sales transactions and determines and displays the total sales for each product. Assume the file has the following fields:

```
transaction # (an integer)
sale date (a date in format yyyy-mm-dd)
product ID (a string)
customer # (an integer)
units sold (an integer)
price per unit (a float)
```

5) Mode of a List

Write a program that prompts the user for a list of numbers and determines the mode of the list. The mode is the value (or values) that occur the most often. For example, the mode of [1, 2, 3, 2] is 2. The mode of [1, 2, 3, 2, 3] is [2, 3]. To solve this problem, use lists and a dictionary.

Here are some suggested functions you could define as part of the solution to this problem:

```
# build a dictionary of values and their counts
def findCounts(alist)
# find the largest value of key/value pairs in a dictionary
def findMaxCount(adict)
# find the keys having a given value
def findValuesWithCount(adict, count)
```

PART II

Python for Data

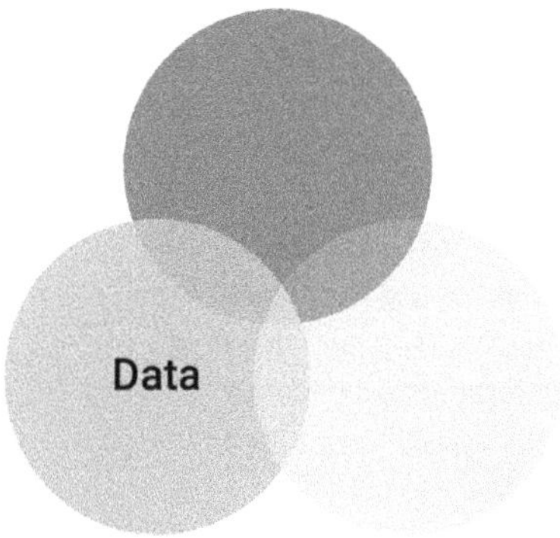

Accessing and organizing data are critical parts of building systems and data analysis. In the opening chapters of Part II, we look at the pandas package in detail for organizing, cleaning, and manipulating lists and tables of data. The pandas package also has concise ways to access tabular data from web pages, databases, and spreadsheets, and these are covered in corresponding chapters in Part II.

Data is available in a variety of forms, and Part II covers the most common of these. Chapter 12 introduces HTML and how to access web page data with pandas (for tabular data) and Beautiful Soup (for more general cases). As an alternative to web pages, web APIs (application programing interfaces) can provide stable programmatic access to data.

Data stored in relational databases can be accessed with SQL in conjunction with Python. A brief introduction to SQL is included in Chapter 14 along with coverage for using Python and SQL together with the SQLite database. The concepts and syntax covered in this chapter are applicable to other types of relational databases as well.

Spreadsheets are another potential source of data, and Python can be used to read Excel workbooks. Chapter 15 covers how pandas as well as the openpyxl package can be used to read Excel spreadsheet data into a Python program.

Chapter 10

pandas DataFrames and Series

Chapter Objectives

In this chapter, you'll learn:

- about the pandas package and its powerful DataFrame for working with data tables
- how to create a DataFrame from code or a data file
- how to use exploratory methods to get an overview of DataFrame data
- how to get DataFrame data by column, row, or an individual cell
- how to use concise methods to sort and filter your DataFrame data
- how missing values are handled in pandas
- operations for maintaining DataFrame data, including adding, changing, and deleting
- how to join two DataFrames into one

Chapter Outline

10.1 About pandas DataFrames and Series

About pandas

In Part I of the book, we've looked at many aspects of the Python language. In Parts II and III, we focus on how you can use Python in concert with third-party packages to access, transform, and analyze data. Python's extensibility makes it possible for others to create great packages that you can leverage to create analyses and systems, where each step can be performed in a few lines of code.

There are thousands of publicly available Python third-party packages. Among these, several have gained widespread adoption due to their power and ease of use. One of the most popular packages is pandas, which provides featureful types for working with tables of data.

The pandas package provides great tools for collecting, transforming, and analyzing tabular data. For example, if you have a database of sales transactions that you need to summarize to seek trends and opportunities, you can use pandas to:

- Read the specific database records of interest.
- Filter out incorrect data.
- Summarize the data according to different dimensions such as product type, region, and period.
- Display different types of charts to visualize the results.

Each of these steps can be done in relatively few lines of code. Further, pandas is not limited to working with databases; it can access tabular data from other sources such as text files, web pages, and spreadsheets. In order to help you leverage pandas, this chapter introduces you to the key pandas types, `DataFrame` and `Series`.

DataFrames and Series

Tables are often a natural way to organize data in programs. However, Python does not have a built-in table data structure. Although you can use lists or dictionaries to store tabular data, these data structures do not have a row/column organization.

The pandas package provides the data type `DataFrame`, a powerful and concise way to load, store, transform, process, and analyze tabular data. The pandas package also provides the `Series` type, which stores a list of items, each of which is named. These data structures, in combination with the more than 200 functions and methods provided by the package, allow you to write efficient, powerful programs with less code.

The pandas DataFrame allows you to store a matrix of data and name each column and row. Figure 10-1 shows the arrangement of a DataFrame.

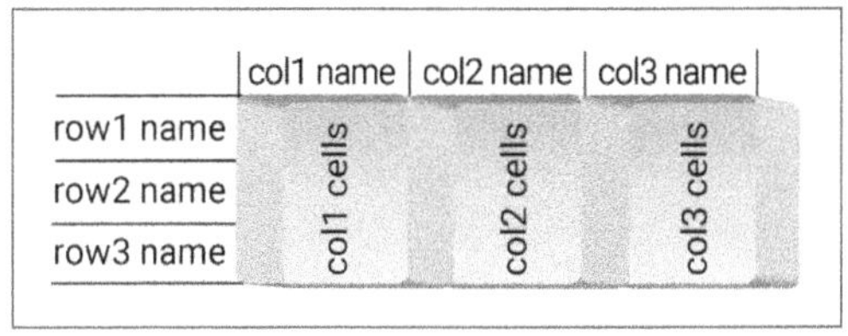

Figure 10-1: DataFrame layout

In pandas, the set of row names is referred to as the DataFrame's ***index***.

Notably, within each column of the DataFrame, the cells generally have the same type. Further, for numeric types (integers and real numbers), pandas uses the NumPy package's types so that a DataFrame is efficient in speed and memory usage. But pandas is not limited to numeric data—a DataFrame cell can contain *any* Python type.

In this chapter, we'll look at how to create a DataFrame in code or by reading a file. In subsequent chapters, we'll see how to populate a DataFrame from a database query, web page, or spreadsheet.

In addition to DataFrame, pandas provides the Series type. A Series is like a one-column DataFrame that optionally has a name. A Series is similar to a Python list, except each item is named instead of numbered. Some DataFrame methods return a Series as their result. Figure 10-2 shows the layout of a Series.

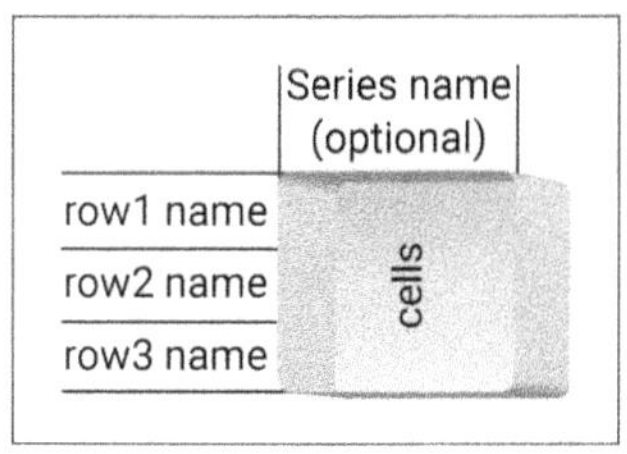

Figure 10-2: Series layout

In addition to covering how to create DataFrames, this chapter focuses on other basics: how to access and maintain their data, how missing values are handled, and how you can join two DataFrames. The next chapter covers features of pandas that allow you to further manipulate and prepare data.

In order to use pandas, you need to have it installed on your computer. Appendix B describes how to install packages. In addition, to use pandas, your program needs to load it into memory by using the `import` statement. A common way to do this is with the statement:

```
import pandas as pd
```

In the above `import` statement, we are indicating that we'd like to use the pandas package. Also, the `as` keyword allows us to use an alternative name when referring to the package, in this case `pd`. As we'll see, this allows certain statements to be a little shorter. For more information about packages and the import statement, see Appendix B.

Spreadsheet Compare/Contrast
DataFrames and Spreadsheet Sheets

A spreadsheet is made up of tabular sheets, which are conceptually similar to DataFrames. However, spreadsheet rows are named alphabetically (A, B, C, . . .) and rows are named numerically (1, 2, 3 . . .), whereas DataFrames allow you to name the rows and columns in more meaningful ways, similar to Python variables.

Another difference is that while DataFrames generally have one type within each of their columns, spreadsheets allow for any type to be placed in any cell. In this sense, spreadsheets can be more free form (but also tabular), whereas DataFrames are always tabular.

Section Summary

The key types in pandas are `DataFrame` and `Series`. Since tabular data are common and Python doesn't provide a table type, pandas fills this gap by providing the `DataFrame` type for storing and manipulating data tables. A Series can be thought of as a table with one column. Notably, DataFrames allow us to name the rows and columns of our data. Having these names, in combination with the many methods pandas provides, allows for concise and powerful statements to manipulate our data, as we'll see in the subsequent sections in this chapter.

10.2 Creating DataFrames

pandas provides a number of ways to create a DataFrame. In this section, we'll see how you can define a DataFrame in code, as well as how to read the contents of a text file into a DataFrame. In subsequent chapters, we'll see how we can use pandas to concisely access tabular data from other sources including web pages, dates, and spreadsheets. In practice, usually DataFrame data are loaded from outside your program, but if you have a small table of fixed data, a DataFrame can be easily constructed in your program directly.

For this chapter, we'll use two datasets, one small and one larger. For the small data, we'll return to the AQI table introduced in Chapter 6, which represents different air quality levels and their associated colors and concerns, as shown in Table 6-1. For the larger dataset, we'll use the list of stocks in the S&P 500 index. Our S&P stock data have a few factual columns about each stock in the S&P 500 that might form the starting point for analyses comparing stocks or sectors in terms of valuation, prices, or other financial measures.

In Chapter 6, we represented the AQI levels using three parallel lists. As an alternative, we can represent the data in one DataFrame, which keeps the related data together. As with lists and dictionaries, we can define a DataFrame and its data in one statement. To do this, we can provide a dictionary where the keys are column names, and dictionary values are in lists. Program 10-1 shows an example for this.

Program 10-1

```
# define AQI levels as a DataFrame of colors, concerns, and ranges
# using a dictionary of parallel lists

import pandas as pd

# define the dictionary of parallel lists
dictAQI = {
    'color': ['Green', 'Yellow', 'Orange', 'Red', 'Purple', 'Maroon'],
    'concern': ['Good', 'Moderate', 'Unhealthy for Sensitive Groups',
                'Unhealthy', 'Very Unhealthy', 'Hazardous'],
    'low': [0,  51,  101, 151, 201, 301],
    'high':[50, 100, 150, 200, 300, 999]
    }

# create a DataFrame from the dictionary
dfAQI = pd.DataFrame(dictAQI)

# display the DataFrame
print(dfAQI)
```

Program 10-1 Output

```
    color                         concern  low  high
0   Green                            Good    0    50
1  Yellow                        Moderate   51   100
2  Orange  Unhealthy for Sensitive Groups  101   150
3     Red                       Unhealthy  151   200
4  Purple                  Very Unhealthy  201   300
5  Maroon                       Hazardous  301   999
```

As an alternative to defining a DataFrame in code, your program can load a text file's data into a DataFrame. The function in pandas for this is `read_csv()`. The function has a lot of intelligence and options—if you have a file with data in a tabular layout, `read_csv()` can read it into a DataFrame. Although its name includes csv (short for comma-separated values), this function can read text files with other types of delimiters such as spaces, tabs, or other characters.

The simplest use of `read_csv()` is to provide it with the name of the file to read data from. For example, the statement:

```
dfAQI = pd.read_csv('aqi.txt')
```

will read a comma-separated text file called aqi.txt, and create a DataFrame variable called `dfAQI`. The `read_csv()` function also assumes by default that the first row of the file contains the column headings of the data, and the DataFrame will use these as the column names.

Tip
Data File Location

In this chapter, the data files are assumed to be in the same directory as the program you are running. This simplifies the file arrangement, but data files can be located in other directories. In fact, it is often preferable to separate the location of code from that of data. If your data file is in a different directory from your program, you can refer to it in `read_csv()` using an absolute path, like /Users/jsmith/data/aqi.txt, or using a path that's relative to your program's directory, like ../data/aqi.txt. Further, to specify path names independent of operating system convention (e.g., forward vs. backward slashes), Python's os.path module has useful functions.

For example, assume we have the AQI data in a file, aqi.txt, that reads as follows:

```
color,concern,low,high
Green,Good,0,50
Yellow,Moderate,51,100
Orange,Unhealthy for Sensitive Groups,101,150
Red,Unhealthy,151,200
Purple,Very Unhealthy,201,300
Maroon,Hazardous,301,999
```

We can read these data into a DataFrame using `read_csv()`, as shown in Program 10-2.

Program 10-2

```
# define AQI levels as a DataFrame of colors, concerns and ranges
# use read_csv to populate a DataFrame from a file

import pandas as pd

dfAQI = pd.read_csv('aqi.txt')

print(dfAQI)
```

Program 10-2 Output

```
    color                         concern  low  high
0   Green                            Good    0    50
1  Yellow                        Moderate   51   100
2  Orange  Unhealthy for Sensitive Groups  101   150
3     Red                       Unhealthy  151   200
4  Purple                  Very Unhealthy  201   300
5  Maroon                       Hazardous  301   999
```

As of pandas version 1.2.2, the `read_csv()` function has 50 named parameters, to help guide the reading of the file and the setup of the DataFrame. This allows you to read tabular data from a text file in a wide variety of cases. For example, the index (row names) of the DataFrame can be set with a column of data by specifying the `index_col` parameter. If your data have a unique key (a column with distinct values), you can use these values as the index to name each row. For example, for our AQI data, the color name is unique for every level.

Another useful parameter is `comment`, which allows you to specify that there are comment lines interspersed with the data lines in the file. By setting `comment`, `read_csv()` will know which lines to skip over when reading the data file and forming the DataFrame. Another useful parameter, `delimiter`, indicates the file has a data separator other than the comma. To demonstrate the use of these parameters, `read_csv()` could be called as follows:

```
dfAQI = pd.read_csv('aqi_pipe.txt', index_col='color', comment='#',
                    delimiter='|')
```

In this case, the DataFrame's index will be set to the data in the color column. Lines beginning with # will be skipped over and not used to populate the DataFrame. Also, the separator between fields is specified as the pipe character `|`, which will be assumed instead of commas.

For example, the file aqi_pipe.txt uses the pipe character | as a delimiter, and has a comment line starting with #, as shown below:

```
color|concern|low|high
# data for Air Quality Index
Green|Good|0|50
Yellow|Moderate|51|100
Orange|Unhealthy for Sensitive Groups|101|150
Red|Unhealthy|151|200
Purple|Very Unhealthy|201|300
Maroon|Hazardous|301|999
```

Program 10-3 shows how to read the aqi_pipe.txt file using named parameters with the `read_csv()` function.

Program 10-3

```
# define AQI levels as a DataFrame of colors, concerns and ranges
# use read_csv() to populate a DataFrame from a file

import pandas as pd

dfAQI = pd.read_csv('aqi_pipe.txt', index_col='color', comment='#',
                    delimiter='|')

print(dfAQI)
```

Program 10-3 Output

```
                              concern  low  high
color
Green                            Good    0    50
Yellow                       Moderate   51   100
Orange  Unhealthy for Sensitive Groups  101   150
Red                         Unhealthy  151   200
Purple                 Very Unhealthy  201   300
Maroon                      Hazardous  301   999
```

Notebooks such as Jupyter Notebook and Google Colab allow you to indicate the name of a DataFrame variable as the last statement in the cell, and provide a more formatted display of the table. Below, we use a code cell in a notebook to read the aqi.txt file shown earlier in this section and display the DataFrame.

```
[1]: import pandas as pd
     dfAQI = pd.read_csv('aqi.txt', index_col='color')
     dfAQI
```

	concern	low	high
color			
Green	Good	0	50
Yellow	Moderate	51	100
Orange	Unhealthy for Sensitive Groups	101	150
Red	Unhealthy	151	200
Purple	Very Unhealthy	201	300
Maroon	Hazardous	301	999

Note that in the above, our code input cell has three lines. The last statement, simply `dfAQI`, asks that the DataFrame dfAQI be displayed as output. Note that the column and row labels are displayed as bold.

In general, pandas is well suited for use in .py programs as well as notebooks, since, as we'll see in the chapter, you can use very concise and few statements to work with DataFrames and Series.

Section Summary

DataFrames can be created in a variety of ways, depending on where your data are. For static data, you can create a DataFrame by defining data in code using a dictionary of data. In this case, the dictionary keys become column names, and the dictionary values are lists of data, which become DataFrame columns. Alternatively, your program can read its data from a text file. (In later chapters, we'll see other ways to read data into a DataFrame, such as from a database, web page, or spreadsheet.). To read a file into a DataFrame, use the `read_csv()` function, indicating the name of the file. `read_csv()` has a wide array of parameters to tailor the read to the specifics of the file. For example, the `delimiter` parameter can be used to specify a field separator other than a comma. Other parameters include `index_col`, which indicates which data will be used to name the rows, and `comment`, which indicates the character that introduces lines in the data file that can be ignored when loading the data.

Test Your Understanding

1) The pandas function that can read a file's data into a DataFrame is:
   a) `load_data()`
   b) `read_csv()`
   c) `read_data()`

2) When reading data from a file into a DataFrame, the data must be separated by commas.
   a) True
   b) False

10.3 Getting Information about DataFrame Data

In the previous section, we looked at a pandas DataFrame for storing a small table of AQI data. As shown, data can be set in code or read from outside the program—for example, from a file. But DataFrames are not limited to small datasets. In fact, given the wide array of methods available, combined with their efficient numeric data storage, DataFrames are well suited for large datasets. And given their concise syntax, DataFrames are suitable as a building block in your programs, as well as for exploring data interactively. In this section, we'll look at some simple ways to get an overview of your DataFrame data.

In this section, we'll look at data about the stocks in the S&P 500. This dataset has certain facts about the stocks and their companies and, in combination with other data, could be the basis for different financial analyses. The stock data are of course a larger dataset than the AQI data previously used. A file of these data is available as SP500_Constituents.csv. To read these data into a DataFrame, assuming the data file is in the same directory as our Notebook or Python session, we can use:

```
[1]: import pandas as pd
     dfsp = pd.read_csv('SP500_Constituents.csv')
```

As seen in the previous section, we can name each row using a distinct data value as the row index. In this file, a distinct data item is in the first column, which has the header "Symbol." To use these data under "Symbol" as our row names, we would use the statement:

```
[2]: dfsp = pd.read_csv('SP500_Constituents.csv', index_col='Symbol')
```

If you try this statement and get an error, make sure the data file is in the same directory as your program or notebook.

head() and tail()

After we've loaded the data into a DataFrame, we typically want to confirm the data were loaded OK. Unlike the AQI data, the file with the stock data has many rows. In other data files, you could have thousands or even millions of rows. Rather than display all the rows, we can initially check if the data were loaded OK by reviewing the first few rows using the head() method:

```
[3]: dfsp.head()
```

Symbol	Security	SEC filings	GICS Sector	GICS Sub-Industry	Headquarters Location	Date first added	CIK	Founded
MMM	3M Company	reports	Industrials	Industrial Conglomerates	St. Paul, Minnesota	1976-08-09	66740	1902
ABT	Abbott Laboratories	reports	Health Care	Health Care Equipment	North Chicago, Illinois	1964-03-31	1800	1888
ABBV	AbbVie Inc.	reports	Health Care	Pharmaceuticals	North Chicago, Illinois	2012-12-31	1551152	2013 (1888)
ABMD	ABIOMED Inc	reports	Health Care	Health Care Equipment	Danvers, Massachusetts	2018-05-31	815094	1981
ACN	Accenture plc	reports	Information Technology	IT Consulting & Other Services	Dublin, Ireland	2011-07-06	1467373	1989

The head() method by default returns the first 5 rows, but you can get more or fewer rows by providing an integer—for example, to get the first 10 rows, use:

```
dfsp.head(10)
```

Since the head() method returns a DataFrame, we can also assign the result to a variable; for example:

```
[4]: dfsp5 = dfsp.head()
```

In the above, the variable dfsp5 is a DataFrame with 5 rows and 8 columns.

Many DataFrame methods return a DataFrame or Series as a result, which makes pandas useful for successive statements as you refine and explore data.

Similar to head(), the tail() method returns the last rows of the DataFrame (as a DataFrame).

In a notebook, indicating a DataFrame variable as the last line in an input cell will by default show the first 30 and last 30 rows of a DataFrame. You can control how many rows are displayed by setting a property—for example, using the following statement:

```
pd.options.display.max_rows = 10
```

To show all rows, this property can be set to None:

```
pd.options.display.max_rows = None
```

Similarly, if your DataFrame has a lot of columns, Jupyter Notebook will display the first few and last few, with . . . in between. The property to control the column count display is options.display.max_columns.

`len()` and `shape`

The `len()` function works with DataFrames to indicate the number of rows. For example:

```
[5]: len(dfsp)
505
```

indicates that there are 505 rows in our table. (Note that the S&P 500 can have more than 500 stocks in its makeup.)

To find the number of rows and columns, use the `shape` property, as in:

```
[6]: dfsp.shape
(505, 8)
```

which indicates that the dataset has 505 rows and 8 columns.

`info()`

Another useful overview method is `info()`, which displays column names, count of data items in each column, and the type of data. For example:

```
[7]: dfsp.info()
<class 'pandas.core.frame.DataFrame'>
Index: 505 entries, MMM to ZTS
Data columns (total 8 columns):
 #   Column                 Non-Null Count  Dtype
---  ------                 --------------  -----
 0   Security               505 non-null    object
 1   SEC filings            505 non-null    object
 2   GICS Sector            505 non-null    object
 3   GICS Sub-Industry      505 non-null    object
 4   Headquarters Location  505 non-null    object
 5   Date first added       450 non-null    object
 6   CIK                    505 non-null    int64
 7   Founded                502 non-null    object
dtypes: int64(1), object(7)
memory usage: 35.5+ KB
```

In the above output, the first column indicates:

- the name of each column in the data (Security, SEC filings, etc.)
- the number of data elements
- the type of the data column

In our example, there are some aspects to highlight. Note that for the "Founded" column, there are 502 non-null items. The term ***non-null*** indicates where there is populated (as opposed to missing) data. For Founded, this indicates that three rows (502 vs. 505 rows overall) are missing data for this data item. Similarly, the "Date first added" column has fewer than 505 non-null data items, so data are missing here as well. All the other columns have 505 data items and so are fully populated. This shows a key use of the `info()` function—we can quickly determine how well populated our dataset is. In a later section, we'll discuss how DataFrames handle missing values.

The last piece of information that `info()` provides is the data type of the column. There is one column marked differently—the CIK column is noted as `int64`. This highlights that pandas uses special types for its numeric values, in particular types provided by the NumPy package. pandas uses NumPy numeric types, since these are more efficient in space and speed relative to Python data types. In our example data, `int64` indicates that the NumPy type `int64` is used for the CIK column. (CIK stands for "Central Index Key," which is used to identify corporations.) Another commonly used numeric type in pandas is the `float64` type, used for floating point numbers. In practice, we can do numeric operations in pandas comfortably without worrying about the distinction between Python `float` and `int` vs. NumPy `float64` and `int64`, as we'll see in this and subsequent chapters. The other columns are noted as `object`, indicating these column values aren't NumPy types; in this case, these columns are Python strings.

This section has so far assumed interactive use to explore a DataFrame's data. Program 10-4 recaps some of the statements presented in this section in a Python .py program that can be used for data overview.

Program 10-4

```
# Display selected information about the S&P 500 stock data
import pandas as pd

filename = 'SP500_Constituents.csv'
indexcol = 'Symbol'

dfsp = pd.read_csv(filename, index_col=indexcol)

print('Rows=', dfsp.shape[0], ';', 'Columns=', dfsp.shape[1], '\n')

dfsp.info()

print(dfsp.head())
```

Program 10-4 Output

```
Rows= 505 ; Columns= 8

<class 'pandas.core.frame.DataFrame'>
Index: 505 entries, MMM to ZTS
Data columns (total 8 columns):
 #   Column                Non-Null Count  Dtype
---  ------                --------------  -----
 0   Security              505 non-null    object
 1   SEC filings           505 non-null    object
 2   GICS Sector           505 non-null    object
 3   GICS Sub-Industry     505 non-null    object
 4   Headquarters Location 505 non-null    object
 5   Date first added      450 non-null    object
 6   CIK                   505 non-null    int64
 7   Founded               502 non-null    object
dtypes: int64(1), object(7)
memory usage: 35.5+ KB
                      Security SEC filings … CIK           Founded
Symbol …
MMM               3M Company     reports … 66740            1902
ABT       Abbott Laboratories     reports … 1800             1888
ABBV             AbbVie Inc.     reports … 1551152   2013 (1888)
ABMD             ABIOMED Inc     reports … 815094           1981
ACN           Accenture plc     reports … 1467373          1989

[5 rows x 8 columns]
```

The output from Program 10-4 includes . . . , which indicates that there are additional columns in the DataFrame that were not displayed. This allows you to see a subset of the columns that fit on the screen, which is useful if there are many columns. As we'll see later, pandas also uses . . . when a subset of rows is shown.

Program 10-4 is written to read and give information about the S&P 500 data, but it can easily be modified to work with any comma-separated value file. Program 10-5 is a variation that will work with other text files.

Program 10-5

```
# Display selected information about a comma-separated value file
import pandas as pd

filename = input('Enter filename: ')
indexcol = input('Enter index column name: ')

df = pd.read_csv(filename, index_col=indexcol)

print('Rows=', df.shape[0], ';', 'Columns=', df.shape[1], '\n')

df.info()

print(df.head())
```

Program 10-5 Output

```
Enter filename: aqi.txt
Enter index column name: color
Rows= 6 ; Columns= 3

<class 'pandas.core.frame.DataFrame'>
Index: 6 entries, Green to Maroon
Data columns (total 3 columns):
 #   Column   Non-Null Count  Dtype
---  ------   --------------  -----
 0   concern  6 non-null      object
 1   low      6 non-null      int64
 2   high     6 non-null      int64
dtypes: int64(2), object(1)
memory usage: 192.0+ bytes
                                 concern  low  high
color
Green                               Good    0    50
Yellow                          Moderate   51   100
Orange  Unhealthy for Sensitive Groups   101   150
Red                            Unhealthy  151   200
Purple                    Very Unhealthy  201   300
```

Section Summary

The pandas package provides a few methods and properties to get an overview of a DataFrame's data. The `head()` and `tail()` methods return the first and last rows of a DataFrame (as a DataFrame). These provide a quick way to see if our DataFrame data have been loaded OK. To determine the number of rows, the `len()` function can be used. A DataFrame also has a `shape` property that provides the number of rows and columns. The `info()` method gives a brief overview of the columns in a DataFrame, including their type and how many values are populated.

Test Your Understanding

1) The `head()` and `tail()` methods can only show one row of a DataFrame's data.
   a) True
   b) False

2) Which can be used to determine the number of columns in a DataFrame?
   a) the `len()` function
   b) the `shape` property
   c) the `info()` method
   d) (b) and (c)

10.4 Getting Data: Columns, Rows, and Cells

In the previous sections, we've looked at how to load data into a DataFrame as well as how to get some information about it. In this section, we'll see how to access one or more columns, as well as how to see a particular row. We can also access an individual cell's value. In the next section, we'll look at ways to see multiple rows through sorting, filtering, and other techniques.

Columns

In this and subsequent sections, we'll work interactively (at the Python >>> prompt, or in a Notebook cell) and assume you've run the following statements to set up our two example DataFrames:

```
[1]: import pandas as pd
     dfAQI = pd.read_csv('aqi.txt', index_col='color')
     dfsp = pd.read_csv('SP500_Constituents.csv', index_col='Symbol')
```

To access a particular column of a DataFrame, we can use square brackets [ ], consistent with how items of Python lists or dictionaries can be accessed. For example:

```
[2]: dfAQI['concern']

     color
     Green                                Good
     Yellow                           Moderate
     Orange    Unhealthy for Sensitive Groups
     Red                             Unhealthy
     Purple                     Very Unhealthy
     Maroon                          Hazardous
     Name: concern, dtype: object
```

Note that the result displays the row names, as well as the values for "Concern." In fact, the result returned is a pandas Series. We can check this by doing the following:

```
[3]: type(dfAQI['concern'])

     pandas.core.series.Series
```

If we'd like a DataFrame's column values as a Python list—for example, if we need to save these to a file in a specific format—we can use the `tolist()` method:

```
[4]:  dfAQI['concern'].tolist()

      ['Good',
       'Moderate',
       'Unhealthy for Sensitive Groups',
       'Unhealthy',
       'Very Unhealthy',
       'Hazardous']
```

Since accessing a DataFrame's column is a common operation, there is a shorthand. If the column name is a valid Python variable name (without invalid characters like spaces), we can access it as a property. For example:

```
dfAQI.concern
```

To access several columns of a DataFrame, we provide a `list` of names inside square brackets. For example, to see the AQI colors and their corresponding numeric ranges, we can use:

```
[5]:  dfAQI[['low', 'high']]
```

	low	high
color		
Green	0	50
Yellow	51	100
Orange	101	150
Red	151	200
Purple	201	300
Maroon	301	999

Note that in the above case, the result is itself a DataFrame.

To access the values in the index (the row names), we can use the index property, like:

```
dfAQI.index
```

which returns a type of `Index`; to get the index values as a `list`, we can use:

```
[6]:  dfAQI.index.tolist()

      ['Green', 'Yellow', 'Orange', 'Red', 'Purple', 'Maroon']
```

The column selection statements we've used with the AQI data would work similarly with the S&P 500 data. For example, to access the names of the securities in the S&P 500, we can use:

```
[7]:  dfsp.Security

      Symbol
      MMM                3M Company
      ABT       Abbott Laboratories
      ABBV              AbbVie Inc.
      ABMD              ABIOMED Inc
      ACN             Accenture plc
      ...
      YUM              ! Brands Inc
      ZBRA      Zebra Technologies
      ZBH             Zimmer Biomet
      ZION            Zions Bancorp
      ZTS                    Zoetis
      Name: Security, Length: 505, dtype: object
```

Or to get the symbols of the S&P 500 in a Python list, we can use:

```
dfsp.index.tolist()
```

Rows

Having looked at variations for accessing a DataFrame's columns, let's look at how we can access a particular row of a DataFrame. We can do this using a row name or a row number. If our DataFrame rows have meaningful row names (the DataFrame's index), the `.loc[]` method is a good way to access a particular row of the DataFrame. For example, from the AQI data, to get the information related to the `'Yellow'` level of concern, use:

```
[8]: dfAQI.loc['Yellow']

concern    Moderate
low              51
high            100
Name: Yellow, dtype: object
```

Note that `.loc[]` returns a pandas Series, where each data value (in the right column above) is labeled with the DataFrame's column name (in the left column above). You can think of `.loc[]` as getting the row's data but also retaining the column headers in the result returned. Another way to think of the Series is as an ordered set of name/value pairs. In the above output, the row orientation in the DataFrame is rotated to a column orientation of the Series. Further in this case, the returned Series itself has a name (noted as Yellow above)—in general, a Series may or may not have a name. In this case, the Series name is taken from the index name of the selected row from the DataFrame.

As an alternative to using named access to a row using `.loc[]`, you can use numbered access to a row using `.iloc[]`. For example, what is the first (0th) row of the AQI data? This can be found by doing the following:

```
[9]: dfAQI.iloc[0]

concern    Good
low           0
high         50
Name: Green, dtype: object
```

As with `.loc[]`, the result is a named Series.

The `.loc[]` method allows for a negative index, which like a Python `list`, counts from the end. For example, the last row of our AQI table, which contains the highest level of concern, can be accessed by doing the following:

```
[10]: dfAQI.iloc[-1]

concern    Hazardous
low              301
high             999
Name: Maroon, dtype: object
```

We've used `.loc[]` and `.iloc[]` on the small AQI data, and of course it works well on larger data, such as our S&P 500 dataset. In that data, we have the security's symbol as the row name. To access the data for 3M, whose symbol is MMM, we can use:

```
[11]: dfsp.loc['MMM']

Security                                    3M Company
SEC filings                                    reports
GICS Sector                                Industrials
GICS Sub-Industry          Industrial Conglomerates
Headquarters Location            St. Paul, Minnesota
Date first added                          1976-08-09
CIK                                            66740
Founded                                         1902
Name: MMM, dtype: object
```

As with the row selection for AQI, our result is a Series, which nicely provides the data values and labels for each value. As a pandas Series is a one-dimensional structure, it can represent a DataFrame's row, as we get using `loc[]`. It can also represent a single DataFrame column, as we saw earlier using `[]`.

Similar to a Python dictionary, if we use a row name with `loc[]` that is not in the DataFrame data, we'll get a `KeyError` as opposed to an empty result such as `None`.

To demonstrate the use of row lookup, Program 10-6 reads the S&P 500 data, allows the user to enter a symbol of interest, and if found, displays the row's data.

Program 10-6

```python
# interactive data lookup program - S&P 500 securities
import pandas as pd
dfsp = pd.read_csv('SP500_Constituents.csv', index_col='Symbol')

symbol = input('Enter security symbol of interest: ')

while symbol != '':
    symbol = symbol.upper()   # symbols are upper case in the data
    if symbol in dfsp.index:  # check if symbol is in data
        row = dfsp.loc[symbol] # locate the data using loc[]
        print(row)
    else:
        print('Not found')

    symbol = input('Enter security symbol of interest (or <return> to end): ')
```

Program 10-6 Example Run

```
Enter security symbol of interest: mmm
Security                                    3M Company
SEC filings                                    reports
GICS Sector                                Industrials
GICS Sub-Industry          Industrial Conglomerates
Headquarters Location            St. Paul, Minnesota
Date first added                          1976-08-09
CIK                                            66740
Founded                                         1902
Name: MMM, dtype: object
Enter security symbol of interest (or <return> to end):
```

Cells

Having looked at how to access a DataFrame's row or column, let's look now at how to get a single cell's value in a DataFrame. In practice, it's common to work with a DataFrame and its rows and columns, but it is easy to access a cell value if needed. We can do this by chaining together the row and column access syntax described earlier. For example, we can access a row by name and immediately get a value from the returned Series using:

```
df.loc['rowname']['colname']
```

For example, to find the concern for the AQI color red, we can use:

```
[12]: dfAQI.loc['Red']['concern']
      'Unhealthy'
```

Alternatively, and more concisely, we can access the column first and then the value according to the row name: `df['colname']['rowname']`

The concern for the AQI color red is also found using:

```
[13]: dfAQI['concern']['Red']
      'Unhealthy'
```

Similarly, for the S&P 500 data, we can find the GICS Sector for the stock symbol MMM using:

```
[14]: dfsp['GICS Sector']['MMM']
      'Industrials'
```

Section Summary

pandas expressions allow for easy access to a DataFrame's columns, rows, and cells. Assuming `df` is a DataFrame and `'colname1'` is a column name in it, `df['colname1']` returns a Series with the column's data as well as the names of the rows. The `tolist()` method allows you to extract a Series' values into a Python list: for example, `df.['colname1'].tolist()`. If `'colname'` is a valid Python variable name, the short-hand `df.colname1` can be used to get a DataFrame column. The index values of a DataFrame are accessible with `df.index`. Multiple columns can be accessed using a `list`, as `df[['colname1', 'colname2']]`.

DataFrame rows can be accessed by name or number. `df.loc['rowname']` returns a Series, with values from the row, and names from the DataFrame column names. If `i` is an integer, `df.iloc[i]` returns a row (also as a Series), using the row number (similar to a Python `list` index).

A cell's value can be accessed by chaining the column and row access described above. For example, `df.loc['rowname']['colname']` or `df['colname']['rowname']` return the value of the cell for a given column and row name.

Test Your Understanding

1) If `df` is a DataFrame, and `'colname'` is one of its column names, the expression `df['colname']` returns:
   a) a DataFrame
   b) a Series
   c) a Python list
2) If `df` is a DataFrame, and `'colname1'` and `'colname2'` are two of its column names, the expression `df[['colname1', 'colname2']]` returns:
   a) a DataFrame
   b) a Series
   c) a Python list
3) The `tolist()` method returns:
   a) a list from a Series
   b) a Series from a list
   c) a Series from a DataFrame
4) If `df` is a DataFrame, and `'rowname'` is a name of one of its rows, which expression returns a row of the DataFrame?
   a) `df.iloc['rowname']`
   b) `df.loc['rowname']`
   c) `df.rowname`

10.5 Getting Data: Sorting, Filtering, Slicing, Looping

In the previous section, we saw how to get data for a particular row of a DataFrame. pandas has a lot of functionality related to working with multiple DataFrame rows. For example, we can sort the DataFrame based on one or more columns. We can also filter to select a subset of rows. In this section, we'll look at techniques that work with rows.

```
[1]: import pandas as pd
     pd.set_option('display.max_rows', 6)
     dfAQI = pd.read_csv('aqi.txt', index_col='color')
     dfsp = pd.read_csv('SP500_Constituents.csv', index_col='Symbol')
```

Sorting

In order to sort a DataFrame, we can use the `sort_values()` method. This method allows for sorting in ascending or descending order. For example, if we'd like to see the AQI levels in descending order, we can use:

```
[2]: dfAQI.sort_values('low', ascending=False)
```

	concern	low	high
color			
Maroon	Hazardous	301	999
Purple	Very Unhealthy	201	300
Red	Unhealthy	151	200
Orange	Unhealthy for Sensitive Groups	101	150
Yellow	Moderate	51	100
Green	Good	0	50

which results in a new DataFrame.

The `sort_values()` method, like several other methods in pandas, allows us to get a new DataFrame result, or change our DataFrame directly. For example, the statement:

```
dfsort = dfAQI.sort_values('low', ascending=False)
```

creates a new DataFrame and leaves our original `dfAQI` as is. If instead we'd like to sort our `dfAQI` DataFrame so that `dfAQI` is updated with the sorted result, we can use the `inplace` parameter:

```
dfAQI.sort_values('low', ascending=False, inplace=True)
```

The parameter `inplace` is available in several pandas methods, and the choice of whether to refine a DataFrame directly or create a new DataFrame to work with is up to you. This depends in part on whether you need to save interim DataFrames, in which case you can omit `inplace`, and assign the method's result to a new variable.

Sorting can also be done based on multiple columns. For example, if we'd like to sort the S&P 500 securities by sector and then by sub-industry, we can use a list of fields, as in:

```
dfsp2.sort_values(['GICS Sector', 'GICS Sub-Industry'])
```

Spreadsheet Compare/Contrast
Sorting

Spreadsheets provide for table sorting, and the reordered data are "in place" in the spreadsheet, replacing the prior order of the data. In contrast, in pandas, we can rearrange the order of the data in a DataFrame, or we can generate a new DataFrame with the sorted data, leaving the original DataFrame as is.

Filtering

Filtering allows us to create a new DataFrame with a subset of rows from an original DataFrame. This powerful pandas feature is much like filtering in a spreadsheet or using WHERE clauses in SQL. For example, to see all AQI levels at or below 100, we can use:

```
[3]: dfAQI[dfAQI.high <= 100]
```

	concern	low	high
color			
Green	Good	0	50
Yellow	Moderate	51	100

which returns the above DataFrame.

How does this work? The expression within the square brackets `[]` is evaluated first by pandas, and returns a Boolean Series. We can see this by evaluating it:

```
[4]: dfAQI.high <= 100
```

```
color
Green      True
Yellow     True
Orange    False
Red       False
Purple    False
Maroon    False
Name: high, dtype: bool
```

In other words, for this filter condition, the first two rows (Green, Yellow) are `True`, and the other rows are `False`. This Series result is then used to include only the rows which evaluate to `True` in the returned result.

pandas filtering allows for a variety of expressions. The usual Python comparison operators are supported (<, >, <=, >=, ==, !=). However, whereas Python provides keywords `and`, `or`, and `not`, pandas uses symbols as shown in Table 10-1.

Python keyword	pandas symbol
and	&
or	|
not	~

Table 10-1: pandas composite operators for filtering

For example, to determine what concern a level of 75 falls under, we can use:

```
[5]: level = 75
     dfAQI[(dfAQI.low <= level) & (level <= dfAQI.high)]
```

	concern	low	high
color			
Yellow	Moderate	51	100

Note that in the above expression, the parentheses are needed to define the parts of the filtering condition.

Spreadsheet Compare/Contrast
Filtering

Spreadsheet filtering is similar in concept to DataFrame filtering. However, as spreadsheets typically support "and-ing" filter conditions across columns, DataFrames provide more flexibility in how composite conditions can be defined.

In the above examples, numeric conditions are used; we can also use string expressions. For example, to find the rows corresponding to Good or Moderate concerns, we can use:

```
[6]: dfAQI[(dfAQI.concern == 'Good') | (dfAQI.concern == 'Moderate')]
```

	concern	low	high
color			
Green	Good	0	50
Yellow	Moderate	51	100

Note that the following gives an error (since == has higher precedence than |):

```
dfAQI[dfAQI.concern == 'Good' | 'Moderate']
```

There are more powerful string comparisons in pandas—for example, you can compare with substrings, ignore case, or use regular expressions. As an example, we can use:

```
dfAQI[dfAQI.concern.str.contains('good|moderate',case=False)]
```

Other `str` methods for Series include `startswith()` and `endswith()`.

Applying filtering to our S&P 500 data, we can find the securities in the Information Technology sector using:

```
dfsp[dfsp['GICS Sector'] == 'Information Technology']
```

We can filter rows and subset the columns by combining these operations into one statement. For example, to see the security name and date founded for stocks in the Information Technology sector, we can use:

```
[7]: dfsp[dfsp['GICS Sector']=='Information Technology'][['Security', 'Founded']]
```

	Security	Founded
Symbol		
ACN	Accenture plc	1989
ADBE	Adobe Inc.	1982
AMD	Advanced Micro Devices Inc	1969
...	...	...
XRX	Xerox	1906
XLNX	Xilinx	1984
ZBRA	Zebra Technologies	1969

73 rows × 2 columns

Slicing

Similar to lists, pandas DataFrames provide slicing as a way to get a part of the data. For example, to get the first three rows of AQI data, we could use:

```
[8]: dfAQI[:3]
```

	concern	low	high
color			
Green	Good	0	50
Yellow	Moderate	51	100
Orange	Unhealthy for Sensitive Groups	101	150

Also, similar to lists, negative indexes can be used in slicing to work backward from the bottom of the DataFrame. Relatedly, there is an optional third skip argument, which can be negative, to move backward through the rows. For example, the following displays the last two levels, in reverse order:

```
[9]: dfAQI[-1:-3:-1]
```

	concern	low	high
color			
Maroon	Hazardous	301	999
Purple	Very Unhealthy	201	300

Looping

In the last part of this section, we'll look at how you can use `for` to loop over the rows of a DataFrame. Since pandas provides many concise methods that operate efficiently on all the rows of a DataFrame, you can often solve a task without resorting to explicit looping. However, there may be cases where you want to explicitly access the data one row at a time. For example, if you needed to create a file in a specific format based on the data in a DataFrame, you could solve this by using explicit looping. The syntax is somewhat like how you would loop over the elements of a Python dictionary.

The `iterrows()` method, in combination with the `for` statement, will loop over the rows of a DataFrame, returning both the row name as well as the row's data as a Series. For example:

```
[10]: for color, row in dfAQI.iterrows():
          print(color+':', row.concern)
```

```
Green: Good
Yellow: Moderate
Orange: Unhealthy for Sensitive Groups
Red: Unhealthy
Purple: Very Unhealthy
Maroon: Hazardous
```

Section Summary

Row-oriented operations of DataFrames include sorting, filtering, slicing, and looping. Sorting can be done with the `sort_values()` method. One of this method's parameters is `inplace`, which indicates if a new sorted DataFrame is created or the existing DataFrame is itself sorted.

Filter expressions allow you to generate a new DataFrame with a subset of the rows, given a filter condition. The filter condition is similar in concept to Python logical expressions, but the syntax is different. For example, symbols such as `&`, `|`, and `~` are used, instead of the keywords `and`, `or`, and `not`. In addition to filtering, and similar to Python list slicing, DataFrame slicing can be used to generate a DataFrame with a subset of rows.

Certain pandas methods let you explicitly loop over a DataFrame's rows. For example, the `iterrows()` method can be used with `for` to iterate over the rows of a DataFrame to access data one row at a time. In practice, it is often more concise and more efficient to use a pandas technique to process the rows of a DataFrame, without explicitly looping over them.

Test Your Understanding

1) The method to sort a DataFrame is:
   a) `sort()`
   b) `sort_values()`
   c) `order()`
2) In order to sort a DataFrame without creating a new DataFrame, the parameter specified is:
   a) `inplace`
   b) `ascending`
   c) `newframe`
3) If `df` is a DataFrame, and `colname` is one of its columns, which is a valid filter expression?
   a) `df[df.colname == 100]`
   b) `df[df['colname'] == 100]`
   c) `df[colname == 100]`
   d) (a) and (b)

10.6 Missing Values

In a dataset, some values may be unknown or not applicable. The handling of missing values is built into pandas. This includes how missing values are represented as well as how methods can identify and handle missing values.

Spreadsheet Compare/Contrast
Missing Values

In spreadsheets, it's natural to represent a missing value as an empty cell. Further, spreadsheet functions such as `sum()`, `average()`, and others know to exclude empty cells in their calculations. Similarly, pandas uses the NumPy value `NaN` to mark a numeric cell as empty, and methods in pandas know to handle a missing value marked this way.

In Python, missing values are sometimes indicated by the special value `None`. In pandas, this can also be done for non-numeric columns. However, as numeric columns are represented using NumPy types (for example, `int64` and `float64`), pandas uses the NumPy value `NaN` as an indicator for a missing numeric value. (`NaN` is short for "not a number.")

We can see this using the AQI data. In prior examples, these data included the high value for the highest concern level as 999. Another approach is to represent this value as missing, since there is no upper bound for this range. The file aqina.txt uses this approach by omitting the last value on the last line of the file:

```
Maroon,Hazardous,301,
```

We can read this file into a DataFrame using:

```
[1]: import pandas as pd
     dfAQI2 = pd.read_csv('aqina.txt', index_col='color')
     dfAQI2
```

	concern	low	high
color			
Green	Good	0	50.0
Yellow	Moderate	51	100.0
Orange	Unhealthy for Sensitive Groups	101	150.0
Red	Unhealthy	151	200.0
Purple	Very Unhealthy	201	300.0
Maroon	Hazardous	301	NaN

Note that the high value for the last row is displayed as `NaN`, reflecting the missing value in the file we imported.

You may want to know which rows have missing values. In the data above, it is easy to see, but for larger datasets, we would want a programmatic alternative to looking for all the data with `NaN`s. We can use the pandas method `isna()` within a DataFrame filter. For example, to see all rows for which the column high has a missing value, we can use this filter expression:

```
[2]: dfAQI2[dfAQI2.high.isna()]
```

	concern	low	high
color			
Maroon	Hazardous	301	NaN

Conversely, we can filter out all the rows that have `NaN` values in a column using `notna()`.

For the S&P 500 data, we found earlier using the `info()` method that certain columns had missing data. For example, the Founded column had 502 non-null values, thus having three missing values. What securities have missing data for this field? We can use the `isna()` method to find out:

```
[3]: dfsp = pd.read_csv('SP500_Constituents.csv', index_col='Symbol')
     dfsp[dfsp.Founded.isna()].Security
```

```
Symbol
L                Loews Corp.
ROP      Roper Technologies
SBAC     SBA Communications
Name: Security, dtype: object
```

This is an interesting data quality problem—if our analysis required complete Founded dates for all the S&P 500 companies, we would need to augment the data (or find another data source). Changing values in DataFrame cells is covered in the next section. We also might want to find all rows that have a missing value in any column. Further, we may want to remove rows with missing values, or fill missing values with a particular value. These and other data quality–related techniques are covered in the next chapter.

Section Summary

A missing value within a pandas numeric column is represented as a NumPy `NaN`. This convention is different from a Python `None`, since NumPy numeric types cannot store a `None` value. pandas provides some methods to identify missing and not-missing values. The `isna()` method can be used to locate missing data. Conversely, the `notna()` method can be used to locate non-missing data.

Test Your Understanding

1) For numeric columns, pandas uses what value to represent missing data?
   a) Python `None`
   b) empty string
   c) NumPy `NaN`

2) What method can be used to locate DataFrame cells with missing data?
   a) `ismissing()`
   b) `isna()`
   c) `novalue()`

10.7 Maintaining DataFrame Data

In the previous sections, we've populated DataFrame data from a text file. In some cases, you may want to adjust the DataFrame's data. Alternatively, you can use DataFrames as you might use Python's lists and dictionaries: as a way to organize data in your program. In either case, pandas supports ways to add, change, and delete information from a DataFrame. In this section, we'll look at how to add columns and rows, delete columns and rows, and change cell values.

For this section, we'll return to the aqi.txt file, loading it with:

```
[1]: import pandas as pd
     dfAQI = pd.read_csv('aqi.txt', index_col='color')
     dfAQI
```

	concern	low	high
color			
Green	Good	0	50
Yellow	Moderate	51	100
Orange	Unhealthy for Sensitive Groups	101	150
Red	Unhealthy	151	200
Purple	Very Unhealthy	201	300
Maroon	Hazardous	301	999

Appending columns

To add list data as a new column to a DataFrame, we can use the following syntax:

```
df['colname'] = list
```

For example, if we'd like to augment our AQI data with a number, we can use the statement:

```
[2]: dfAQI['number'] = [1, 2, 3, 4, 5, 6]
```

which adds a new column in place, so that `dfAQI` becomes:

```
[3]: dfAQI
```

	concern	low	high	number
color				
Green	Good	0	50	1
Yellow	Moderate	51	100	2
Orange	Unhealthy for Sensitive Groups	101	150	3
Red	Unhealthy	151	200	4
Purple	Very Unhealthy	201	300	5
Maroon	Hazardous	301	999	6

Appending rows

We can append one or more rows to a DataFrame using the `append()` method. We can provide this method either a Series or a DataFrame. For example, if we'd like to add a new level of concern to the AQI data, we could use:

```
[4]: newrow = \
      pd.DataFrame({'concern':['Worst'],'low':[1000],'high':[9999],'number':7},
                   index=['Brown'])
     dfAQI = dfAQI.append(newrow)
```

As `append()` generates a new DataFrame, the above statement reassigns the result to the `dfAQI` variable. (Note that `append()` doesn't have an `inplace` parameter like `sort()` and some other methods have.)

```
[5]: dfAQI
```

	concern	low	high	number
Green	Good	0	50	1
Yellow	Moderate	51	100	2
Orange	Unhealthy for Sensitive Groups	101	150	3
Red	Unhealthy	151	200	4
Purple	Very Unhealthy	201	300	5
Maroon	Hazardous	301	999	6
Brown	Worst	1000	9999	7

If you need to append a single row, you can provide `append()` with a named Series argument instead of a one-row DataFrame as above. For example:

```
newrow = pd.Series(name='Brown',
                  ['Worst', 1000, 9999, 7],
                  index=['concern', 'low', 'high'])
dfAQI = dfAQI.append(newrow)
```

Deleting columns and rows

The `drop()` method can be used to remove a row or column from a DataFrame. This method takes a parameter named `axis` to indicate whether we are dropping a row or column. By default, `drop()` assumes we are dropping a row. For example, we can drop the number column we added earlier with:

```
[6]: dfAQI.drop('number', axis='columns', inplace=True)
     dfAQI
```

	concern	low	high
Green	Good	0	50
Yellow	Moderate	51	100
Orange	Unhealthy for Sensitive Groups	101	150
Red	Unhealthy	151	200
Purple	Very Unhealthy	201	300
Maroon	Hazardous	301	999
Brown	Worst	1000	9999

In the above statement, since we include `inplace=True`, `drop()` doesn't create a new DataFrame.

Similarly, we can use `drop()` to remove rows from a DataFrame. So, for example, we can remove the last row of our AQI data with:

```
[7]:  dfAQI.drop('Brown', inplace=True)
      dfAQI
```

	concern	low	high
color			
Green	Good	0	50
Yellow	Moderate	51	100
Orange	Unhealthy for Sensitive Groups	101	150
Red	Unhealthy	151	200
Purple	Very Unhealthy	201	300
Maroon	Hazardous	301	999

Updating a cell's value

To change a DataFrame cell, we can use `at[]`. The syntax is:

```
df.at['rowname', 'colname'] = value
```

For example, to change a label in the AQI data, we could use:

```
[8]:  dfAQI.at['Orange', 'concern'] = 'Unhealthy For Some'
      dfAQI
```

	concern	low	high
Green	Good	0	50
Yellow	Moderate	51	100
Orange	Unheathly For Some	101	150
Red	Unhealthy	151	200
Purple	Very Unhealthy	201	300
Maroon	Hazardous	301	999

Section Summary

DataFrame data can be revised in a number of ways. To add a column, use `df['colname'] = list`. To add a row, you can use the `append()` method.

Deleting rows or columns is done with the `drop()` method. A cell's value can be updated using `df.at['rowname','colname']=value`.

10.8 Joining Two DataFrames

In this last section, we'll see how you can combine two or more DataFrames side by side. For example, you may have two files with related data and would like to read and combine these into one DataFrame.

In this section, we'll continue to use the original AQI data, loaded as:

```
[1]: import pandas as pd
     dfAQI = pd.read_csv('aqi.txt', index_col='color')
     dfAQI
```

	concern	low	high
color			
Green	Good	0	50
Yellow	Moderate	51	100
Orange	Unhealthy for Sensitive Groups	101	150
Red	Unhealthy	151	200
Purple	Very Unhealthy	201	300
Maroon	Hazardous	301	999

There are alternative methods in pandas for combining DataFrames side by side. In this section, we'll assume that the DataFrames to join have the same row names (index values). We can do this with the `join()` method.

For example, assume for our AQI data that we have another file, aqidays.txt, that contains the number of days in a year the air quality was at each concern level for a certain location. We can read in this new file with the statement:

```
[2]: dfAQIdays = pd.read_csv('aqidays.txt', index_col='color')
     dfAQIdays
```

	days
color	
Green	320
Yellow	30
Orange	15
Red	0
Purple	0
Maroon	0

and then join the original data with this new DataFrame using:

```
[3]: dfAQI.join(dfAQIdays)
```

	concern	low	high	days
color				
Green	Good	0	50	320
Yellow	Moderate	51	100	30
Orange	Unhealthy for Sensitive Groups	101	150	15
Red	Unhealthy	151	200	0
Purple	Very Unhealthy	201	300	0
Maroon	Hazardous	301	999	0

Note that `join()` returns a new DataFrame as opposed to modifying `dfAQI`.

In the above join, the DataFrames each have six corresponding rows. However, in other cases, not all the rows may match between the two DataFrames. For example, we have a file aqidays2.txt that omits colors with 0 days. We can load it with:

```
[4]: dfAQIdays2 = pd.read_csv('aqidays2.txt', index_col='color')
     dfAQIdays2
```

color	days
Green	320
Yellow	30
Orange	15

and we can join this smaller DataFrame to the original AQI data:

```
[5]: dfAQI.join(dfAQIdays2)
```

color	concern	low	high	days
Green	Good	0	50	320.0
Yellow	Moderate	51	100	30.0
Orange	Unhealthy for Sensitive Groups	101	150	15.0
Red	Unhealthy	151	200	NaN
Purple	Very Unhealthy	201	300	NaN
Maroon	Hazardous	301	999	NaN

Since `dfAQI2` does not have rows for Red, Purple, and Maroon colors, the joined DataFrame shows missing values for days for these rows.

Alternatively, if we want the `join()` result to include rows based on `dfAQI2`, we can specify the `how` parameter as:

```
[6]: dfAQI.join(dfAQIdays2, how='right')
```

color	concern	low	high	days
Green	Good	0	50	320
Yellow	Moderate	51	100	30
Orange	Unhealthy for Sensitive Groups	101	150	15

Here, `how='right'` indicates that the DataFrame argument (within the parentheses) controls what rows are returned in the result.

The `join()` method has four options for the `how` parameter. Table 10-2 shows these settings.

join type	Included rows
`left`	Include all rows from `dfLeft` and matching rows from `dfRight` (default)
`right`	Include all rows from `dfRight` and matching rows from `dfLeft`
`outer`	Include the union of all rows in `dfLeft` and `dfRight`
`inner`	Include rows where index values are in both `dfLeft` and `dfRight`

Table 10-2: Included rows in `dfLeft.join(dfRight)`

Section Summary

In order to combine two DataFrames side by side, assuming they have matching row names, you can use the `join()` method. The syntax is `df1.join(df2)`. A few named parameters for this method are available, including `how`, which directs `join()` regarding how to handle rows that are in one DataFrame but don't correspond to rows in the other.

Chapter Review

The pandas package provides the DataFrame and Series types to allow you to write more powerful, more concise programs. A DataFrame is a matrix with named columns and rows, and has a large number of methods for working with tabular data. Series is a list of values, each of which has a name. DataFrames can be created in code or read from outside the program—for example, from a text file. Several methods provide a quick overview of a DataFrame, such as `head()`, `tail()`, and `info()`.

Rows, columns, and individual cells can be accessed in several concise ways. Row-oriented operations on DataFrames include sorting, filtering, and slicing. DataFrames can be maintained using syntax to create additional columns or rows, remove columns or rows, and update a cell's value. Numeric missing values are represented as NumPy `NaN`, and pandas provides methods related to these. To combine two DataFrames side by side, the `join()` method can be used.

Key Terms

column	A DataFrame is made up of named columns.
column name	Each column of a DataFrame has a name.
DataFrame	A primary type provided in the pandas package for representing a matrix of data.
filter condition	A way to select a subset of a DataFrame's rows.
index	The rows of a DataFrame and Series are named with an index.
join	The join method allows for side-by-side merging of two DataFrames.
NaN	Numeric missing values are represented by NumPy `NaN` in pandas.
NumPy	The NumPy package provides numeric types that pandas uses to more efficiently store numbers in DataFrames and Series.
pandas	A third-party package that provides the key types DataFrame and Series.
row	DataFrames are made up of named rows.
row name	Each row of a DataFrame and Series is named.
Series	A primary type provided in the pandas package, for representing a list, where each item is named.

Programming Problems

1) Contacts

For the following questions, the file contacts.csv is provided, with the following fields: email, first, last, phone. Data in the email field are unique for each row and can be used as the index in a DataFrame.

Loading and getting information

a) Use the pandas `read_csv()` function to load the data in contacts.csv into a DataFrame. Use the email column as the DataFrame index. Display the data in the DataFrame.
b) Use the `info()` method to determine how many non-null values there are for each column and what each column's type is.
c) Determine the number of rows and columns using the `len()` function as well as the `shape` property.
d) Display the first two and last two rows using `head()` and `tail()`.

Accessing rows, columns, and cells

e) Display the phone column.
f) Display the columns named first and last in a DataFrame.
g) Display the phone number for the row with email=smsmith@yahoo.com.

Filtering, sorting

h) Display the data sorted by last name then first name.
i) Display rows where the first name is Dan.

Missing values

j) Display rows where the phone number is missing.

Updating

k) For the row where email=tessbijon@bijon.com, set the value of phone to '555-1233'.
l) Drop the row where email=tessbijon@gmail.com.
m) Append a row with the following data:
   email-stefsmith@gmail.com
   First: Stef
   Last: Smith
   Phone: 801-555-9876

Joining

n) The file contact-states.csv has additional data, including email and state. Use `join()` to augment the DataFrame with the state data.

2) Dow Jones Industrial Average

For the following questions, the file djia.csv is provided, with the following fields: Symbol, Company, Exchange, Industry, Data added, Notes, and Index weighting. Data in the Symbol field are unique for each row and can be used as the index in a DataFrame.

Loading and getting information

a) Use the pandas `read_csv()` function to load the data in djia.csv into a DataFrame. Use the Symbol column as the DataFrame index. Display the data in the DataFrame.
b) Use the `info()` method to determine how many nonnull values there are for each column and what each column's type is.
c) Determine the number of rows and columns using the `len()` function as well as the `shape` property.
d) Display the first and last rows using `head()` and `tail()`.

Accessing rows, columns, and cells

e) Display the Company column.
f) Display the columns named Company and Industry in a DataFrame.
g) Display the data for the row with the Symbol AAPL.
h) Display the Exchange for the row with the Symbol AAPL.

Filtering, sorting

i) Display the data sorted by Industry, then Company.
j) Display the rows where Industry is Information Technology.

Missing values

k) Display rows where Notes has a value (other than missing).

Dropping

l) Create and display a DataFrame from the full DataFrame but excluding the Notes column.

Joining

m) The file djia-prices.csv has additional data, including Symbol and Price. Use `join()` to augment the DataFrame with the Price data.

Chapter 11

pandas for Data Preparation

Chapter Objectives

In this chapter, you'll learn:

- how to discover and address data quality issues with numeric and date data
- how to work with category data for representing fields with a particular set of values
- how to use certain pandas methods for working with string data
- about key fields for capturing unique identifiers
- how to check for consistency between fields
- how to check if there are any missing rows between two related DataFrames
- how to rearrange data layout

Chapter Outline

11.1 About Data Preparation

In the previous chapter, we looked at how you can load data into a pandas DataFrame and some basic ways you can work with these data. In this chapter, we look further at how you can refine your data—for example, to ready them for analysis. In particular, we'll focus on two aspects: data cleaning, to find and repair or filter issues in your data, and data reorganization, to change the layout of your data. Data cleaning is covered in most of the sections in this chapter, and data reorganization is covered in the last two sections.

In contrast with spreadsheets, Python and pandas are great tools for data preparation. Python and pandas can be used interactively and/or in scripts in order to prepare data. This allows for you to define concise logic and generate new data columns and data layouts from an original source of data.

One pattern used in this chapter involves creating new data from original data. This is as opposed to replacing original data. This has the benefit of allowing you to retain the original data and compare them to newly generated data.

There are a variety of types of issues you may encounter in your data. This chapter covers data issues across several data types including numbers, dates, categories, and strings. Several small datasets are provided in files to demonstrate the techniques; however, the approaches are applicable to large datasets as well.

Spreadsheet Compare/Contrast
Data Cleaning

Spreadsheets provide convenient ways to filter and sort data, allowing you to visually review data quality. However, repairing data in general is more manual with a spreadsheet than with a programming language, since you can use a language to define rules for refining data. For this reason, Python in combination with pandas is well suited for finding and cleaning data issues, either interactively or implemented in an automated system.

The dataset

For most of this chapter, we'll work with hypothetical data assumed to be collected from a survey of respondents about their demographics (age, gender, number of children) and their job (state working in, annual salary, income tax rate). In practice, this could be a large dataset, but for this chapter, we'll look at a few responses. Intentionally our data have a lot of issues, so we can see how to programmatically detect and repair these. For example, among the several issues in our data, our age data have an invalid value (-1), our gender data have several values for male (*m*, *male*), and our state data have invalid state abbreviations and missing data.

Category	Field	Description	Variable Type
Identification	id	respondent #	
	date	date of survey response	
	email	email address of respondent	
	address	home address	
Demographic	age	age of respondent (years)	numerical
	gender	gender (male, female)	categorical
	children	children? (yes, no)	categorical
	children_num	number of children	numerical
Work-related	state	state of work address	categorical
	salary	annual salary ($)	numerical
	taxrate	federal income tax rate (%)	numerical

Table 11-1: Sample dataset fields

Table 11-1 summarizes the fields in the survey data we'll work with in this chapter. It distinguishes between numerical variables (e.g., age, salary) vs. categorical variables that can take on a particular set of values (e.g., gender, state).

11.2 Numbers

In this section, we'll look at numeric data and how pandas can be used to find issues and address them. There are a variety of potential issues you may encounter with numeric data fields, such as data values that aren't numeric, cases where the data are out of expected bounds, missing values, and decimal values where integers are expected.

An example dataset

To discuss a variety of numeric-related features in pandas, we'll look at a subset of the fields described in the prior section, as listed in Table 11-2.

Field	Description
id	respondent #
age	age of respondent (years)
children_num	number of children
salary	annual salary ($)
taxrate	federal income tax rate (%)

Table 11-2: Sample dataset numeric fields

Here is our data, in comma-separated format, in the file DQ-Example-Numeric.csv:

```
id,age,children_num,salary,taxrate
1,21,2,40000,22
2,29.4,0,50000,25
3,40,1,60,28
4,6,2,45000,21
,-1,-1,52500,25
six,28,1,47515,100
4,26,,,105
```

Let's load the above file into a DataFrame:

```
[1]: import pandas as pd
     df = pd.read_csv('DQ-Example-Numeric.csv')
     df
```

	id	age	children_num	salary	taxrate
0	1	21.0	2.0	40000.0	22
1	2	29.4	0.0	50000.0	25
2	3	40.0	1.0	60.0	28
3	4	6.0	2.0	45000.0	21
4	NaN	-1.0	-1.0	52500.0	25
5	six	28.0	1.0	47515.0	100
6	4	26.0	NaN	NaN	105

At a glance, you may see several issues with the data—this is intentional in order to review how we can use pandas to uncover and repair our data. In practice, you may work with large datasets and their issues will not be apparent at a glance. But the techniques we discuss in this chapter are applicable to any size dataset.

In each of the following sections, we'll look at each of the columns in this DataFrame.

Missing values

As discussed in the prior chapter, pandas uses NumPy types to represent numbers. This is for efficiency in both space and speed, as NumPy can store a sequence of numbers in contiguous memory. Further, pandas uses the NumPy special value `NaN` to represent missing values. (`NaN` is an abbreviation for "not a number.")

In the output above, we see `NaN` displayed in a few cells. This corresponds to places in the CSV file where there was no value. The use of `NaN`, as opposed to a particular value like 0, is an important feature of pandas so that its calculations can exclude these cells. As an example, children_num is 0 in the second row, as opposed to missing in the last row. If we were interested in the average number of children of the respondents, we would want to include zeros but exclude missing values. We also have an errant value of -1.0, which we will cover later.

Float vs. integer vs. other

In the DataFrame above, some columns show integers (id, taxrate), while others show one decimal place (age, children_num, salary). The pandas `read_csv()` function follows these rules regarding numbers:

- if any value of a column is non-numeric, the column's values are represented as Python types as opposed to NumPy values
- if all values in a column are numeric and any value has decimal places, the column is represented as a NumPy float type
- if any value in a column is missing, even if all the non-missing values are integers, the column is stored as a NumPy float type

What is a column's type?

Given the rules noted above, for a dataset, you may want to know what data types pandas used to store the data it read. The simplest way to do this is with the `info()` method:

```
[2]:  df.info()
```

```
<class 'pandas.core.frame.DataFrame'>
RangeIndex: 7 entries, 0 to 6
Data columns (total 5 columns):
id                6 non-null object
age               7 non-null float64
children_num      6 non-null float64
salary            6 non-null float64
taxrate           7 non-null int64
dtypes: float64(3), int64(1), object(1)
memory usage: 360.0+ bytes
```

For each column, `info()`'s output indicates the data type. In this case, `float64` is a NumPy float type, and `int64` is a NumPy integer type. Further, object indicates that other types are used, in this case Python built-in types are used.

As an alternative to `info()`, a Series has a `dtype` property, and this tells the type used for that column. For example:

```
[3]:  df.taxrate.dtype
```

```
dtype('int64')
```

An example column: Age

As seen above, in our small dataset, age is a `float64` column. There are a few issues with this column:

a) The value -1 is not valid, perhaps it was intended as a special value to represent missing data.
b) There is a noninteger value, 29.4—perhaps we want to have integer values for age.
c) There is a small value, 6—let's assume this survey was expected to be given to respondents age 18 and older, meaning this value is invalid.

Although these issues are easy to see in a small dataset, the techniques we present here and elsewhere are useful in datasets of any size, including very large ones.

For invalid values, we can use missing values instead. This is useful if we are calculating descriptive statistics such as mean or median, as pandas will omit the missing values from these calculations. To derive a new column from another, we can write a utility function and use the `apply()` method to find and handle invalid values.

```
def lowValuesToNone(value):
   if value < 18:
      return None
   else:
      return value
```

or alternatively, we can use the terse `if/else` statement:

```
def lowValuesToNone(value):
   return None if value < 18 else value
```

Once defined, our `lowValuesToNone()` function can be used with the pandas `apply()` method to create a new column:

```
df['age_clean'] = df['age'].apply(lowValuesToNone)
```

In the above statement, the `apply()` method takes the name of our function and will call that function for each value in the age column. Further note that we are creating a new column, age_clean, which allows us to retain the initial data for comparison purposes, or if we need to research original values later.

Tip
Adding vs. Overriding Columns

In this chapter, we'll often add a new column with clean values based on an original column. This pattern allows you to retain the original column's dataset for subsequent reference as well as compare it with a cleaner version of the data that you can create using data quality rule-based transformations.

Because our function is one line, we can further streamline the above example. For this, we introduce Python's ***lambda*** functionality. Lambda functions are one-statement functions that don't have a name and are not defined with the def keyword. For example, the above statement can be rewritten as:

```
df['age_clean'] = df['age'].apply(lambda x: None if x < 18 else x)
```

In the above statement, the `lambda` keyword is followed by `x`, which indicates the function has one parameter called `x`. (Here we use `x` for the variable name, but we could choose any valid Python variable name—for example, `age`.) In this case, after the colon, the logic follows, our terse `if/else` statement replaces values less than 18 with None.

Another potential issue in our age column is the value 29.6, assuming we would like integer (discrete) values for this field. We could use the `apply()` method, with the Python `int()` or `round()` function:

```
df['age_clean] = df['age_clean].apply(lambda x: round(x))
```

But more concisely, we can use the `round()` method available in a pandas Series:

```
df['age_clean'] = df['age_clean'].round()
```

Combining our concise statements, we have:

```
[4]: df['age_clean'] = df['age'].apply(lambda x: None if x < 18 else x)
     df['age_clean'] = df['age_clean'].round()
     df[['age', 'age_clean']]
```

	age	age_clean
0	21.0	21.0
1	29.4	29.0
2	40.0	40.0
3	6.0	NaN
4	-1.0	NaN
5	28.0	28.0
6	26.0	26.0

Note that in the above result, age_clean has NaNs to indicate missing values, as desired. Also, age_clean, despite having rounded values, is displayed as float with trailing .0. This is because if any number in a column is missing (`NaN`), pandas will use a float representation for all values in the column, since NumPy integer types do not have a special missing value.

Another integer column: children_num

Similar to the age column, children_num is an integer column with a missing value. And as with age, there is an invalid value of -1. Using the methods presented in the prior section, we can create a cleaner column representing the number of children:

```
[5]: df['children_num_clean'] = \
         df['children_num'].apply(lambda x: None if x < 0 else x)
     df[['children_num', 'children_num_clean']]
```

	children_num	children_num_clean
0	2.0	2.0
1	0.0	0.0
2	1.0	1.0
3	2.0	2.0
4	-1.0	NaN
5	1.0	1.0
6	NaN	NaN

As one further refinement, we can optimize the space required for the children_num column. The default float type used by pandas is the NumPy `float64`, which takes eight bytes of space for each value and can represent very large values. However, children_num would have small values, so we can use a smaller data type to conserve memory. In particular, for children_num, we can get by with the smaller `float16` type. The column can be switched to this type using the `astype()` method:

```
[6]: df['children_num_clean'] = df['children_num_clean'].astype('float16')
     df.info()
```

```
<class 'pandas.core.frame.DataFrame'>
RangeIndex: 7 entries, 0 to 6
Data columns (total 7 columns):
 #   Column              Non-Null Count  Dtype
---  ------              --------------  -----
 0   id                  6 non-null      object
 1   age                 7 non-null      float64
 2   children_num        6 non-null      float64
 3   salary              6 non-null      float64
 4   taxrate             7 non-null      int64
 5   age_clean           5 non-null      float64
 6   children_num_clean  5 non-null      float16
dtypes: float16(1), float64(4), int64(1), object(1)
memory usage: 478.0+ bytes
```

In practice, you may not have memory constraints, but `astype()` can be used to "right-size" the numeric type if you need to. The NumPy documentation shows the variety of NumPy types and the range of values they can store.

Marking outliers and transforming values: The taxrate column

Let's look at some other numeric techniques, focusing on the taxrate column. From the data, we can see some outliers, in particular tax rates at or above 100%. As with age, we could decide to use `NaN` to represent missing values for these fields. Another approach is to create a new column that marks potential outliers as True in case we want to do follow-up research on selected rows. Using `apply()`, we could do this like so:

```
[7]: df['taxrate_outlier'] = \
        df['taxrate'].apply(lambda x: True if x >= 100 else False)
     df[['taxrate', 'taxrate_outlier']]
```

	taxrate	taxrate_outlier
0	22	False
1	25	False
2	28	False
3	21	False
4	25	False
5	100	True
6	105	True

This new column can be used easily to find all potential outliers using a filter expression:

```
[8]: df[df.taxrate_outlier]['taxrate']
```

```
5    100
6    105
Name: taxrate, dtype: int64
```

Further, if we want to have a tax rate column as decimals, as opposed to percentages, we can simply use an arithmetic expression, like:

```
[9]: df['taxrate_decimal'] = df['taxrate'] / 100
     df[['taxrate', 'taxrate_decimal']]
```

	taxrate	taxrate_decimal
0	22	0.22
1	25	0.25
2	28	0.28
3	21	0.21
4	25	0.25
5	100	1.00
6	105	1.05

Non-numeric values in numeric columns

In the previous sections, we've looked at valid values for numeric data: integers, floats, and missing values. What if there are stray non-numeric data in what should be a numeric column? Our small dataset has an example of this, in particular the `id` column. This column is intended to be positive integers and fully populated (no missing values). However, there is a stray value of six in the data, as well as a missing value.

In a small dataset, these issues are easy to spot, but in a larger dataset, you may need to programmatically find and fix or filter out non-numeric data. The initial clue that there is an issue is in the result of the `info()` method. The `id` column is noted as having an object type as opposed to `int64` or `float64`. This is an indication that `read_csv()` detected non-numeric data in this column. We'll look further at the `id` field later in the chapter in the section on key fields.

Note that pandas provides an additional method, `to_numeric()`, to convert values to float or integer types. This method has a named parameter, `errors`, to control what will happen for non-numeric values. There is also the named parameter `downcast` to control what size NumPy type is used for the result.

Section Summary

This section covered a variety of potential issues and approaches for working with numeric data. pandas uses NumPy types for numbers, as well as the NumPy special value `NaN` to represent a missing value. Since `NaN` is a float type, pandas will use a float type in a numeric column that has any missing values, even if all the non-missing values are integers. The type of a DataFrame column can be checked with the `info()` method or with the dtype property.

To derive a clean column from a dirty one, the `apply()` method can be used. This can work with a function you write that can be defined with the `def` keyword. For simple functions, `apply()` can more concisely be used with a lambda function, which defines a one-statement unnamed function.

The `astype()` method was introduced, which can create a column of a desired type. For numeric columns, this can be used to conserve space by indicating a smaller type that takes less memory.

Test Your Understanding

1) Which can be used to review the datatype of a DataFrame column?
   a) the dtype property
   b) the `info()` method
   c) the Python `type()` function
   d) (a) and (b)

2) Which method can be used to process each value of a given column?
   a) `each()`
   b) `apply()`
   c) `lambda()`

11.3 Dates

Dates can be represented in a wide variety of ways. For this reason, date data can be challenging to get into a consistent format. Fortunately pandas provide some tools to help you work with dates.

In our example dataset, we have a date column, which has a variety of formats as well as an erroneous value and missing data. Here is the data:

```
id,date
1,3/1/21
2,3/2/2021
3,03/03/2021
4,2021-03-4
5,2021-03
6,x3/6/21
7,3/32/21
```

We can read these data into a DataFrame as:

```
df = pd.read_csv('DQ-Example-Dates.csv', index_col='id')
```

```
[1]: import pandas as pd
     df = pd.read_csv("DQ-Example-Dates.csv", index_col='id')
     df
```

id	date
1	3/1/21
2	3/2/2021
3	03/03/2021
4	2021-03-4
5	2021-03
6	x3/6/21
7	3/32/21

A call to `df.info()` will show that the data column is indicated as object. In fact, all the values for the data are Python strings.

Reviewing the data, we see a few issues:

- There is a mix of separators: some dates use / and others use -.
- The order of the month, day, and year varies.
- The value on row 5 omits the day of the month.
- The value on row 6 has a stray x.
- Row 7 is an invalid date (March 32nd).

To clean these data, we'll use two techniques. First, the `replace()` method can be used with a ***regular expression*** to remove errant characters. Although regular expressions are beyond the scope of this book, note that they are a flexible and powerful way to find patterns in strings. Second, the pandas `to_dateime()` function can be used, which will attempt to convert each value to a pandas `DateTime` type.

The pandas `replace()` method can be used to remove certain characters. As shown below, `replace()` is used together with `str.`, so that cell values are accessed as strings. In this example, we'd like to remove characters that are not digits, /, or -. We can do this by defining a regular expression as a string preceded by an r. In particular, `r'[^0-9/-]'` is used below to identify characters that are not digits, a slash, or a dash. For our data, we can remove invalid data characters with:

```
[2]:  df['date_clean'] = df['date'].str.replace(r'[^0-9/-]','', regex=True)
      df[['date', 'date_clean']]
```

id	date	date_clean
1	3/1/21	3/1/21
2	3/2/2021	3/2/2021
3	03/03/2021	03/03/2021
4	2021-03-4	2021-03-4
5	2021-03	2021-03
6	x3/6/21	3/6/21
7	3/32/21	3/32/21

The above expression will remove the stray *x* from the data in row 6.

The other tool we can use is the pandas `to_datetime()` function, which has built-in logic to try to convert strings to the pandas `Timestamp` type. We can use the statement:

```
[3]:  df['date_clean'] = pd.to_datetime(df['date_clean'], errors='coerce')
      df
```

id	date	date_clean
1	3/1/21	2021-03-01
2	3/2/2021	2021-03-02
3	03/03/2021	2021-03-03
4	2021-03-4	2021-03-04
5	2021-03	2021-03-01
6	x3/6/21	2021-03-06
7	3/32/21	NaT

In the statement, the setting `errors='coerce'` ensures that invalid values will not generate an error but instead result in a `NaT` (not a time) value.

For a last step, we can filter out rows that do not have a value for date_clean. The `dropna()` method lets you filter out rows with missing values in a variety of ways. For our data, we can indicate that date_clean is a critical column, and we'd like to filter out rows where date_clean is NaT by using:

```
[4]: dfnona = df.dropna(subset=['date_clean'])
     dfnona
```

id	date	date_clean
1	3/1/21	2021-03-01
2	3/2/2021	2021-03-02
3	03/03/2021	2021-03-03
4	2021-03-4	2021-03-04
5	2021-03	2021-03-01
6	x3/6/21	2021-03-06

In the above statement, we use the `subset` parameter to indicate that date_clean is the column to check for missing values, and to filter out those rows. Although we didn't use `dropna()` in Section 11.2, the `dropna()` method works for numeric columns as well, in order to filter out rows with missing values denoted by `NaN`.

Section Summary

Dates can be represented in a variety of formats. If your data includes date data that have a mixture of formats, you can use string methods such as `replace()`, as well as the `to_datetime()` function, to transform your dates into a consistent format.

11.4 Categories

In some datasets, you may have fields that can have a particular limited set of values. These are referred to as categorical data. Some examples include:

- binary (yes or no) choices such as marital status (single, married)
- multiple choice types such as size (small, medium, large)
- ranges of values, such as age range: for example, less than 18, 18–24, 25–34, and so on.

In some cases, these values are a single character (n, y), a string (no, yes), or indicated by an integer code (1, 2, . . .). Further, these types of fields may be unordered (marital status), or have an order (small, medium, large).

Pandas provides the `category` data type to work with categorical fields. In this section, we'll look at how to work with this type of data, including identifying unique values, cleaning a column's values to conform to a desired set of values, and using ***binning*** to create values corresponding to ranges of numeric values. As we've seen previously, missing values may be important to represent, and we'll look at this as well.

For this section we'll look at the small dataset in DQ-Example-Category.csv. It can be loaded into a DataFrame using:

```
[1]: import pandas as pd
     df = pd.read_csv('DQ-Example-Category.csv', index_col='id')
     df
```

id	age	gender	children	state
1	21.0	m	y	NY
2	29.4	male	n	NJ
3	40.0	F	Y	IL
4	6.0	f	N	CA
5	-1.0	female	Y	CT
6	28.0	female	N	NY
7	26.0	NaN	NaN	NaN

Unique values: The children field

Let's first look at the children field in the above data. This is intended to be a binary field indicating if the respondent has children. Since this is a small set of data, we can see we have lower and uppercase values of a single letter, as well as a missing value. In a larger dataset, for a category field, we usually want to see what its unique values are. You can use the `unique()` method to see distinct values for a column:

```
[2]: df.children.unique()
```

```
array(['y', 'n', 'Y', 'N', nan], dtype=object)
```

Reviewing these, we see the mixture of lower- and uppercase values. Assuming we want to have all lowercase values for consistency, we can use the `str.lower()` method available in pandas:

```
[3]: df['children_clean'] = df.children.str.lower()
     df[['children', 'children_clean']]
```

id	children	children_clean
1	y	y
2	n	n
3	Y	y
4	N	n
5	Y	y
6	N	n
7	NaN	NaN

The above statement creates a new column called children_clean, by taking the lowercase of all values in the children column.

The pandas `str.lower()` method is similar to the Python string `lower()` method, and it also handles missing values as shown on the last row of the table above. There are many other pandas `str` methods, and several are covered later in this chapter.

As mentioned earlier, pandas provides the notion of the category type for a column. You can apply this to a column using the `astype()` method:

```
df.children_clean.astype('category')
```

The category designation is more memory-efficient, and in cases where a category field has an ordering, you can indicate the order as well.

Our children_clean column may be sufficient for use now. However, there may be cases where you'd like to represent a category using integers. For example, if the subsequent use of the data is for input to a model that needs numeric instead of string values, we need to further transform the data. For this case, we can use the pandas `map()` method to convert strings to associated numbers. The `map()` method can take a dictionary as its input argument. For example:

```
[4]: df['children_clean'] = df.children_clean.map({'y':1, 'n':0})
     df[['children', 'children_clean']]
```

id	children	children_clean
1	y	1.0
2	n	0.0
3	Y	1.0
4	N	0.0
5	Y	1.0
6	N	0.0
7	NaN	NaN

Note that in the above for children_clean, pandas is using floating point values, since there is a missing value in the column. To conserve space, we may want to use the `astype()` method, to indicate a smaller size float:

```
df.children_clean.astype('float16')
```

Another category field: Gender

Next, from the above dataset, we'll review the Gender column. In our data, it has a variety of values, though we'd like to map the data to a few consistent values. As with the Children field, we can first review the distinct values:

```
[5]: df.gender.unique()
     array(['m', 'male', 'F', 'f', 'female', nan], dtype=object)
```

By reviewing these values, we can transform this column to a consistent set of values by taking the lowercase of the first character. We can also indicate to pandas that this is a category column. The statements for this are:

```
[6]:  df['gender_clean'] = df.gender.str.lower()
      df['gender_clean'] = df.gender_clean.str[0]
      df.gender_clean.astype('category')
      df[['gender', 'gender_clean']]
```

	gender	gender_clean
id		
1	m	m
2	male	m
3	F	f
4	f	f
5	female	f
6	female	f
7	NaN	NaN

Category by numeric ranges: The age field

Another type of category involves bracketing a numeric value, such as age. The different brackets (also referred to as ***bins***) give rise to category values that can be numbered or named. In order to generate category bracket values, pandas provides the `cut()` function.

For example, if we'd like to define certain age ranges associated with our data, we could use the `cut()` function:

```
[7]:  df['age_range'] = pd.cut(df.age, [0, 18, 21, 30, 100],
                                labels=['<18', '18-21', '22-30', '31+'])
      df[['age','age_range']]
```

	age	age_range
0	21.0	18-21
1	29.4	22-30
2	40.0	31+
3	6.0	<18
4	-1.0	NaN
5	28.0	22-30
6	26.0	22-30

In the above statement, the first argument is the column to partition, the second argument provides the dividing values for the bins, and the named `labels` parameter provides the values that will name each bin. Note that the dividing values list has five items, and the labels list has four items. This is because the first item in the dividing list is the *lower* bound of the first bin, and the last item of the dividing list is the *upper* bound of the last bin. Also note that `cut()` assigns the age of -1 an age_range of `NaN` (denoting missing), since -1 falls outside of any bin—in other words, this age is less than or equal to 0, which is the lowest bin value.

In our example, we defined the last bin's upper bound as 100. If you'd like to use an open-ended upper bound, you can use infinity—for example, if you've imported numpy as np, you can use `np.inf`. Similarly, you can use `-np.inf` for an open-ended lower bound for the first bin.

Mapping fine to coarse labels: States and regions

For category columns that have many values, you may want to derive a new column that groups values into fewer labels. This can be thought of as fine-to-coarse label mapping. For our data, we have a state column, and perhaps we would like to group states into regions. This can be done by creating a DataFrame with state-to-region mapping and merging it with the master dataset.

For our data, we can define a state-to-region mapping file, regions.csv:

```
state,region
CA,West
CT,East
NY,East
NJ,East
IL,Midwest
```

(In practice, our file would have all states.) The states and regions can be read into a DataFrame using:

```
[8]: regions = pd.read_csv('regions.csv')
     regions
```

	state	region
0	CA	West
1	CT	East
2	NY	East
3	NJ	East
4	IL	Midwest

Then a new column can be appended to the main DataFrame using the `merge()` method:

```
[9]: df = df.merge(regions, on='state', how='left')
     df[['state', 'region']]
```

	state	region
0	NY	East
1	NJ	East
2	IL	Midwest
3	CA	West
4	CT	East
5	NY	East
6	NaN	NaN

In the above statement, the `on` parameter indicates the common field to connect between the two DataFrames. The `how` parameter indicates that all the rows from the df DataFrame will be included in the result.

Section Summary

pandas provides features to work with data fields whose values are drawn from a particular set of choices. This includes the `unique()` method to show the unique values of a column. String methods, such as `str.lower()`, `str.upper()`, and `str[]` may be useful to transform data into a targeted set of values. Alternatively, to narrow the set of values in a column to a desired set, the `map()` method can be used. The `astype()` method with the argument of `category` will create a category type in pandas.

A category column can be generated by bucketing ranges of numeric data using the `cut()` function. Relatedly, a category that has many distinct values can be used to derive a coarser data column using the `merge()` method.

Test Your Understanding

1) The method to display the set of values occurring in a DataFrame column is:
   a) `occurring()`
   b) `unique()`
   c) `tolist()`

2) A category column can be created using which method?
   a) `category()`
   b) `astype()`
   c) `unique()`

11.5 Strings

To work with string data in a DataFrame, you can choose from the many string methods available in pandas that function similarly to Python's string methods. We looked at some of these in the previous section, such as `str.lower()` and `str[]`. In this section, we'll look at a few more string manipulations. These are meant to show a few more approaches to strings—in general, if you need to process strings there is usually a concise way to do it in pandas.

To demonstrate some string-related techniques with pandas, we'll load the following data:

```
id|address
1|Bronx,NY
2|New Haven,CT
3|Chicago,IL
4|Monterey, CA
5|Washington
6|Nashville,XX
7|
```

Note that the fields are separated by the pipe character | as opposed to by commas.

In order to read this file into a DataFrame, we can use:

```
[1]: import pandas as pd
     df = pd.read_csv("DQ-Example-Strings.txt", sep='|', index_col='id')
     df
```

	address
id	
1	Bronx,NY
2	New Haven,CT
3	Chicago,IL
4	Monterey, CA
5	Washington
6	Nashville,XX
7	NaN

If we'd like to have city and state in different columns, we can use the `split()` method. The following statement adds new columns to our DataFrame by splitting the data in the existing address column:

```
[2]: df[['city', 'state']] = df.address.str.split(',', expand=True)
     df
```

	address	city	state
id			
1	Bronx,NY	Bronx	NY
2	New Haven,CT	New Haven	CT
3	Chicago,IL	Chicago	IL
4	Monterey, CA	Monterey	CA
5	Washington	Washington	None
6	Nashville,XX	Nashville	XX
7	NaN	NaN	NaN

The first argument is a comma, indicating the data should be split where the comma is found. The second argument, `expand`, is set to True to indicate that we'd like to create new columns.

Tip

More Involved Splitting

Note that the splitting works fine in our example, but your data may have less clean cases requiring you to write more involved logic. For example, you could write a function that detects and handles strings with an incorrect number of delimiters, as well as uses the `split()` method presented in this section. This function could then be used with the `apply()` method.

There are a few data-quality issues with the above DataFrame:

- The state in row 4 has a leading blank (though this is not apparent, it is in the original data).
- The state XX in row 6 is an invalid state.
- We are missing data on row 7.

Let's address each of these issues.

For the leading blank in row 4, we can use another pandas `str` method. The pandas `strip()` method works like the Python string `strip()` method. By default, we can use it to trim leading and trailing whitespace such as blanks:

```
[3]: df['state'] = df['state'].str.strip()
     for index, row in df.iterrows():
         print(row['state'])
```

```
NY
CT
IL
CA
None
XX
nan
```

In the above, since the leading space is not apparent when displaying the DataFrame, we use `iterrows()` and `print()` to see each cell's value.

The next issue is an invalid state—in particular, the XX in row 6. If we have a list of valid states—for example, in a file—we can check for invalid state abbreviations, and replace these with a missing value. The following statements read a list of state abbreviations from a file, and create a Python list from these:

```
[4]: dfstates = pd.read_csv('states.txt')
     stateList = dfstates.state.tolist()
     stateList
```

```
['CA', 'CT', 'IL', 'NY']
```

Once we have this list of valid states, we can create a new column indicating whether the state abbreviation is valid:

```
[5]: df['state_ok'] = df.state.apply(lambda x: x in stateList)
     df
```

id	address	city	state	state_ok
1	Bronx,NY	Bronx	NY	True
2	New Haven,CT	New Haven	CT	True
3	Chicago,IL	Chicago	IL	True
4	Monterey, CA	Monterey	CA	True
5	Washington	Washington	None	False
6	Nashville,XX	Nashville	XX	False
7	NaN	NaN	NaN	False

We can also create a new column, state_clean, that has valid state abbreviations (or `None`):

```
[6]: df['state_clean'] = \
       df.state.apply(lambda x: x if x in stateList else None)
     df
```

id	address	city	state	state_ok	state_clean
1	Bronx,NY	Bronx	NY	True	NY
2	New Haven,CT	New Haven	CT	True	CT
3	Chicago,IL	Chicago	IL	True	IL
4	Monterey, CA	Monterey	CA	True	CA
5	Washington	Washington	None	False	None
6	Nashville,XX	Nashville	XX	False	None
7	NaN	NaN	NaN	False	None

Section Summary

Similar to Python's string methods, pandas provides many string-related features. In this section, we focused on two: `str.split()` and `str.strip()`. As with Python string methods, there is often a concise way to work with strings to achieve the desired result for a string column in a DataFrame.

Test Your Understanding

1) New string columns can be created from a single string DataFrame column using which method?
   a) `multicol()`
   b) `split()`
   c) `astype()`

2) Assume a Python list has valid values for a given DataFrame column. To create a new DataFrame column indicating which rows have valid values, we can use:
   a) the `split()` method
   b) a lambda function with an in expression
   c) list slicing

11.6 Key Fields

In this section, we'll look at *key fields*—those columns whose values are intended to identify each row. Key fields can be of different types, such as strings (e.g., email addresses) or integers (e.g., generated numbers). In some datasets, more than one column is needed to uniquely identify each row, but in this section, we'll focus on data with single-key fields. Once again, we'll use a small dataset to demonstrate some useful techniques. The file DQ-Example-IDs.csv has the following data:

```
id,email
1,obraund@stateu.edu
2,jcummings@mymail.com
3,lheikkinen@redletter.com
4,jfutrelle@epost.org
,wallen@fivegee.net
six,jmoran@mymail.com
4,wallen@fivegee.net
```

This can be loaded into a DataFrame using:

```
[1]: import pandas as pd
     df = pd.read_csv("DQ-Example-IDs.csv")
     df
```

	id	email
0	1	obraund@stateu.edu
1	2	jcummings@mymail.com
2	3	lheikkinen@redletter.com
3	4	jfutrelle@epost.org
4	NaN	wallen@fivegee.net
5	six	jmoran@mymail.com
6	4	wallen@fivegee.net

Checking for uniqueness: The email field

To check if a column's values are different from one another, we can use the pandas Series `is_unique` property, as in:

```
[2]: df.email.is_unique
```

```
False
```

For our dataset, the result is False, since rows 4 and 6 have the same email value. To see which rows have non-unique values in a given column, we can use the `duplicated()` method:

```
[3]: df['email_dup'] = df.email.duplicated(keep=False)
     df
```

	email	email_dup
0	obraund@stateu.edu	False
1	jcummings@mymail.com	False
2	lheikkinen@redletter.com	False
3	jfutrelle@epost.org	False
4	wallen@fivegee.net	True
5	jmoran@mymail.com	False
6	wallen@fivegee.net	True

Our DataFrame is now augmented by a new column, email_dup, that tells us if the row has an email value that is duplicated somewhere else in the data. We can locate all the rows with duplicates by simply using:

```
[4]: df[df.email_dup]
```

	id	email	email_dup
4	NaN	wallen@fivegee.net	True
6	4	wallen@fivegee.net	True

which shows the subset of the DataFrame with duplicate email values:

Checking for integral, unique, not missing: The ID field

Similar to the email field, the ID field is intended to be unique. But in contrast with the string email field, the ID field is intended to have integer values. We'd also like to check if it has any missing values.

We'll start by checking for missing values. This can be done with the `isna()` and `sum()` methods:

```
[5]: df.id.isna().sum()
```

```
1
```

The resulting value of 1 indicates we have one missing value. The filter expression to show us this row is:

```
[6]: df[df.id.isna()][['id', 'email']]
```

	id	email
4	NaN	wallen@fivegee.net

Next we'd like to find rows where the ID value is not an integer. For this, the pandas `str.isdigit()` method can be used, which is similar to the Python string `isdigit()` method. However, `str.isdigit()` doesn't handle NumPy `NaN`s. To work around this, we can first ensure all values in the ID column are strings using `astype('str')`. We can create a new True/False column, id_isdigit, that indicates for each row if the value in the `id` column has only digits or not:

```
[7]: df['id'] = df.id.astype('str')
     df['id_isdigit'] = df.id.str.isdigit()
     df
```

	id	email	email_dup	id_isdigit
0	1	obraund@stateu.edu	False	True
1	2	jcummings@mymail.com	False	True
2	3	lheikkinen@redletter.com	False	True
3	4	jfutrelle@epost.org	False	True
4	nan	wallen@fivegee.net	True	False
5	six	jmoran@mymail.com	False	False
6	4	wallen@fivegee.net	True	True

We can then see all the non-digit IDs using:

```
[8]: df[['id']][~df.id_isdigit]
```

	id
4	nan
5	six

Last, we can check for duplicate IDs, as we did for emails, using the `duplicated()` method:

```
[9]: df['id_dup'] = df.id.duplicated(keep=False)
     df[df.id_dup][['id','email']]
```

	id	email
3	4	jfutrelle@epost.org
6	4	wallen@fivegee.net

For duplicate ID values, you might decide to review the data closer to determine the best way to repair the issue. Alternatively, pandas provides the `drop_duplicates()` method. For example, if we want to remove all rows with duplicate values for the `id` field, we can use:

```
[10]: df.drop_duplicates(subset=['id'], keep=False)
```

	id	email	email_dup	id_isdigit	id_dup
0	1	obraund@stateu.edu	False	True	False
1	2	jcummings@mymail.com	False	True	False
2	3	lheikkinen@redletter.com	False	True	False
4	nan	wallen@fivegee.net	True	False	False
5	six	jmoran@mymail.com	False	False	False

In the statement above, `subset=['id']` indicates that duplicates should be considered based on the `id` column alone. Also, `keep=False` indicates to remove the duplicated rows. The `keep` parameter could instead be set to `'first'` (keep the first of the duplicates), or `'last'` (keep the last of the duplicates).

Section Summary

We usually want key fields to have distinct, non-missing values. pandas provides the `is_unique` property to check if values in a column are unique. If they are not, we can use the `duplicated()` method to locate the duplicate values. As well, in order to find missing values, you can use the `isna()` method. To programmatically remove rows with duplicate data, you can use the `drop_duplicates()` method. Further, if a column's values are expected to be integers, the `str.isdigit()` method can be used to locate invalid values.

Test Your Understanding

1) Which can be used to see if a column has unique values?
   a) The is_unique property
   b) the `duplicated()` method
   c) the `info()` method

2) Which can be used to see those rows that have duplicate values within a certain column?
   a) the is_unique property
   b) the `duplicated()` method
   c) the `isna()` method

11.7 Inter-field Checking

In the previous sections, we've focused on handling data issues for one column at a time. In some cases, you may want to perform data quality checks for consistency between two (or more) columns. In this section, we'll look at a few techniques for inter-field checking. For the code examples, we'll look at a small dataset that indicates whether a respondent has children, as well as how many children they have. Given data in a file DQ-Example-CrossField.csv, we can load these data into a DataFrame using:

```
[1]: import pandas as pd
     df = pd.read_csv('DQ-Example-CrossCheck.csv', index_col='id')
     df
```

id	children	children_num
1	yes	2.0
2	n	0.0
3	Y	1.0
4	N	2.0
5	Y	-1.0
6	x	1.0
7	NaN	NaN

The cross-field check we are interested in is whether the binary value for children is consistent with the numeric value for children_num. In other words, we would expect:

1) If children is no, children_num should be zero.
2) If children is yes, children_num should be positive.

Before checking these conditions, we should have clean versions of these columns. Using background from earlier sections in this chapter, we can create clean columns using the following:

```
[2]: # clean the children field
     df['children_clean'] = df.children.str[0].str.lower()
     df['children_clean'] = \
        df.children_clean.apply(lambda x: x if x in ('y','n') else None)
     # clean the children_num field
     df['children_num_clean'] = \
        df.children_num.apply(lambda x: x if x>=0 else None)
     df
```

id	children	children_num	children_clean	children_num_clean
1	yes	2.0	y	2.0
2	n	0.0	n	0.0
3	Y	1.0	y	1.0
4	N	2.0	n	2.0
5	Y	-1.0	y	NaN
6	x	1.0	None	1.0
7	NaN	NaN	None	NaN

Now we can check our inter-field conditions. We can check rows not meeting condition 1 using a filter expression:

[3]:
```
df[(df.children_clean=='n') & (df.children_num_clean > 0)]
```

	children	children_num	children_clean	children_num_clean
id				
4	N	2.0	n	2.0

This above expression identifies the row as having a data conflict between the two fields.

Note that in the above filter expression, we use & instead of the Python keyword `and`. Also importantly, we need to use parentheses to surround each individual part of the condition.

As with Python, pandas lets us use comparison operators == , != , > , < , >= , <=. Also, in addition to & to indicate *and*, we can also use | for *or* as well as ~ for *not*.

If we'd like to identify both conditions 1 and 2, we can write a more involved filter condition. However, as the filter expression becomes more involved, it can become harder to read. As an alternative, we can write a function to check for inter-field consistency and use the DataFrame `apply()` method to check each row. A function to check both conditions 1 and 2 is:

[4]:
```
# a helper function to find a set of cross-field issues
def checkChildren(row):
    check1 = row.children_clean=='n' and row.children_num_clean>0
    check2 = row.children_clean=='y' and row.children_num_clean==0
    return not (check1 or check2)
```

We can use this function to identify rows that don't meet condition #1 or #2 using:

[5]:
```
df['children_cross_ok'] = df.apply(checkChildren, axis=1)
df['children_cross_ok']
```

```
id
1     True
2     True
3     True
4    False
5     True
6     True
7     True
Name: children_cross_ok, dtype: bool
```

This creates a new column, children_cross_ok, whose values are True if the inter-field check is OK or False if it is not. The `axis=1` argument indicates that the program should process each row (as opposed to each column). We can then find all the rows that have inter-field issues using:

[6]:
```
df[~df.children_cross_ok][['children_clean', 'children_num_clean']]
```

	children_clean	children_num_clean
id		
4	n	2.0

We have looked at one simple inter-field check. In practice, your dataset may require checking across multiple fields or need checks with more involved logic. The choice of filter conditions and Python functions gives you a lot of flexibility to verify your data have consistency between fields. Table 11-3 highlights a few pandas True/False methods that can be helpful in checking for issues.

Category	Method
numeric	`str.isdigit()`
	`str.isdecimal()`
	`str.isnumeric()`
string	`str.isalpha()`
	`str.isspace()`
	`str.isalnum()`
substring	`str.contains()`
	`str.endswith()`
	`str.startswith()`
membership	`isin()`
missing	`isna()`
	`notna()`

Table 11-3: Some pandas methods returning True/False results

Section Summary

In some datasets, you may want to check rules that relate field values within each record. One approach is to use a pandas filter expression to check inter-field consistency. Another approach is to write a function that can be applied to the DataFrame using the `apply()` method, to find inconsistencies between field values.

11.8 Finding Missing Related Rows between Tables

In the previous sections, we looked at addressing a variety of data quality issues within a DataFrame. In this section, we'll look at joining two tables into one DataFrame to see if data in one DataFrame are missing related data that appear in the other DataFrame. We'll assume the two tables have a common key field on which to do the join. Note that joining DataFrames was covered in Chapter 10.

The data for this section involve names and phone numbers. The names are in one file, contacts.csv; the phone numbers are in another file, phone.csv. Both files have an integer `id` field on which we can connect the names with the phone numbers.

We can load the names data using:

```
[1]: import pandas as pd
     dfContacts = pd.read_csv('contacts.csv', index_col='id')
     dfContacts
```

	name
id	
1	Alan
2	Bonnie
3	Chris
4	Deborah
5	Edward
6	Florence

We can load the phone number data using:

```
[2]: dfPhones = pd.read_csv('phones.csv', index_col='id')
     dfPhones
```

	phone
id	
1	212-555-1212
1	212-555-1234
2	212-555-3434
3	212-555-4545
4	212-555-5656
5	212-555-6767
7	212-555-9876

To combine these DataFrames into one DataFrame, we can use the `join()` method:

```
[3]: dfJoin = dfContacts.join(dfPhones)
     dfJoin
```

	name	phone
id		
1	Alan	212-555-1212
1	Alan	212-555-1234
2	Bonnie	212-555-3434
3	Chris	212-555-4545
4	Deborah	212-555-5656
5	Edward	212-555-6767
6	Florence	NaN

By default, the `join()` method will match the indexes of the two DataFrames. If the data to match are in a column as opposed to in the index, you can use the `on` parameter. For example, if `id` was a column, the statement would be:

```
dfJoin = dfContacts.join(dfPhones, on='id')
```

Note that in the `dfJoin` table above, there is a row for Florence with no phone number. This is because by default, the `join()` method result includes at least one row for each row in the `dfContacts` table. This is referred to as a `'left'` join, as the `dfJoin` variable is to the left of the `join()` method. There are in fact four variations to control which rows are included from the two tables. The `how` parameter lets you select one of these four variations.

Another join variation is to include at least one row in the result corresponding to all rows in both tables being joined. This is done using `how='outer'`:

```
[4]: dfJoinOuter = dfContacts.join(dfPhones, how='outer')
     dfJoinOuter
```

id	name	phone
1	Alan	212-555-1212
1	Alan	212-555-1234
2	Bonnie	212-555-3434
3	Chris	212-555-4545
4	Deborah	212-555-5656
5	Edward	212-555-6767
6	Florence	NaN
7	NaN	212-555-9876

Note that in the above table, the last two rows have missing values. This is because a row in one DataFrame does not have any corresponding row in the other DataFrame.

The missing values may be of note if you expect your data to have matches between all rows in the two DataFrames. With the pandas `isna()` method, we can easily find the unmatched cases. For example, those contacts who do not have a phone number is given by:

```
[5]: dfJoinOuter[dfJoinOuter['phone'].isna()]
```

id	name	phone
6	Florence	NaN

And conversely, the phone numbers that do not have a corresponding contact is given by:

```
[6]: dfJoinOuter[dfJoinOuter['name'].isna()]
```

id	name	phone
7	NaN	212-555-9876

Section Summary

To check if related data are in one DataFrame but not another, we can use the `join()` method with its named parameter `how`. The result of `join()`, in combination with `isna()`, can be used to find data quality issues where rows are expected but not found in the data.

11.9 Reorganizing Table Layout

In this section, we'll look at how you can transform tables so that the data are organized as you need them. Using pandas, you can reorganize the layout of your data in a number of ways. For this, we'll use a small dataset related to population and GDP (gross domestic product). The goal is to see how GDP per capita has compared between a few countries over a few years.

To start, let's load data in a file popgdp.csv:

```
[1]: import pandas as pd
     df = pd.read_csv('popgdp.csv')
     df
```

	country	year	population	gdp
0	us	2017	325.1	19.485
1	us	2018	327.1	20.612
2	us	2019	329.1	21.433
3	ca	2017	36.7	1.650
4	ca	2018	37.1	1.716
5	ca	2019	37.4	1.736
6	mx	2017	124.8	1.159
7	mx	2018	126.2	1.222
8	mx	2019	127.6	1.269

The data above include population (in millions) and GDP (in trillions) for the US, Canada, and Mexico for 2017–2019. If we'd like to calculate GDP per capita, we can add a column with:

```
[2]: df['gdppercap'] = round(df.gdp / df.population * 1000000)
     df
```

	country	year	population	gdp	gdppercap
0	us	2017	325.1	19.485	59935.0
1	us	2018	327.1	20.612	63014.0
2	us	2019	329.1	21.433	65126.0
3	ca	2017	36.7	1.650	44959.0
4	ca	2018	37.1	1.716	46253.0
5	ca	2019	37.4	1.736	46417.0
6	mx	2017	124.8	1.159	9287.0
7	mx	2018	126.2	1.222	9683.0
8	mx	2019	127.6	1.269	9945.0

To compare GDP per capita among countries and over time, it may be useful to use a different layout, such as with years as rows and countries as columns. We can do this using the `pivot()` method:

[3]:
```
gdppercap = \
    df.pivot(index='year', columns='country', values='gdppercap')
gdppercap
```

country	ca	mx	us
year			
2017	44959.0	9287.0	59935.0
2018	46253.0	9683.0	63014.0
2019	46417.0	9945.0	65126.0

If you instead want to have countries as rows and years as columns, you could adjust the `index` and `columns` arguments for the `pivot()` method. Alternatively, we can transpose the matrix above:

[4]:
```
gdppercap.transpose()
```

year	2017	2018	2019
country			
ca	44959.0	46253.0	46417.0
mx	9287.0	9683.0	9945.0
us	59935.0	63014.0	65126.0

The use of `pivot()` allows you to take a normalized dataset, where each row has one or more keys—in our example the keys are country and year—and reshape the data. Conversely, you may have a dataset that has a key whose values are spread across several columns, and you would like to normalize the data so that these key values are in one column. This is essentially the opposite of `pivot()`. For this case, we'll look at a file, gdp.csv, which can be read using:

[5]:
```
df = pd.read_csv('gdp.csv')
df
```

	year	us	ca	mx
0	2017	19.485	1.650	1.159
1	2018	20.612	1.716	1.222
2	2019	21.433	1.736	1.269

In the above data, we have each country as its own column. If you want instead for the country names to be in a common column, you can use the `melt()` method:

```
[6]:  df.melt(id_vars='year', var_name='country')
```

	year	country	value
0	2017	us	19.485
1	2018	us	20.612
2	2019	us	21.433
3	2017	ca	1.650
4	2018	ca	1.716
5	2019	ca	1.736
6	2017	mx	1.159
7	2018	mx	1.222
8	2019	mx	1.269

Section Summary

pandas provides a number of ways to reorganize the layout of your data. This includes using `pivot()` to take values that are initially in one column and generate a result where each of these values are in their own column. Conversely, if you have data where a set of columns are related to one dimension, these can be consolidated into a column using the `melt()` method. Another method, `transpose()`, allows you to exchange the row and column dimensions.

Chapter Review

Data preparation involves identifying and addressing data quality issues, as well as potentially joining and rearranging your data. Python and pandas provide a wide variety of powerful, concise ways to achieve these data goals.

For numeric data, pandas lets you handle integers and floats, as well as missing values. The `info()` method and `dtype` property provide ways to confirm the types of DataFrame columns. To derive clean data from raw data, you can use the `apply()` method in combination with a function. For date data, the pandas `to_datetime()` function can generate dates in a consistent format.

A category column has values that are limited to a fixed set of values. The `unique()` method lets you identify the unique values in a column. This is useful if you want to check that values fit into a desired set and, if not, help you see how you might transform the data. Transformation can be done with string methods, as well as the `map()` method.

pandas provides you with many string methods that are analogous to most of Python's string methods. For example, the pandas `split()` method can create multiple columns based on a single column of string data.

A particular column in your data may be a key field that is intended to be unique with no missing values. The `is_unique` property can confirm if the data have unique values. If not, the `duplicated()` method allows you to find where values are duplicated. Rows with missing values in a particular column can be found with `isna()`.

In some cases, you may need to check if two fields within a row are consistent in their values. pandas filter expressions can be used for simple checks. For more involved checks, the DataFrame `apply()` method can be used to check for and flag inter-field inconsistencies.

Two DataFrames with a common key column can be combined using the `join()` method. There are several alternatives for this (left, right, outer, inner), depending on what subset of rows you'd like to be included in the result.

From a source DataFrame, you can generate alternative layouts for your data. The `pivot()` method takes values in a single column and generates a result that will have these in multiple columns. Conversely, the `melt()` method will take related columns and generate a result where these are consolidated into a single column. Another useful layout method is `transpose()`, which will rotate the DataTable in order to display rows as columns and columns as rows.

Key Terms

apply()	Method that can apply a function to all rows of a Series or DataFrame.
astype()	Method to create a column of a given type from an existing column.
category data	A data column whose values are drawn from a fixed set.
cut()	Function that can create a new column representing a bracketing of the data.
data cleaning	Developing a clean set of data from an initial dataset with potential data quality issues.
data reorganization	Generating a new representation of data with a different layout with respect to rows and column organization.
duplicated()	Method to indicate where data values are duplicated.
field	A particular named column of data within a dataset.
is_unique	Property indicated if all the values are different from one another.
isna()	Method to determine if a value is missing.
join()	Method to combine two DataFrames.
key field	A data field that can be used to uniquely identify a row.
lambda function	A one-statement unnamed function.
map()	Method to generate new values based on a mapping from old values.
melt()	Method to generate a new data organization by combining related columns into one column.
NaN	"Not a number," a NumPy special value used to indicate a missing value.
pivot()	Method to generate a new data organization by splitting out data from one column into several columns.
to_datetime()	Function to convert date data into a pandas datetime.
transpose()	Method to generate a new data organization by exchanging the rows and columns.
unique()	Method to indicate if a particular value is unique for a given row.

Programming Problems

1) S&P 500 Constituents

For the following questions, the file sp500-constituents.csv is provided, which contains a list of stocks in the S&P 500 and certain data about each. Use pandas `read_csv()` to load the file into a DataFrame, and for each part below, develop statements to further refine the data.

a) SEC filings: Use the `drop()` method to take out the SEC filings column from the DataFrame.
b) Headquarters location: Use the `split()` method to create two new fields, one for city and one for state.
c) Date first added missing values: Create a DataFrame that has the symbols with those stocks where the Date first added field is missing a value.
d) Date first added (multivalued): Find rows where "Date first added" is different from a single date, then use statements to create a new column that determines and populates the date in first position if more than one date is indicated for this field.
e) Founded (missing values): Create a DataFrame that has the symbols with those stocks where the Founded column is missing a value.
f) Founded (multivalued): Find rows where "Date first added" is different from a single date, then use statements to create a new column that determines and populates the date in first position if more than one date is indicated for this field.

2) S&P 500 Price Data

For the following questions, the files sp500-constituents.csv and sp500-prices.csv are provided, which contain a list of stocks in the S&P 500 and certain data about each. For the constituent data, Symbol is a unique key column. For the prices data, Symbol and Date together uniquely define a row.

a) Use `read_csv()` to load two files, sp500-constituent.csv and sp500-prices.csv, each into their own DataFrame.
b) Rotate data: Transform the data organization for the price DataFrame, so prices for each date are each in their own dated column: for example, price-20201031, price-20200930, and so on.
c) Join data: Use `join()` to combine the constituents data and rotated price data from part (b) into one DataFrame. The resulting DataFrame should have one row for each row in the constituents data.
c) Augment data: Add a column that is the difference between the price on the last available date and the price on the prior available date.
d) Missing values: Determine which stocks do not have price data.

Chapter 12

Reading Web Pages with pandas and Beautiful Soup

Chapter Objectives

In this chapter, you'll learn:

- about HTML and some of its tags
- how you can use pandas to easily access table data from an HTML page
- how you can use the Beautiful Soup package to access any data from an HTML page

Chapter Outline

12.1 About HTML, pandas, and Beautiful Soup

As of this writing, there are over 50 billion web pages on the internet. Within these is a wealth of publicly available data, some of which may be useful for you in a future project. Although web pages were originally designed to be viewed in a browser, the data within them can often be accessed programmatically. Programmatic access to web pages is useful if we need data for a one-time study, or for an automated system running periodically. And Python together with certain packages is ideal for writing concise programs to collect data from web pages. In this chapter, we'll see how you can write Python programs that can collect data from web pages.

Web pages are defined in the HTML language, which we'll introduce in the next section. An HTML file may be locally stored on your computer or available as a web page on the internet. In this chapter, we'll mostly use an HTML file stored on your computer, but the same concepts and syntax are used for working with HTML files stored on the internet (web pages).

In cases where you'd like to programmatically access the data on an HTML page, there are a few approaches. In this chapter, we'll look at how pandas can be used to easily access data in HTML tables, as covered in Section 12.3. If you need HTML data that fall outside of a table, you can use the Beautiful Soup package to locate the data of interest. This is covered in the last section of the chapter.

HTML was designed for viewing content in browsers. In contrast, web APIs have been designed with programmatic (as opposed to browser) access to data and calculations available on the web. In cases when you can use a web API, instead of programmatically reading a web page, this can be a more stable, reliable, and efficient approach. We'll cover web APIs in the next chapter.

12.2 HTML

HTML basics

At the heart of a web browser is the ability to read an HTML file and display these data in a formatted way. In order to write Python that reads and processes data from an HTML page, it is helpful to know about HTML's format and structure, which we'll cover in this section. Note that the HTML that is stored in a file looks very different from what ultimately is displayed in a web browser.

For this chapter, we'll use a very small HTML page to show the approaches for reading its data of interest into a Python program. The page is from a hypothetical bank that would like to advertise its certificate of deposit rates to potential customers. As displayed in a browser, the page looks like Figure 12-1.

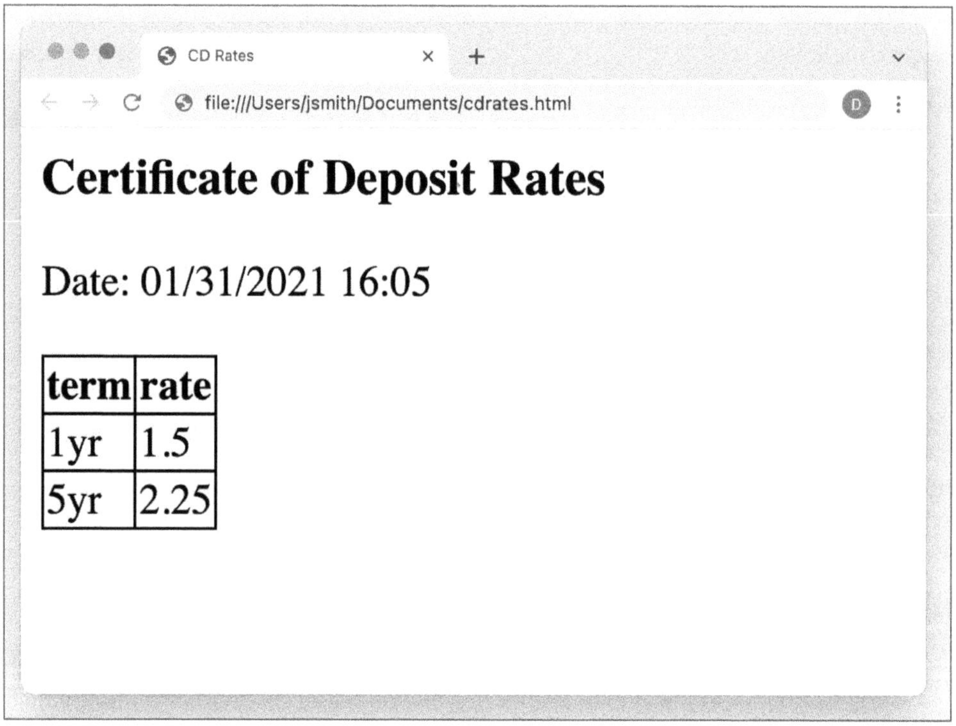

Figure 12-1: Example HTML page displayed in a web browser

As displayed in a browser, this example page has some formatting: different size fonts, some words bolded, and values in a table. However, the file that is used by the browser to display the above is a text file containing several HTML keywords to achieve the desired display.

The HTML file corresponding to the display in Figure 12-1 is shown in Figure 12-2:

```
<html>
  <head>
    <title>CD Rates</title>
    <style>
table, th, td {
  border: 1px solid black;
}
table {
  border-collapse: collapse;
}
    </style>
  </head>
  <body>
    <h3>Certificate of Deposit Rates</h3>
    <div id='asofdatetime'>
       Date: <span id='asofdate'>01/31/2021</span> <span>16:05</span>
    </div>
    <br>
    <div>
      <table id='rates'>
        <tr>
          <th>term</th><th>rate</th>
        </tr>
        <tr>
          <td>1yr</td><td>1.5</td>
        </tr>
        <tr>
          <td>5yr</td><td>2.25</td>
        </tr>
      </table>
    </div>
  </body>
</html>
```

Figure 12-2: Example HTML file for certificate of deposit rates page (cdrates.html)

In this section, we'll explain the parts of the HTML file in Figure 12-2. As mentioned, in order to write Python programs that can read web pages, it will be helpful to have a basic understanding of HTML.

In the example HTML file, the first thing that stands out is the use of angle brackets < >. The angle brackets aren't displayed in the page we see through the browser; instead, they contain the HTML keywords such as head, body, and table to give the browser information about the organization and style of the HTML file. In particular, keywords within angle brackets are called ***tags***. In our sample HTML file, `<head>`, `<body>`, `<table>`, and others are the tags.

Further, we see some tags have a forward slash, such as `</table>`. These are ***closing tags***, and each closing tag has a corresponding opening tag before it (without the /). A corresponding pair of opening and closing tags denote a start and end point, or section, in the HTML file.

Before discussing further, let's simplify our HTML page by removing the table. The smaller HTML file is shown in Figure 12-3. We'll come back to the table section later.

```
<html>
  <head>
    <title>CD Rates</title>
  </head>
  <body>
    <h3>Certificate of Deposit Rates</h3>
    <div id='asofdatetime'>
       Date: <span id='asofdate'>01/31/2021</span> <span>16:05</span>
    </div>
  </body>
</html>
```

Figure 12-3: Example HTML without a table (cdrates-notable.html)

In our HTML examples, note that the initial tag is `<html>`, signifying that the file is following the HTML format. At the bottom of the file is the matching closing `</html>` tag.

Between the opening and closing `<html>` and `</html>` tags are tags denoting the two major sections of all HTML files: <head> and <body>. The <head> section provides information about the document, and the <body> section provides the contents of the document to be displayed. Within the <head> section is the `<title>` section; between the title opening and closing tags is the phrase "CD Rates." The `<title>` section lets the browser know what to display on the tab for the displayed page.

By considering how each tag is contained within other opening and closing tags, we can think of the HTML file as a tree of tags. Figure 12-4 shows the tree corresponding to our smaller (no-table) HTML file.

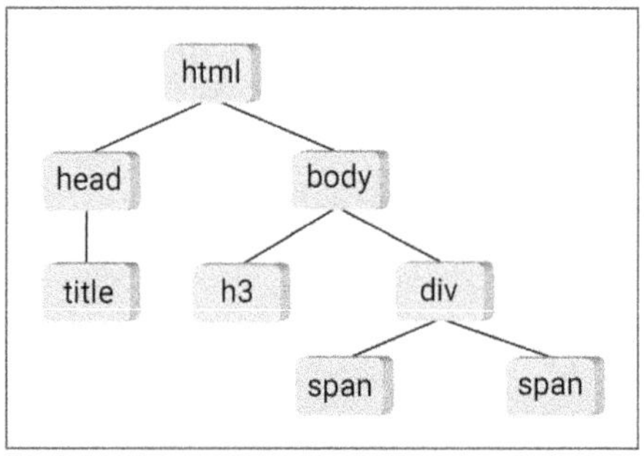

Figure 12-4: Tree representation of example HTML without a table (cdrates-notable.html)

Based on the tree representation, we note a few related terms and examples in Table 12-1.

Term	Definition	Examples
parent	The tag immediately above a tag is called the parent tag; all tags (other than `<html>`) have a parent tag	`<title>`'s parent is <head>
child	A tag can be thought of as the child of its parent tag	`<title>` is a child of <head>
sibling	Tags with the same parent can be thought of as ***sibling*** tags	<head> and <body> are siblings

Table 12-1: Relationships between HTML tags

Additionally note that a tag may have no children, one child, or several children. As well, the HTML tree may have multiple tags of the same type—for example, we have two `<span>` tags.

One note on terminology: as related to HTML, an ***element*** refers to a start tag, some content within, and an end tag. In general we'll use the word *tag* in this chapter.

Spreadsheet Compare/Contrast
Tables vs. Trees

While spreadsheets provide a tabular layout for data, this is in contrast with the tree representation of data in HTML. Trees allow for a wide variety of data organization and are ultimately more flexible than a tabular representation of data. A table of data can be represented as a tree in HTML, which we'll see later in this chapter.

Some additional HTML tags

Let's describe a few more HTML tags. Although there are over 100 different types of tags, in practice, you need to know only a few to become familiar with HTML and ultimately learn how to access HTML data in Python.

Our page has an overall heading. To capture this, we surround the heading text with one of the heading tags. There are these six heading tags: `<h1>`, `<h2>`, `<h3>`, `<h4>`, `<h5>`, and `<h6>`. The number indicates the relative size: `<h1>` is the largest heading and `<h6>` the smallest. In our example, we use `<h3>`:

```
<h3>Certificate Deposit Rates</h3>
```

Next, let's consider the date and time information within our page's data. As the page includes two distinct sections (data and time, as well as the rates table), we've marked the date and time data as being part of their own division using the `<div>` tag.

We'll see later that organizing our data within `<div>` and `<span>` tags can make it easier for programs to access particular data in our page. The `<div>` tag is generally used to identify particular blocks of the page and by default may introduce a line break when the page is displayed in the browser. Similarly, the `<span>` tag identifies a section of the page, but does not introduce a line break and is generally used for a small section of the page—for example, to surround a phrase or field. In our example, we use the `<span>` tag to surround the date and time fields.

HTML attributes

At this point, it will be helpful to introduce the HTML feature of attributes. HTML allows attributes as a way of adding information to a tag. A tag can optionally have one or more attributes specified inside the angle brackets. Each attribute has a name and associated value. A special attribute, `id`, is used to uniquely identify tags within a document. For example, we've used the `id` attribute for the `<div>` tag as:

```
<div id='asofdatetime'>
```

In the above, the `<div>` tag has the `id` attribute with a value of `'asofdatetime'`. The setting won't have an effect on what's displayed in the browser, but it will make the date and time fields easier to access programmatically, as we'll see later.

HTML tables

Having covered the basics of the small HTML page, we'd like to return back to the expanded example in Figure 12-1 that includes a table of rates data. Recall that the table within our page is displayed as:

Term	Rate
1yr	1.5
5yr	2.25

The corresponding HTML for the above table is:

```
<table id='rates'>
   <tr>
      <th>term</th><th>rate</th>
   </tr>
   <tr>
      <td>1yr</td><td>1.5</td>
   </tr>
   <tr>
      <td>5yr</td><td>2.25</td>
   </tr>
</table>
```

To define a table in HTML, we use the `<table>` tag. Within a `<table>`, we use the `<tr>` tag to identify each row, and the `<td>` tag to identify each cell of the row. As well, the `<th>` tag identifies the table's header row.

Given the `<table>`, `<th>`, `<tr>`, and `<td>` tags, we can consider our table as like a tree, as shown in Figure 12-5.

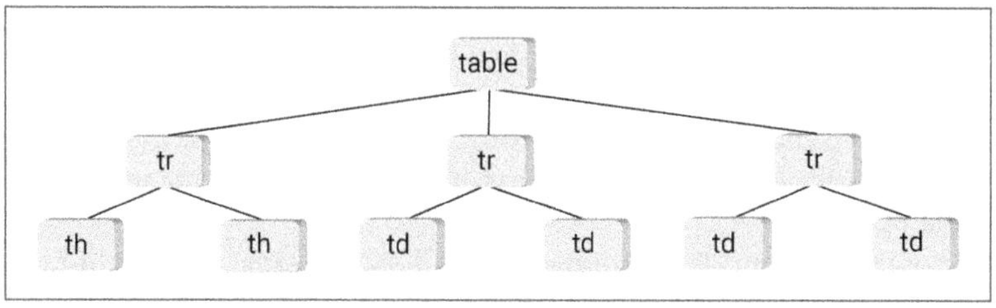

Figure 12-5: Tree representation of an HTML table

Section Summary

Given the huge amount of data available on the web in HTML format, we may be able to find data we need in HTML pages. An HTML document can be thought of as a tree of tags. In an HTML file, elements have an opening tag in angle brackets < > and a corresponding closing tag. Each tag may have one or more attributes. The `id` attribute can be used to uniquely identify an element in a document. There may be text within tags as well.

The highest level tags of an HTML document are `<html>`, `<head>`, and `<body>`. In this chapter, we looked at a number of important tags, including `<div>`, `<span>`, and `<table>` and its related tags.

Test Your Understanding

1) An HTML document can be thought of as a tree of:
   a) attributes
   b) tags
   c) text
   d) none of the above
2) An HTML tag can have at most one attribute.
   a) True
   b) False
3) The `id` attribute can be used to uniquely identify a tag within an HTML page.
   a) True
   b) False

12.3 pandas `read_html()`

If we find data of interest in web pages, they may well be stored in tables. And pandas provides a way—in one line of code—to read tables from a web page into pandas DataFrames. This lets us write very concise Python code to collect tabular data from web pages.

As we saw in the previous section, tables can be represented in an HTML page using the `<table>` and related tags. To read an HTML page's tables into a list of `DataFrames`, pandas provides the `read_html()` function. This function accepts an argument that can be a filename, the contents of an HTML page, or a URL. For example, to read an example HTML document stored locally, we can simply write:

```
[1]: import pandas as pd
     dfs = pd.read_html('cdrates.html')
```

In the above, dfs is a list of DataFrames, even if the HTML page has only one table. We can confirm this with:

```
[2]: type(dfs)
```
```
list
```

In this example, there is one table on the HTML page, so there is one DataFrame item in the list dfs:

```
[3]: len(dfs)
```
```
1
```

We can view this DataFrame as:

```
[4]: dfs[0]
```

	term	rate
0	1yr	1.50
1	5yr	2.25

Note that if there are no tables in the HTML page, then read_html() will generate an error, as opposed to returning an empty list. To handle this error, you can use try/except logic, as introduced in Chapter 7 and covered further in Appendix D.

In the previous example, read_html() is used to read an HTML from your computer. But more powerful is that read_html() can also be used to read a web page from the internet. The details of communication across the internet are handled by pandas so that we can still use one line of code to access and read a web page stored remotely on the internet.

In order to read a web page's tables into a list of pandas DataFrames, we simply pass a URL as an argument to read_html(). For example, to access a web page of the S&P 500 stocks from Wikipedia and get a list of DataFrames corresponding to the web page's tables, we can use:

```
[5]: url = 'https://en.wikipedia.org/wiki/List_of_S%26P_500_companies'
     dfs = pd.read_html(url)
```

Let's see what was returned in the DataFrame list dfs. We can see how many DataFrames are in the list:

```
[6]: len(dfs)
```
```
2
```

Since there are two tables returned by read_html(), we can interactively explore what was returned in dfs using the DataFrame head() method:

```
[7]: dfs[0].head()
```

	Symbol	Security	SEC filings	GICS Sector	GICS Sub-Industry	Headquarters Location	Date first added	CIK	Founded
0	MMM	3M	reports	Industrials	Industrial Conglomerates	Saint Paul, Minnesota	1976-08-09	66740	1902
1	AOS	A. O. Smith	reports	Industrials	Building Products	Milwaukee, Wisconsin	2017-07-26	91142	1916
2	ABT	Abbott	reports	Health Care	Health Care Equipment	North Chicago, Illinois	1964-03-31	1800	1888
3	ABBV	AbbVie	reports	Health Care	Pharmaceuticals	North Chicago, Illinois	2012-12-31	1551152	2013 (1888)
4	ABMD	Abiomed	reports	Health Care	Health Care Equipment	Danvers, Massachusetts	2018-05-31	815094	1981

Based on the above, we see the first returned DataFrame in `dfs` is the table of stocks we're looking for.

We can refine the call to `read_html()` to indicate which column should be used as the row index. The named parameter for this is `index_col`, similar to `read_csv()`. We can indicate that the first column of the data should be used for the DataFrame index using:

```
[8]: dfs = pd.read_html(url, index_col=0)
     dfs[0].head()
```

	Security	SEC filings	GICS Sector	GICS Sub-Industry	Headquarters Location	Date first added	CIK	Founded
Symbol								
MMM	3M	reports	Industrials	Industrial Conglomerates	Saint Paul, Minnesota	1976-08-09	66740	1902
AOS	A. O. Smith	reports	Industrials	Building Products	Milwaukee, Wisconsin	2017-07-26	91142	1916
ABT	Abbott	reports	Health Care	Health Care Equipment	North Chicago, Illinois	1964-03-31	1800	1888
ABBV	AbbVie	reports	Health Care	Pharmaceuticals	North Chicago, Illinois	2012-12-31	1551152	2013 (1888)
ABMD	Abiomed	reports	Health Care	Health Care Equipment	Danvers, Massachusetts	2018-05-31	815094	1981

Since we've located the DataFrame of interest as returned from `read_html()`, we can process it further as desired. For example, if we'd like to save the data in a text file, we can use:

```
[9]: dfs[0].to_csv('sp500-constituents.txt', sep='|')
```

In the above, we set `sep='|'` to override the default separator of comma, since there may be commas in the data.

There are a few more named parameters available in `read_html()`; several are similar to those available in `read_csv()`. The official pandas documentation describes each of these parameters.

One additional parameter worth noting is called `attrs`, which is short for attributes. Recall that an HTML tag can have attribute/value pairs. If there is a particular HTML table that we are interested in, we can use the `attrs` parameter to have `read_html()` locate and return it, and omit the other tables from the result. In this way, our result from `read_html()` just contains the table or tables of interest. For example, if our HTML page has a table tag with an `id` attribute set to `'rates'`, this approach is a good way to get only that table returned in our list of DataFrames. Returning to our local CD rates HTML file, this following statement uses the `attrs` property to locate and load a specific table:

```
dfs = pd.read_html('cd_rates.html', attrs={'id':'rates'})
```

As our rates.html file has only one table, the inclusion of `attrs` will not change our result. However, in HTML pages with several tables, the use of the `attrs` parameter may allow you to better identify the table of interest.

If your HTML page of interest is locally on your computer, you can look at the file's contents directly—for example, in a text editor. However, there is a better alternative in a browser that lets you view and navigate an HTML document's elements for both local HTML and web pages. For example, in Chrome, you can right-click on an item in the page (control-click on a Mac), and choose Inspect. Below shows the side-by-side view Chrome provides when inspecting a page.

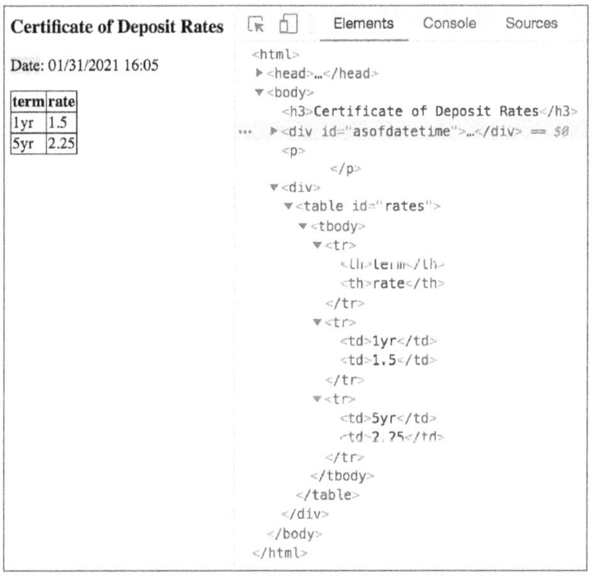

The Inspect view is an excellent way to reverse engineer the organization of an HTML page. For example, this can be helpful for finding `id` attributes for use in `read_html()` with pandas. More importantly, you will need to analyze an HTML page's tree structure to navigate it when working with the Beautiful Soup package, covered in the next section.

Tip

Web Page Changes

For one-time data collection, you can create code to access web page data and save the results to a file for subsequent analysis. If you would like to develop a system to periodically read a web page, you need to take care, since web pages are subject to change by others at any time. Changes to the organization of a web page may break code that you originally had working. For more reliable approaches, consider sourcing the data from databases or web APIs, which can be more stable in layout than web pages. These topics are covered in subsequent chapters in this book.

Section Summary

With one function call, you can use the pandas function `read_html()` to get the tables from an HTML page into a list of `DataFrames`. This function works with local HTML pages as well as web pages. There are a number of named parameters available in `read_html()`, including `attr` to locate specific tables on an HTML page.

Test Your Understanding

1) The pandas `read_html()` function can be used to read a comma-separated (CSV) file.
   a) True
   b) False
2) The pandas `read_html()` function can be used to read all the data on an HTML page.
   a) True
   b) False
3) The pandas `read_html()` function returns an empty list if there are no table elements on the HTML page.
   a) True
   b) False

12.4 Beautiful Soup

Beautiful Soup basics

In the previous section, we used the pandas `read_html()` function to read tables from an HTML page into a list of DataFrames. If all the data of interest are in a page's tables, this is sufficient for getting the data. However, if there are data you need in HTML that aren't in a table, `read_html()` can't be used to access them. Instead, another package can be used. In this section, we'll look at how the Beautiful Soup package can be used to access any data on an HTML page and bring the data into your Python program.

As described in Section 12.2, HTML documents can be thought of as a tree of tags. Beautiful Soup can read HTML and parse its contents to create a tree representation of the page. You can then programmatically seek data through the methods that Beautiful Soup provides.

Let's look at a specific case with the CD rates page shown in Figures 12-1 and 12-2. Let's assume you'd like to programmatically read the date within the page. Recall that the date is within a `<span>` element that has an `id` of `'asofdate'`.

In order to use Beautiful Soup, we first need to get the HTML data of interest. Since we're starting with a local HTML page, we can use the following to read the cdrates-notable.html file's contents into the string variable `contents`:

```
[1]: filename = 'cdrates-notable.html'
     f = open(filename, 'r')
     contents = f.read()
```

Since we are working with a small file, we can check that the file was read OK by displaying `contents`:

```
[2]: contents
```

```
"<html>\n    <head>\n      <title>CD
Rates</title>\n    </head>\n    <body>\n       <h3>Certificate of
Deposit Rates</h3>\n       <div id='asofdatetime'>\n          Date:
<span id='asofdate'>01/31/2021</span>
<span>16:05</span>\n       </div>\n    </body>\n</html>\n"
```

Next in order to use the Beautiful Soup package (also called bs4), we can import the `BeautifulSoup` type using:

```
[3]: from bs4 import BeautifulSoup
```

Using the `contents` data as previously read from the HTML file, we can create a `BeautifulSoup` instance using:

```
[4]: soup = BeautifulSoup(contents, 'html.parser')
     type(soup)

     bs4.BeautifulSoup
```

The first statement above creates a `BeautifulSoup` instance using the contents previously read from the cdrate-notable.html file. The second argument, `'html.parser'`, indicates to `BeautifulSoup` to use its parser for HTML formatted data. In the above, we also check and see that the type of the `soup` variable is `BeautifulSoup`.

At this point, variable `soup` contains the HTML page data represented as a tree. We can do a quick check of what was loaded by checking the `title` property:

```
[5]: soup.title

     <title>CD Rates</title>
```

Above, we see the `title` property including the opening and closing `<title>` tags as well as the text contained within. Here `soup.title` is of the type `Tag`. A tag represents a node in the tree. A `Tag` has a `text` property, and we can get the `<title>` text with:

```
[6]: soup.title.text

     'CD Rates'
```

Displaying the tree

We can display the HTML tree, for example:

```
[7]: soup

     <html>
     <head>
     <title>CD Rates</title>
     </head>
     <body>
     <h3>Certificate of Deposit Rates</h3>
     <div id="asofdatetime">
             Date: <span id="asofdate">01/31/2021</span> <span>16:05</span>
     </div>
     </body>
     </html>
```

This confirms that the HTML has been loaded, though the output has lost the indenting of our page. For easier reading, we can use the `prettify()` method. For example:

```
[8]:  print(soup.prettify())
```

```
<html>
 <head>
  <title>
   CD Rates
  </title>
 </head>
 <body>
  <h3>
   Certificate of Deposit Rates
  </h3>
  <div id="asofdatetime">
   Date:
   <span id="asofdate">
    01/31/2021
   </span>
   <span>
    16:05
   </span>
  </div>
 </body>
</html>
```

Navigating the tree

Having loaded the file's contents into the `BeautifulSoup` variable `soup`, let's return to our earlier question of how to get the date field. We can use the same syntax we used to get the title; for example:

```
[9]:  soup.span.text
```

```
'01/31/2021'
```

Although there are two `<span>` elements in our data, this works because the property access `BeautifulSoup` uses will locate the first element of that type in the tree. In our case, that happens to be the date. Conversely, if there were no `<span>` elements in our tree, `soup.span` would return `None`, and `soup.span.text` would return an error.

The use of an expression like `soup.span` is shorthand for using `soup.find('span')`. The `find()` method is a concise and flexible way to search down the HTML tree to find a matching element. Here is an example that uses `find()`:

```
[10]: spanElement = soup.find('span')
      if spanElement != None:
         print(spanElement.text)
```

```
'01/31/2021'
```

In its simplest use, the `find()` method will find the first occurrence of an element of a given type. Notably, `find()` will return a `Tag` corresponding to the first element match it finds, even if there are multiple matching elements. If there is not a match, `find()` will return `None`.

We can also provide `find()` with one or more HTML attribute name/value pairs. Notably, this works well with the `id` attribute, as this will locate a specific element in the tree. In general, if your data of interest are located at or near an element with an `id` attribute, you can use simple code to access the data. The `find()` method uses a Python dictionary in order to restrict its match to attribute/values; for example:

```
[11]: soup.find('span', {'id':'asofdate'})
      <span id="asofdate">01/31/2021</span>
```

As a shorthand for the above, we can use the name of the HTML attribute as a parameter for the `find()` method. For example, we can use an `id` parameter in the `find()` method, as in:

```
[12]: soup.find('span', id='asofdate')
      <span id="asofdate">01/31/2021</span>
```

In cases where we want to collect data located at multiple elements in the HTML tree, the `find_all()` method can be used. This is in contrast with `find()`, which will return at most one tag. For example, for our page, we can use:

```
[13]: spanElements = soup.find_all('span')
```

In this case, `spanElements` is a `ResultSet` type. We can loop over the items in the `ResultSet` using the `for` statement; for example:

```
[14]: for element in spanElements:
          text = element.text
          if text != None:
              print(text)
      01/31/2021
      16:05
```

The items of a `ResultSet` can also be accessed using square brackets `[]` similar to a list. For our example, we know the time field is the second `<span>`, so we can access it like:

```
[15]: spanElements[1].text
      '16:05'
```

Combining the aspects above, we can write a program that finds the date and time data within our page. This is shown in Program 12-1.

Program 12-1

```
from bs4 import BeautifulSoup

f = open('cdrates.html', 'r')
contents = f.read()

soup = BeautifulSoup(contents, 'html.parser')

dateElement = soup.find(id='asofdate')
date = dateElement.text
spans = soup.find_all('span')
time = spans[1].text
print(date, time)
```

Program 12-1 Output

```
01/31/2021 16:05
```

Since our example is small and has `id` attributes, the `find()` and `find_all()` methods are sufficient to locate the date and time fields from the top of the tree. In a larger example, or one without distinct attributes like `id`, you may need to navigate the HTML tree. For this, you can use a sequence of `find()` and/or `find_all()` methods to navigate to the data of interest. For example:

```
[16]: datetimeDiv = soup.find('div', id='asofdatetime')
      dateSpan = datetimeDiv.find('span')
      print(dateSpan)

      <span id="asofdate">01/31/2021</span>
```

In the above statements, we first find the `<div>` element of interest using `find()` and its `id` attribute. From there we navigate further to the date `<span>` using `find()` with the `datetimeDiv` variable, which is of the type `Tag`. In general, you can use `find()` and `find_all()` not only from the top level of the tree but from any element (`Tag`) within the tree. This allows you to locate an element and continue to navigate from there to find a lower down element of interest.

The Beautiful Soup package also provides methods to move across from one element to another. In our example, we can find the date `<span>` and then move sideways across the tree to its sibling `<span>` to get the time. For example:

```
[17]: datetimeDiv = soup.find('div', id='asofdatetime')
      dateSpan = datetimeDiv.find('span')
      timeSpan = dateSpan.find_next_sibling('span')
      print(dateSpan.text, timeSpan.text)

      01/31/2021 16:05
```

Once you know the layout of an HTML document, you can use Beautiful Soup methods to "walk" the tree of data. Table 12-2 summarizes these methods.

search direction	method	return type
down	`find()`	`Tag` or `None`
	`find_all()`	`ResultSet`
up	`find_parent()`	`Tag` or `None`
	`find_parents()`	`ResultSet`
across	`find_next_sibling()`	`Tag` or `None`
	`find_next_siblings()`	`ResultSet`
	`find_previous_sibling()`	`Tag` or `None`
	`find_previous_siblings()`	`ResultSet`

Table 12-2: Beautiful Soup navigation methods

Note that the find methods in Table 12-2 have some named parameters that may be useful. To limit results to matches of the elements' text, the string parameter can be used. This parameter has a few variations—it can be given a string, a regular expression,* a list, a function, or the value True. To limit the number of results,

* Regular expressions allow for string pattern matching and are supported by the Requests package in Python.

the limit parameter can be set to a number: for example, `limit=2`. As well, to limit the search to an element's children as opposed to all its descendants, `recursive=False` can be used. Combining all these, an example is:

```
children = node.find_all('div', string=['rate', 'price'],
                         recursive=False, limit=2)
```

The above statement will find the first two children elements of the type `<div>`, which include the text of either `'rate'` or `'price'`.

Accessing web pages

In this section so far, we've used Beautiful Soup to read a local HTML file—in other words, a file that you have on your computer. The more common case is to access an HTML page on the web. Though Beautiful Soup itself can't access a web page itself, it can work easily with another package that can get the web page for it. There are a few packages that allow us to access a web page; in this section, we'll introduce and use the Requests package.

The Requests package has a number of capabilities; in this chapter, we'll focus on the essentials for getting a web page. Assuming you have the URL in a variable called `url`, we only need these statements to retrieve a web page:

```
import requests
page = requests.get(url)
```

In the above, the `get()` function returns a `Response` type. The contents of the web page are available in the `Response`'s `content` property.

Returning to the interest rate web page introduced earlier in the chapter, Program 12-2 shows how to access that page with the Requests package, and provide the page's contents to Beautiful Soup.

Program 12-2

```
import requests
from bs4 import BeautifulSoup
url = 'https://en.wikipedia.org/wiki/List_of_S%26P_500_companies'

# get the page of interest from the web
page = requests.get(url)

# provide the page's content to Beautiful Soup, for it to parse
soup = BeautifulSoup(page.content, 'html.parser')

# display the title of page
print(soup.title.text.strip())
```

Program 12-2 Output

```
List of S&P 500 companies - Wikipedia
```

We can add further to Program 12-2 to navigate to the table of contents within the page. We can inspect the web page in our browser to see where the table of contents is. From this, we find the table of contents in this page is within a `<div>` tag that has an `id` attribute of `'toc'`. As `id` is a commonly used attribute in HTML pages, Beautiful Soup provides the `id` property that we can use with the `find()` method, to locate a particular tag. For example:

```
tableOfContents = soup.find('div', id='toc')
```

Program 12-3 uses the above statement, as well as the `find_all()` method, to locate and display the table of contents for this page.

Program 12-3

```
import requests
from bs4 import BeautifulSoup
url = 'https://en.wikipedia.org/wiki/List_of_S%26P_500_companies'

# get the page of interest from the web
page = requests.get(url)

# provide the page's content to Beautiful Soup, for it to parse
soup = BeautifulSoup(page.content, 'html.parser')

# display the table of contents
toc = soup.find('div', id='toc')
tocItems = toc.find_all('span', {'class':'toctext'})
for tocItem in tocItems:
    print(tocItem.text)
```

Program 12-3 Output

```
S&P 500 component stocks
Selected changes to the list of S&P 500 components
See also
References
External links
```

In Program 12-3, `find_all()` is a key method used for finding all the desired matching elements on the HTML tree. The result of `find_all()` is a Beautiful Soup `ResultSet`, which allows us to access its items using the `for` statement.

Notes on JavaScript and CSS

So far, we've looked at static HTML pages—that is, those pages whose content is defined directly within the HTML document. HTML also works with the JavaScript language, allowing for interaction, activity, and dynamic content in a web page. If your page of interest has dynamic content of interest, pandas or Beautiful Soup may not be able to access it, and you will need another approach. Dynamic coverage is beyond the scope of this chapter, but it is possible to programmatically access dynamic content with other packages.

In addition to working with JavaScript, HTML works with CSS (Cascading Style Sheets), a language that defines the style and layout of the page. In general, you will not use CSS when programmatically accessing data, since CSS code does not contain data. However, the style attribute of an HTML element may have CSS directions. Although you can use the style attribute programmatically to locate an element, this may not be reliable, as web page authors may choose to change a page's style. In general, you should look for the `id` attribute if available on a page to locate an element.

Section Summary

In the prior section, we saw how pandas can be used to get tabular data from HTML. However, if your data of interest are not in a table, the Beautiful Soup package allows your program to parse an HTML document so that you can locate any data of interest in an HTML page. The methods it provides allows your code to navigate to and extract the data of interest from a page. In conjunction with the Requests package, Beautiful Soup can be used to access a web page's data. The primary types provided are `BeautifulSoup`, which represents the whole page; `Tag`, which corresponds to an HTML tag; and `ResultSet`, which can represent a list of tags.

Test Your Understanding

1) Beautiful Soup can access HTML data, but not data within a `<table>`.
   a) True
   b) False
2) Beautiful Soup's main type is called an "element."
   a) True
   b) False
3) The `BeautifulSoup find()` method will find at most one occurrence of a matching element.
   a) True
   b) False

Chapter Review

The web has a wealth of data in HTML pages, and much of it can be accessed programmatically. This is useful for one-time data collection as well as systems that collect data from the web.

HTML models a document as a tree of elements. Each element may have one or more attributes. The `id` attribute can be used to uniquely identify an element in a document. The main elements of an HTML document are `<html>`, `<head>`, and `<body>`. The element types `<div>` and `<span>` divide the document into sections. The `<table>` element—and others, including `<tr>`, `<td>`, and `<th>`—provides a structure for defining tables in HTML.

If your data of interest are in an HTML `<table>`, you can use the pandas function `read_html()` to get the tables from an HTML page into a list of `DataFrames`. There are a number of named parameters available in `read_html()`, including `attr` to locate specific tables on an HTML page.

The Beautiful Soup package allows your program to parse an HTML document, so that you can locate data of interest in an HTML page. The methods it provides allows your code to navigate to and extract the data of interest from a page. In conjunction with the Requests package, Beautiful Soup can be used to access a web page's data.

Key Terms

attribute	An HTML element may have attributes, which are name/value pairs.
element	An HTML document is composed of tags that are arranged in a tree.
HTML	Hypertext markup language, which includes a set of valid tag names that can be included in a page.
id	The `id` attribute in HTML associates a unique identifier with a tag in the page.
read_html()	A pandas function for finding tables on an HTML page and returning these as a list of DataFrames.
ResultSet	A set of results that is returned from certain Beautiful Soup methods—for example, from `find_all()`.
tag	A keyword enclosed in < > brackets, denoting a position an HTML document.
Tag	A Beautiful Soup type that corresponds with an element in an HTML document.
tree	An HTML document can be represented as a tree of elements.
web page	An HTML file that is on a web server and accessible through the internet.

Programming Problems

1) Reading a Local HTML Page

Write a program that uses pandas `read_html()` to read the cdrates.html file, and displays the data in the table contained within that page.

2) Reading a Web Page—Yield Curve Difference—pandas

For the interest rate web page at:

https://home.treasury.gov/resource-center/data-chart-center/interest-rates/TextView?type=daily_treasury_yield_curve&field_tdr_date_value=2022

a) Write a program that uses pandas `read_html()` to read the matrix of yields.
b) Determine the difference between the last two yield curves in the yield matrix. This is the difference between the corresponding rates within each column of the matrix.

3) Reading a Web Page—Yield Curve Difference—Beautiful Soup

Solve the problem described in problem 2, but instead of using pandas, use the Requests and bs4 packages to read the yield matrix.

4) Reading a Web Page—Data from Yahoo Finance

Yahoo Finance provides pages containing stock prices and other data for companies and their stocks. The form of the URL is:

https://finance.yahoo.com/quote/<symbol>

where <symbol> is a particular stock symbol.

a) Using a web browser, access a particular stock's page from Yahoo Finance and inspect it to determine where the latest stock price is located on the HTML tree.
b) Use the Requests and bs4 packages to access a Yahoo Finance page for a given stock.
c) Using Beautiful Soup, navigate to and display the latest stock price.
d) Determine where the stock's Market Cap value is on the HTML tree, and use Beautiful Soup to navigate to and display the Market Cap.

5) Reading a Web Page of Interest—pandas

Choose a web page of interest that has table data, and use pandas to read and display this table.

6) Reading a Web Page of Interest—Beautiful Soup

Choose a web page of interest that has data not in a table, and use Beautiful Soup to locate and display these data.

7) Project: Collect Top Charting Songs from Several Pages into a csv File

The top 100 songs for each year are available from Wikipedia at:

https://en.wikipedia.org/wiki/Billboard_Year-End_Hot_100_singles_of_<year>

where <year> is a four-digit year—for example, 2020.

a) Write a program that accesses the data from the top songs (#, song title, artist) from Wikipedia and writes these data to a .csv file.
b) Enhance your program from part (a) to read pages from a range of years (by accessing each page corresponding to each year), and save the data to one .csv file. In addition to #, title, and artist, include the year in the .csv file.

8) Project: Collect the Last Price and Market Cap of All Stocks in the DJIA

The list of stocks in the Dow Jones Industrial Average (DJIA) can be found at:

https://en.wikipedia.org/wiki/Dow_Jones_Industrial_Average

a) Use this page to determine the list of symbols of the stocks in the DJIA. With these symbols, access the corresponding pages in Yahoo Finance (see question 4) to determine the latest stock price for each stock. Create a .csv file with symbol and latest stock price.

9) Project: Portfolio Valuation

Using work in problem 8 as a starting point, read a portfolio file containing stock symbols and their number of shares, then determine and report the valuation of each stock and for the portfolio as a whole.

Chapter 13

Accessing Web APIs with Requests and JSON

Chapter Objectives

In this chapter, you'll learn:

- about web APIs, which give you an avenue to a wide variety of data and calculations across the web
- how web APIs can be called with a URL request using the Python Requests package and a few lines of code
- about JSON for representing data that can be used in web API requests and responses
- how query parameters can be included in a web API request to tailor requests to your needs
- about web API status codes used to indicate success or issues with a request
- some alternatives for indicating an API access key as part of a request that is necessary for some web APIs
- how a specific Python package, if available, can simplify how you access a web API

Chapter Outline

13.1 About Web APIs

In the previous chapter, we looked at how you can programmatically access web page data. Given the number of web pages available, this can help you get data you need. However, web pages are primarily designed for viewing, and their layout can be changed. If you need ongoing access to data, a change to a web page layout can break something you've developed that reads a given page. As an alternative to reading web page data, ***web APIs*** provide a more stable approach to accessing data available on the web. In fact, API stands for application *programming* interface—indicating that APIs are provided for programs (such as Python) to access. Further, web APIs can support not just data access but data processing, such as language translation or best routing, allowing you to integrate features available in web services.

The World Wide Web's original design allowed browsers to interact with web servers to view and maintain files across the internet. Standards such as HTML and HTTP made it possible to view pages in a web browser and click on hyperlinks to move to and view another page. The World Wide Web's original usage was visual, and in the previous chapter, we looked at how you can programmatically access data on web pages.

This "visual web" is familiar and tangible, as we experience it through web browsers. There is also functionality based on the original web standards as well as subsequent standards and approaches that allow nonbrowser programs to interact across the web—for example, for data retrieval, image access, and calculations. This "invisible web" is based on web APIs, allowing client software such as Python to make requests to, and receive results from, web servers. This arrangement allows you to access data and other features maintained by others using relatively few lines of code in your program.

In a client/server interaction using a web API, the client program makes a request using a URL and possible supplemental data, which is routed to a web server. The server in turn processes the request and sends back a reply to the client program. Figure 13-1 shows the steps involved.

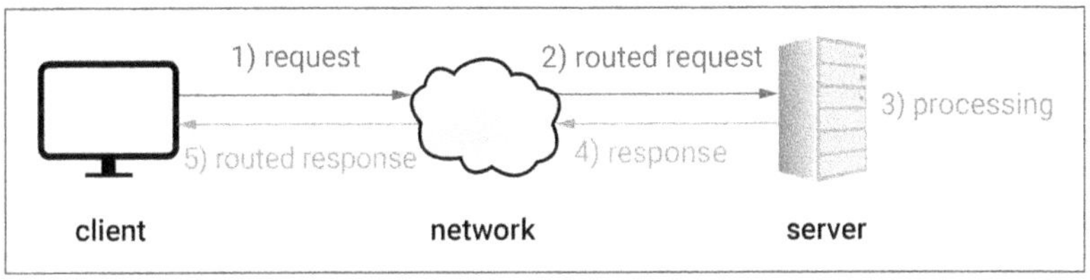

Figure 13-1: Steps in a web API call

In a web API call, the request and reply conform to HTTP messages, which is the same standard used for web page access. In this way, web APIs rely on existing standards used on the internet. For the request, the message involves three parts: a start line that includes a URL, a headers section (technical information about the message), and possibly a body section. If the information in the URL is sufficient, there is no need for a body section in the message. What's required in the URL and message body is based on the design of the web API.

A web API response also uses the ***HTTP*** message format, which includes a ***status code*** in the first line; a ***header section***; and a ***body section***. The status code is an integer giving an overall indication of whether the processing was successful and, if not, then why not. The body of the message is the main result returned to the client program. The data in the message body may be formatted in different ways; often it is in either ***JSON*** or ***XML*** format.

In this chapter, we'll cover JSON, as it is a very commonly used data format in web APIs. JSON provides a simple way to represent data using lists and/or trees of numeric, string, and Boolean values. The acronym JSON is short for JavaScript Object Notation because it was originally designed for use with JavaScript, but JSON is data-oriented and programming language independent and so can be used with Python and other programming languages as well.

XML is short for eXtensible Markup Language. Like JSON, it is also a standard approach to representing data and can cover lists and trees. Some web APIs use JSON, some use XML, while others may support returning results in another format.

In the previous chapter, we used the Requests package to make a request across the internet. In this chapter, we'll also use the Requests package and explore some additional features it provides to make web API requests very concise. Though the Requests package makes it simple to send a request to a web API, it has no knowledge of the format of the response. For this reason, when using this approach, you need to understand the layout of the data being returned and programmatically navigate it to get the results of interest. This is similar in concept to navigating an HTML tree and getting the results of interest from a web page.

As an alternative to using the Requests package, if a specific package for a web API is available, this can provide a simpler way to access that web API, since the package can navigate and organize the returned response. Though the majority of the chapter uses the Requests package, the last section shows how a specific package might be used to simplify web API calls and access their results.

Section Summary

Web APIs provide a more stable alternative to web pages for accessing data across the internet. Further, web APIs can support calculations for specific powerful services—for example, language translation, fastest routing, and many others. As we'll see in this chapter, the Requests package and JSON let us leverage web APIs with a few lines of code.

A web API allows clients to make requests to a service, which in turn provides its answer as a reply. The client program forms its request as a URL in a certain form, possibly including additional data that are often in either JSON or XML format. The service then processes the request as needed, which may involve data retrieval, data update, calculation, or some combination. The response from the service back to the client is then sent. Python clients can use the Requests package to send requests and receive the response. In some cases, a specific package might be available to make web API calls in a more concise way.

Test Your Understanding

1) Requests to a web API are typically made from a web browser.
   a) True
   b) False

2) Calls to a web API are made with:
   a) a URL
   b) possibly supplemental data as JSON or XML
   c) (a) and (b)
   d) none of the above

13.2 A Simple Example: FX Rate

In this section, we'll look at a hypothetical web API that lets you programmatically request a foreign exchange rate between two currencies. We use a hypothetical API in order to build it step-by-step, introducing concepts one section at a time. (Note that this is a hypothetical API; if you run these programs, you'll get a ConnectionError exception.) In the last section of the chapter, we'll introduce a real web API that uses all the concepts discussed in the hypothetical web API.

Our sample programs will make a request to the web API and simply display the result to the screen. These simple programs will allow us to focus on the key steps and concepts for calling a web API. In practice, a real web API like the one we're presenting could be used for real-time access to foreign exchange rates as part of a trading application or payment processing system.

Our hypothetical foreign exchange API accepts requests of the form:

```
api.p4dabook.online/fxrates/v1/<from ccy>/<to ccy>
```

where the `<from ccy>` and the `<to ccy>` are standard three-character currency codes. Note that the request is a URL, similar to what you might use or see as the address in a web browser. However, this URL is not used by a browser; instead, it would be sent from our program.

If the service has the data for the requested exchange rate, the response back to our program will be in the form:

```
{"rate": <fx rate>}
```

For example, in order to request the current FX rate from USD to EUR, the request would be:

```
api.p4dabook.online/fxrates/v1/USD/EUR
```

and the response might be:

```
{"rate": .83}
```

In order to make this request from Python, the Requests package can be used. Program 13-1 shows an example.

Program 13-1

```
1  # request an FX rate
2  import requests
3
4  # make the request to the service
5  url = 'http://api.p4dabook.online/fxrates/v1/USD/EUR'
6  response = requests.get(url)
7
8  # convert the returned JSON string result to a Python dictionary
9  result = response.json()
10 print('Rate=', result['rate'])
```

Program 13-1 Output

```
Rate= .83
```

In Program 13-1, the variable `response` is of type `Response`, which allows access to several aspects of what the service returns. On line 9, the `json()` method accesses the core returned result (the response body), which, since it is enclosed in curly braces { }, converts it to a Python dictionary. Given the JSON result layout of this API, the dictionary will have a key of `'rate'`, which line 10 uses to access the numeric result (the exchange rate).

Section Summary

In this section, we looked at a fictitious web API for retrieving a foreign exchange rate. This service expects a URL of a certain form that includes the from and to currencies of interest. If the rate for this requested pair of currencies is available, the service will provide an FX rate in its response. The package we used in the example code is Requests, which includes a `json()` method that converts the returned response into a Python dictionary.

Test Your Understanding

1) For the hypothetical FX web service described in this section, what is a valid URL request?
   a) GET:USD/EUR
   b) http://api.p4dabook.online/fxrates/v1/CAD/JPY
   c) fxrates/GBP/EUR
   d) all of the above

2) For the hypothetical FX web service described in this section, what is a valid response body?
   a) 1.12
   b) CAD/USD/1.12
   c) {"rate": 1.12}
   d) all of the above

13.3 JSON

In the previous section, we looked at a simple result that returned one value, an FX rate. In practice, APIs return composite results, allowing our programs to get much more information from a single reply. JSON is well suited to handle more complicated results, since it supports lists and trees of data. Further, a JSON string can be converted to an equivalent Python dictionary or list, allowing our program to easily navigate to the parts of the result of interest. In this section, we'll expand on our FX rate service to look at more involved results.

JSON can be thought of as a string representation of a data structure. JSON design is simple but allows it to represent almost any kind of data layout. Like a dictionary, JSON can contain key/value pairs. JSON keys are strings in quotation marks. JSON values can be either simple or composite. A JSON simple value is a string, a number, a Boolean constant (`True` or `False`), or the word `none`. A JSON composite value can be a list (in square brackets [ ], with its items separated by commas), or it can be a JSON object (in curly braces { }).

Multiple FX rates

Our FX service allows requests asking for more than one rate at a time. This is done by using an asterisk for the currency. For example, to request all the available FX rates from USD to other currencies, we could use in the URL:

```
fxrates/v1/USD/*
```

Program 13-2 shows an example of this multi-currency request:

Program 13-2

```
1  # request FX rates
2  import requests
3
4  # make the request to the service
5  url = 'http://api.p4dabook.online/fxrates/v1/USD/*'
6  response = requests.get(url)
7
8  # access and display the core returned result
9  print(response.text)
10
11 # convert the returned JSON string result to a Python dictionary
12 result = response.json()
13 print(result)
14 for currency, rate in result.items():
15     print(currency, format(rate, '6.2f'))
```

Program 13-2 Output

```
{"USD":1, "AUD":1.30, "CAD":1.22, "CHF":0.90, "CNH":6.38, "EUR":0.83, "GBP":0.71, "HKD":7.76,
"JPY": 109.67, "NZD": 1.40, "ZAR": 13.71}
USD   1.00
AUD   1.30
CAD   1.22
CHF   0.90
CNH   6.38
EUR   0.83
GBP   0.71
HKD   7.76
JPY 109.67
NZD   1.40
ZAR  13.71
```

In Program 13-2, the service returns multiple key/value pairs in JSON, which are converted to the equivalent Python dictionary representation. (Getting multiple rates in one request is more efficient than making several requests, one for each rate.) Line 9 introduces the Response type's text property, which in this case shows the JSON string returned for this request. Lines 12 converts this to a Python dictionary, and lines 14–15 iterate over the dictionary's key/value pairs to display a table of FX rates.

> **Tip**
> **Converting JSON to Python Dictionaries**
>
> Note that a JSON string enclosed in curly braces { } can always be converted into a Python dictionary. For example, JSON keys are always quoted strings, and these keys are converted to Python dictionary string keys. However, if the JSON contains duplicate keys (e.g., two FX rates for USD/EUR), the conversion from JSON to the Python dictionary will retain the last key/value pair. In practice, this is not typically an issue, as API responses will avoid duplicate keys in their JSON.

FX metadata: A data tree

Web APIs often arrange their returned data in a tree structure. This provides flexibility in how the results are returned, but requires our program to have code to navigate the data to the values of interest. In this section, we look at an enhanced FX web API that provides a tree of results. In particular, it will provide information about the results (what time the results were as of), as well as the FX rates themselves.

Program 13-3 demonstrates an enhanced API that includes the as-of time in the results. For this new version, we include V2 as part of the URL, and the result returned is of the form:

```
{
  "metadata": {"asoftime": <as-of time>},
  "data": { <currency rates> }
}
```

Program 13-3

```python
# request FX rates - version 2 API including metadata
import requests

# make the request to the service
url = 'http://api.p4dabook.online/fxrates/v2/USD/*'
response = requests.get(url)

# access and display the core returned result
print(response.text)

# convert the returned JSON string result to a Python dictionary
result = response.json()
print('time', result['metadata']['asoftime']) # access the metadata
for currency, rate in result['data'].items(): # access the data
    print(currency, format(rate, '6.2f'))
```

Program 13-3 Output

```
{"metadata":{"asoftime":"2021-06-11T15:05:10"},"data":{"USD":1, "AUD":1.30, "CAD":1.22,
"CHF":0.90, "CNH":6.38, "EUR":0.83, "GBP":0.71, "HKD":7.76, "JPY": 109.67, "NZD": 1.40, "ZAR":
13.71}}
time: 2021-06-11T15:05:10
USD   1.00
AUD   1.30
CAD   1.22
CHF   0.90
CNH   6.38
EUR   0.83
GBP   0.71
HKD   7.76
JPY 109.67
NZD   1.40
ZAR  13.71
```

Spreadsheet Compare/Contrast
Tables vs. Trees

Web APIs cover a diversity of requests and replies. The replies are often organized as a tree of data, which can handle a wide variety of data. Since JSON and Python dictionaries can represent data trees, these provide a more natural fit for accessing web APIs as compared with the tabular data representation that is used by spreadsheets.

Section Summary

In this section, we refined our FX service to handle requests with more than one pair of FX rates. Getting multiple values in one response is more efficient than making individual requests for each rate. Further, as JSON is a common format for returning web API results, we revised our service to return its results in JSON format. Since JSON can represent lists and/or trees or data, it is a flexible way for web APIs to return their result.

Test Your Understanding

1) Which are valid JSON strings?
   a) `{"Rate": 1.23}`
   b) `{"CAD":1.23, "EUR":.97}`
   c) `{"Base":"USD", "Rates": {"CAD":1.23, "EUR":.97} }`
   d) all of the above

2) Any JSON string can be converted into a Python dictionary.
   a) True
   b) False

13.4 Query Parameters

Web APIs may allow or require their request URLs to include ***query parameters*** that provide more information to the API about the request. Query parameters let you tailor your request to your needs—for example, if you need data between two dates, these dates might be indicated as query parameters. The specific query parameters are determined by the web API designer and described in the documentation for it. The URL string approach is to specify query parameters after a question mark. If there are multiple parameters, these are separated by ampersands.

For example, perhaps our FX service allows us to optionally indicate a historical as-of time of interest for the data. This could be indicated in a query parameter called `asoftime`. As well, perhaps we can indicate a provider of the rates. A URL might include:

```
fxrates/v2/USD/*?asoftime=2021-06-11T16:00&provider=ABC
```

This type of URL could be used in a program similar to Program 13-3. The result could add the provider to the metadata that is returned.

In the above URL, the query parameters are included in the URL string. Since query parameters are commonly used, the Requests package provides an alternative way to provide their names and values, as a dictionary. In some cases, building query parameters through a dictionary may be simpler and more readable in code than adding query parameters to a URL string. Program 13-4 shows an example of this.

Program 13-4

```
# request FX rates - version 2 API including metadata and query parameters
import requests

# make the request to the service
url = 'http://api.p4dabook.online/fxrates/v2/USD/*'
queryParams = {'asoftime':'2021-06-11T16:00', 'provider':'ABC'}
response = requests.get(url, params=queryParams)

# access and display the core returned result
print(response.text)
```

Program 13-4 Output


```
{"metadata":{"asoftime":"2021-06-11T15:05:10", "provider":"ABC"}, "data":{"USD":1, "AUD":1.30,
"CAD":1.22, "CHF":0.90, "CNH":6.38, "EUR":0.83, "GBP":0.71, "HKD":7.76, "JPY": 109.67, "NZD":
1.40, "ZAR": 13.71}}
```


Above in Program 13-4, although `queryParams` contains a datetime (the asoftime), it is represented in the dictionary as a string (`'2021-06-11T16:00'`). Other types of data could be included in the URL if supported by the web API. For example, our service might offer to provide results from a secondary alternative provider if the requested provider does not have data. We might request this as:

```
fxrates/v2/USD/*?asoftime=2021-06-11T16:00&provider=ABC&alt=y
```

or specified in the queryParams as a Python Boolean True:

```
queryParams = {'asoftime':'2021-06-11T16:00', 'provider':'ABC', 'alt':True}
```

Section Summary

URLs can contain supplemental data as query parameters, which are placed after a question mark in the URL. Since query parameters are commonly used, the Requests package provides a way to specify them as a dictionary of key/value pairs. To demonstrate this, we augmented our FX service to take query parameters for the as of time and rate provider of interest.

Test Your Understanding

1) A query parameter can be indicated in a URL after a question mark.
   a) True
   b) False

2) The Python Requests package allows code to indicate query parameters as a list.
   a) True
   b) False

13.5 JSON in the Request

Some web APIs require request information in the body of the request to supplement the URL. For this information, the Requests `get()` function provides a `data` parameter that can be set to a dictionary. For example, if our FX service required `asoftime` and `provider` to be provided in the body of the request (as opposed to as query parameters), this could be done as:

```
url = 'api.p4dabook.online/fxrates/v2/USD/*'
dataParams = {'asoftime':'2021-06-11T16:00', 'provider':'ABC'}
response = request.get(url, data=dataParams)
```

The last statement above highlights some of the key parts in an HTTP request:

- the request verb, in this case get (implied by the use of the `get()` method)
- the URL of the request
- the optional body of the request, in this case in the `dataParams` variable

Though not shown here, the request also has a header section, which contains keyword/value pairs that control the technical aspects of the request. We won't cover the details of the header, but if needed, you can set the specific header settings using the `headers` named parameter to more finely control the specifics of the request.

13.6 Status Codes and Errors

Web APIs use the conventions of HTTP, which include the overall numeric status code that is set by the server in the response. The status code provides the program with a concise summary of whether the request succeeded or failed and why. This status code is available in the `Response`'s status_code property. For example, the status code of 200 indicates success. In general, your code should check the status code before accessing the specific results of the response. Table 13-1 indicates the different types of standard status codes that might be returned.

Status Code Range	Meaning
100–199	Informational response
200–299	Successful response
300–399	Redirect
400–499	Client error
500–599	Server error

Table 13-1: HTTP status codes

For our FX service, a status code of 200 indicates the request was successful, and all other codes indicate there was an issue. The code 200 is standard HTTP convention for a successful response. Further assume that if there was an issue, the JSON result returned would be of the form:

```
{"error": <error message>}
```

Program 13-5 shows a status code check included in the program.

Program 13-5

```python
# request FX rates - handling status code
import requests

# make the request to the service
url = 'http://api.p4dabook.online/fxrates/v2/XYZ/*'
response = requests.get(url)

if response.status_code != 200:
    print('Status code:', response.status_code)
    print('Error processing request:', response.json()['error'])
else:
    # access and display the core returned result
    result = response.json()
    for ccy, rate in result['data'].items():
        print(ccy, format(rate, '6.2f'))
```

Program 13-5 Output

```
Status code: 404
Error processing request: Unknown currency 'XYZ'
```

Section Summary

The HTTP standard includes a convention for an overall status code as part of the response to a given request. The response package captures the status code in the status_code property of the Response type. In this section, we demonstrated this by accessing the status code of our hypothetical FX service.

Test Your Understanding

1) An HTTP status code is a string sentence.
   a) True
   b) False

2) The Requests package Response type provides the HTTP status code in a property called:
   a) status
   b) code
   c) statusCode
   d) none of the above

13.7 Access Control

Many web services require your program to provide a username and password, or an API key, or another mechanism indicating authorization to use the service. This allows the service provider to keep track of each client's usage and potentially limit usage to keep performance high for all users. For example, to get an API key (a string), you may need to sign up for the service and then get the API key from the provider's site at a particular web page. The API key can then be included when making requests to the service.

Let's assume our FX service requires an API key in order to make requests, and we've registered and received an API key of FX122334. Further, the FX service lets us pass this key in the request as a query parameter using the name `apikey`. A request URL might be:

```
api.p4dabook.online/fxrates/v1/USD/EUR?apikey=FX122334
```

As an alternative to providing a key in the URL, a service may require the key to be placed in the request header. The Requests package supports setting header values using the headers parameter. For example:

```
apikey = 'FX122334'
response = requests.get(url, headers={'Authorization:': 'Bearer ' + apikey})
```

There are several other approaches to access control, but those are beyond the scope of the chapter.

API keys via a .env file

For development or ad hoc programs, it may be sufficient to hardcode authentication information such as API keys in your code. However, for applications in production, a better practice is to separate API keys, usernames, passwords, and other configuration information from your code so they can be kept private and/or can be changed without redeploying code. One approach is to keep this information in an ***environment file***, which can be read by your code. Further, this information can be loaded and accessed using a Python package such as python-dotenv.

The python-dotenv package provides the function `load_dotenv()` to read a file of names and values. These are placed into the operating system's environment variables, which can be accessed using the os package's `getenv()` function.

For example, we could define a file called apikeys.env, which includes a line:

```
FX_API_KEY=FX122334
```

Program 13-6 demonstrates how to get our API key from the .env file, and include it in the request header in order to call the FX service.

Program 13-6

```
# request an FX rate using an API key
import requests
import os
from dotenv import load_dotenv

# load the name/value pairs into the operating system environment
load_dotenv('api_keys.env')

# get the value associated with FX_API_KEY
apiKey = os.getenv('FX_API_KEY')

# make the request to the service, including the apiKey in the header
url = 'http://api.p4dabook.online/fxrates/v3/USD/EUR'
response = requests.get(url, headers={'Authorization': 'Bearer ' + apiKey})

# convert the returned JSON string result to a Python dictionary
result = response.json()
print('Rate=', result['rate'])
```

Program 13-6 Output

```
Rate= .83
```

Section Summary

Web APIs may require an API key or other access control approach. In this section, we introduced an API key requirement as part of our FX service. We looked at two approaches to providing this key: as a query parameter and as part of the request header. We also introduced the idea of storing the API key in a file separate from the program code and reading the key using the dotenv package.

Test Your Understanding

1) A good practice is to place API keys in your code.
   a) True
   b) False

2) Alternatives for providing an API key as part of a request to a service are:
   a) as a query parameter in the URL
   b) as part of the request's header
   c) (a) or (b)
   d) none of the above

13.8 Python Package Layer over an API

In the previous sections in this chapter, we've used the Requests package to make requests of our FX service. This approach can be used with other web APIs that accept a get request with a URL, and possibly additional data in the body of the requests and/or data in the header of the request.

Another approach is to use a specific Python package, if one is available, that lets us form requests and receive replies in simpler code. In this arrangement, the package translates our request into a URL, possibly including query parameters, a request body, and settings in a request header.

For example, we might provide a specific package called p4dabookapiV3. This package could have one primary function, `get()`, and return its results in a pandas DataFrame. Program 13-7 shows an example use of our hypothetical p4dabookapiV3 package.

Program 13-7

```
# request an FX rate using a specific package
import p4dabookapiV3 as fxapi

# set the API in the required variable
fxapi.apikey = 'FX122334'

# make the request to the service
df = fxapi.get(fromccys=['USD'], toccys=['EUR', 'GBP'],
               format='table', asoftime='2021-06-11:16:00')
print(df)
```

In the above example, the result could be returned in a matrix with currencies as rows as well as columns. Note that error handling would need to be implemented as well—for example, by defining Exceptions that correspond to different error status codes that might be returned by the service.

Section Summary

In prior sections, we demonstrated calling the FX web API service using the Python Requests package. An alternative approach to calling a web API is to use a specific package if it is available. Because a specific package can be tailored to a specific web API, the code we need to write using a specific package is typically simpler than code needed using a general package like Requests. In this section, we introduced a hypothetical specific package called p4dabookapiV3 and showed how it might be used to more concisely call our FX service.

13.9 An Example Web API: Alpha Vantage Data

Alpha Vantage provides a web API for financial data; its home page is alphavantage.co. Its datasets are divided into several categories, one of which is Stock Time Series. Data are available daily as well as intraday and less frequently (weekly, monthly). In this section, we'll look at this site's intraday data.

In order to make a request for intraday stock data to the Alpha Vantage web API, use a URL in the form:

> https://alphavantage.co/query?function=TIME_SERIES_INTRADAY&symbol=
> <symbol>&interval=<interval>&apikey=<apikey>

where:

`<symbol>` is the stock symbol of interest
`<interval>` can be 1min, 5min, 15min, 30min, or 60min
`<apikey>` is your API key, which you can get through the Alpha Vantage website

Since the above URL requires no additional data, you can test Alpha Vantage in a web browser to see the JSON response it would return. Alternatively, Program 13-8 is a brief Python program to submit a request for a particular stock and see the JSON returned as a Python dictionary.

Program 13-8

```
1  # example call to Alpha Vantage
2  import requests
3
4  symbol = input('Enter stock symbol of interest: ')
5  interval = input('Enter the interval of interest: ')
6
7  # set the API key, replace as your own key
8  apikey = 'akey'
9
10 # form the URL
11 url = 'https://alphavantage.co/query?function=TIME_SERIES_INTRADAY'
12 url += '&symbol='+symbol
13 url += '&interval='+interval
14 url += '&apikey='+apikey
15
16 # make the request to the service
17 response = requests.get(url)
18 data = response.json()
19
20 print(data)
```

Program 13-8 builds the URL request on lines 11–14, and sends this request to Alpha Vantage using the Requests `get()` function on line 17. Line 18 converts the response body into a Python dictionary that is displayed on line 20. In practice, you would review the format of the returned result, navigate the dictionary tree to access the data of interest, and process them further as needed.

Section Summary

The Alpha Vantage web service is one example of a data retrieval web API that provides a variety of financial data at different frequencies. To access the data, you can form a URL with the required query parameters and submit the request using the Requests package's `get()` function. Results are returned as JSON, which can easily be converted to a Python data dictionary and from which you can get and further process data of interest.

Chapter Review

Web APIs provide a framework for accessing data, images, and calculations across the internet. As opposed to reading web pages programmatically, we can write programs that leverage web APIs for a more stable system. The approach underlying web APIs is based on standards such as HTTP and JSON for making requests to remote servers and receiving their responses. The Requests package can be used to submit a request to a web API and receive the corresponding response. This package provides convenient methods and functions—for example, the ability to set query parameters in the request, set header properties, and convert JSON replies to Python types (`int`, float, `str`, bool, dict, and list).

A web API may use access control, requiring the client program to include additional information such as an API key in its request. Last, as an alternative to the Requests package, if available, you may be able to use a specific package designed to work with a corresponding web API to further streamline accessing that API.

Key Terms

API key	A string issued by a service provider and included as part of an API request.
body	A response from an API typically includes data in its body; an API request may have supplemental data in its body.
environment file	A file containing environment variable names and their values.
environment variable	A variable with a value known to the operating system.
header	Requests and responses to web API contain headers that contain keywords and values.
JSON	A language for representing data that supports lists and trees of data.
query parameter	A URL may have one or more query parameters at its end, after a question mark, to define aspects of the request.
Response	A type defined in the Requests package that represents the information returned from a web server.
status code	A numeric code included in the response from a web server.
URL	A string representing a request to a server.

Programming Problems

1) Alpha Vantage—Daily Stock Prices

a) Write a program that calls the Alpha Vantage web API to request daily stock prices for a symbol of a given stock of interest. For the returned result, write the dates and prices to a CSV file.

b) Enhance your program in part (a) to detect errors—for example, an unknown stock symbol or invalid API key—and display the error received.

2) Alpha Vantage—Intraday Stock Prices—One Stock

a) Write a program that calls the Alpha Vantage web API to request intraday stock prices for a symbol of a given stock of interest. For the returned result, write the dates, times, and prices to a CSV file.

b) Enhance your program in part (a) to detect errors—for example, an unknown stock symbol or invalid API key—and display the error received.

3) Alpha Vantage—Intraday Stock Prices—Multiple Stocks

a) Write a program that calls the Alpha Vantage web API to request intraday stock prices for two or more symbols of given stocks of interest. For the returned result, write the dates, times, and prices to a CSV file.

b) Enhance your program in part A to detect errors—for example, an unknown stock symbol or invalid API key—and display the error received.

4) *New York Times* API

The *New York Times* provides several web APIs for their content; see:

https://developer.nytimes.com/

Using the Article Search API, write a program that requests articles for a given keyword. Based on the returned results, display the headlines of the articles returned by the API.

5) Find and Access a Web API

Find a web API of interest, and write a Python program to make a call to the API and display its results.

Chapter 14

Querying Databases with Python and SQL

Chapter Objectives

In this chapter, you'll learn:

- how Python can be used with SQL to query relational database data
- a basic introduction to SQL, including statements for querying database tables
- about the primary Python database types, `Connection` and `Cursor`
- alternative ways for processing data returned from a Python SQL query
- how Python handles SQL data types and SQL errors
- how Python can be used with database servers

Chapter Outline

14.1 Introduction to SQL, SQLite, and Python for SQL
  Overview of working with Python and SQL databases
  Tables
  SQL queries in brief
  About SQLite
14.2 Using Python to Query a SQLite Table
  Steps for querying a database table
  Importing the database package
  Establishing a connection to the database
  Getting and using a cursor for submitting database requests
  Getting the database results
  A simple Python query program
14.3 Variations on Processing SQL Results
  Displaying each row
  Accessing and displaying one row at a time
  Getting query results into a pandas DataFrame
14.4 More about using Python with SQL
  Getting the SQL result's field names
  Querying multiple tables
  Handling database errors with exceptions
  Updating database data with SQL insert, update, and delete
14.5 About SQL Data Types and Python Data Types
14.6 Working with Database Servers and Python
  About database servers
  Using the right package for a particular database server with Python
  Connecting to database servers with Python
  Closing connections and cursors
  Working with transactions
  Connection parameters via the dotenv package
Chapter Review
Key Terms
Programming Problems

14.1 Introduction to SQL, SQLite, and Python for SQL

Overview of working with Python and SQL databases

Organizations keep a wealth of data in databases, and relational databases provide a standard way of keeping data in collections of tables. SQL is the standard language for working with relational databases, and has been in use for over 40 years. Python can be used in combination with the SQL language to access and update relational database data. As you'll see in this chapter, the combination of Python and SQL is both concise and powerful.

SQL is its own language, separate from Python. SQL statements can be represented in Python strings, and Python statements can send these SQL strings to a database. Based on the SQL request, the database forms a reply that is sent back to the Python program. Python can then access and process the results, in order to provide a user or system with data and information.

This next section provides a brief introduction to SQL queries. This chapter focuses on relatively simple SQL examples, though the Python statements covered can handle complicated queries as well. SQLite, a database system that comes with Python, is used for most of the examples in this chapter. Python can also work with other relational databases, and using Python with these database servers is discussed at the end of the chapter.

SQL (Structured Query Language) is a language for working with relational databases. As its name implies, it can be used to ***query*** databases, by sending a request and getting a table reply.

In addition to querying data, SQL can be used for many other database activities, including:

- creating and deleting database tables and other database items
- updating database data by adding, changing, or deleting data
- managing user access to data
- grouping database actions together to ensure data quality

In this chapter, we will focus on the query aspects of SQL.

Tables

Data in relational databases is stored in tables. Each table can have one or more named columns, and each column (also known as a field) has a particular data type. Each row of data in a table is a record. The arrangement is similar to a sheet in a spreadsheet that has a table with column headings.

For example, in this chapter, we'll use a small database with two tables that store data about companies and their stock's daily prices and trading volume. These tables and their columns are summarized in Tables 14-1 and 14-2.

Column Name	Column Type	Description	Example Value
id	`int`	integer to identify each stock	1
ticker	text	Ticker symbol identifying the stock	M
name	text	The name of the company	Macy's Inc.

Table 14-1: Example database table: company table

Column Name	Column Type	Description	Example Value
ticker_id	`int`	integer to identify each stock	1
date	date	date for stock prices and volume	2017-12-29
opening	numeric(12,6)	opening stock price	51.48
high	numeric(12,6)	highest stock price	51.88
low	numeric(12,6)	lowest stock price	50.36
closing	numeric(12,6)	closing stock price	50.38
volume	`int`	number of shares traded	13599056

Table 14-2: Example database table: daily_measures table*

* Prices and volumes are not actual values, and are for demonstration purposes.

SQL queries in brief

The SQL command to query data is SELECT. There are several ways that SELECT can be used. The simplest way is to query data from a single table. For example, we can use the SQL statement:

```
SELECT * FROM company
```

The above statement is a query to get all data from the company table. In this statement, the * indicates that all the fields (columns) of the table should be returned in the result.

In order to request only certain fields of interest, you can provide a list of fields in the SELECT command. For example:

```
SELECT id, ticker, name
FROM company
```

indicates that the three fields (id, ticker, name) are of interest and should be included in the result.

In the prior example SELECT commands, all the rows of the table will be returned. In order to filter the records of interest, the SQL WHERE clause is used. For example:

```
SELECT id, ticker, name
FROM company
WHERE ticker = 'AAPL'
```

indicates that only those records with the ticker of AAPL should be returned in the result.

SQL can handle more involved queries. For example, multiple conditions can be combined with the words AND and OR. Also, a single query can request data from multiple tables, using the concept of joining tables. However, for the purposes of this chapter, we'll focus mostly on simple queries.

Tip

About Multi-line SQL

You may have noticed above the SQL statements are sometimes shown in one line or in multiple lines. For readability, it is good practice to use multiple lines when defining SQL queries. As we'll see later, SQL queries are defined as Python strings, and you can define these in one or several lines in your program.

About SQLite

SQLite is a unique type of SQL database for several reasons. First, it is included as part of the standard Python distribution. This means if you have Python, you have SQLite. The second unique aspect of SQLite is that it requires no setup. Third, SQLite does not require a special process or server to run in. Finally, a SQLite database is stored in one file. These aspects in combination mean that it is very easy to start using Python with SQLite.

In addition to being easy to get started with, SQLite also uses the standard Python interface for SQL databases. You can learn or prototype with SQLite, and then easily migrate your program to use another SQL type of database.

In order to get started working with SQL queries in Python with SQLite, you need to have a SQLite database. There are a few options, including:

1) Creating your own database using the sqlite3 tool
2) Using a Python program to create a SQLite database
3) Copy a SQLite database file that has already been created

For this chapter, you can use the SQLite database, dj.db, provided with this textbook. For use with this chapter, you can copy dj.db to any folder on your computer where you would like.

In this chapter, you'll learn how to use Python to query a SQLite database. In addition, though beyond the scope of this chapter, you can use the sqlite3 tool to interactively enter queries and see the results. You can run sqlite3 from the Windows Command Prompt or the Mac Terminal application.

Spreadsheet Compare/Contrast
SQL Where and Spreadsheet Filtering

A single table in a relational database is similar to a table of data in Excel. In Excel, you may have used Excel filters—these allow you to see only the rows that meet conditions you set in the filter. This filtering is very much like the WHERE condition in a SQL query. Further, Excel's approach for filters is to combine multiple conditions, similar to how a SQL statement would use AND to combine conditions.

Also note that a SQL SELECT statement allows you to specify which columns are of interest. Similarly, in Excel, you could hide columns to see only those columns you'd like to view.

Section Summary

Python can be used with SQL to query relational database data. A particular Python package needs to be used to work with a particular database brand. The interaction between Python and the database is done using SQL strings. The SQL SELECT command is used to query a database. The SQLite database system is provided with Python and provides an easy way to get started using Python and SQL.

Test Your Understanding

1) SQL can be used to query databases, but not update them.
   a) True
   b) False
2) Relational databases store their data in:
   a) web pages
   b) tables
   c) encrypted files
3) SQLite is a type of relational database that:
   a) stores each of its databases in a file
   b) requires no specific setup
   c) can run on your computer as opposed to requiring a dedicated computer
   d) all of the above

14.2 Using Python to Query a SQLite Table

In the previous section, we introduced relational databases and SQL queries. In this section, we'll see how to use Python and SQL to query a relational database table. Note that our examples will be small, to focus on the steps and syntax needed to query our database and get the results. In practice, SQL statements can have more complex logic and work with a large number of tables and records. Further, the Python programs you write with SQL will typically do more than display the results—for example, you may need to further process the data and save the results.

Steps for querying a database table

The steps needed to query a SQL table with a Python program are:

1) Import the package for working with the database.
2) Establish a ***Connection*** between your program and the database. The `Connection` variable acts as a "handshake" between your program and the database. The `Connection` variable is similar to the variable returned when you open a file, in that it allows you to work with the database.
3) Get a ***Cursor*** variable. The `Cursor` lets you submit requests to the database and access the results of those requests.
4) Using the `Cursor` variable, submit a request to the database.
5) Using the `Cursor` variable, access and process the returned results.

Importing the database package

In order to work with a database in Python, you need to import an appropriate database package that knows how to work with that database type. For SQLite, the package sqlite3 can be used. You can import this package in your Python program as:

```
import sqlite3
```

As with any Python import, you can provide a nickname for the package. For example:

```
import sqlite3 as db
```

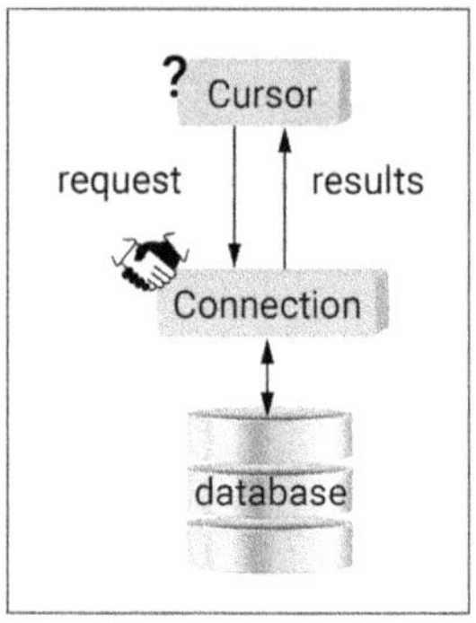

Figure 14-1: Connection and Cursor types

Establishing a connection to the database

In order to work with a SQL database, you need to establish a `Connection` between your program and the database. The `Connection` acts as a "handshake" between your program and the database. Getting a `Connection` to the database is similar to how you get a file variable in order to work with a file. In Chapter 7, you've seen a statement that uses the open function to allow working with a file. For example:

```
file_var = open('myfile.txt', 'r')
```

You use a statement like the following to get ready to work with a database:

```
connection = db.connect('dj.db')
```

Getting and using a cursor for submitting database requests

Once you have a `Connection` variable, you can use its `cursor()` method to get a `Cursor` variable. The `Cursor` variable's methods allow you to submit requests to the database and get the corresponding replies. For example:

```
cursor = connection.cursor()
cursor.execute('SELECT * FROM company')
```

will create a `cursor` variable to use with the database, and submit the query to the database.

> **Tip**
> **About SQL as Strings**
>
> In the above `execute()` method, the SQL command is represented as a Python string. This allows you to form SQL commands in your Python program easily and flexibly. However, if you have an error in your SQL syntax, Python will not detect it until the program is run. It is good practice to carefully review your SQL statements for correctness before running your Python programs containing SQL statements.

Getting the database results

Once you've used a `Cursor`'s `execute()` method to submit a query to the database, there are some alternatives for getting the results. The simplest is to use the `Cursor`'s `fetchall()` method, which returns a matrix of results. As all SQL queries return tabular results, a matrix is suitable for representing the query result.

```
result = cursor.fetchall()
```

The result matrix is composed of a list of tuples, where each item in the list corresponds to a row of the result. (Recall from Chapter 6 that a tuple is like a list with an unchangeable set of items.) There are a few choices for processing this matrix; the simplest is to display it on the screen with:

```
print(result)
```

A simple Python query program

Given the individual steps as described above, here is a complete program for querying a database table and displaying the results. Program 14-1 is a Python program that connects to a SQLite database, submits a query for all fields and records of a table, and displays the results.

Program 14-1

```
1  # a minimal Python + SQL example, using SQlite
2
3  # the sqlite3 package is included in Python 3 distributions
4  import sqlite3 as db
5
6  # establish a Connection to a SQLite file, indicating the filename
7  connection = db.connect('dj.db')
8
9  # the Cursor will be used to submit database requests and get results
10 cursor = connection.cursor()
11
12 # submit a SQL request to the database
13 cursor.execute('SELECT * FROM company')
14
15 # get the full result set as a matrix represented by a list of tuples
16 rows = cursor.fetchall()
17
18 # display the rows returned from the database
19 print(rows)
```

Program 14-1 Output

```
[(1, 'M', "Macy's Inc"), (2, 'AXP', 'American Express Company'), (3, 'AAPL', 'Apple Inc.'), (4,
'BA', 'The Boeing Company'), (5, 'CAT', 'Caterpillar Inc.'), (6, 'CVX', 'Chevron Corporation'),
(7, 'CSCO', 'Cisco Systems Inc.'), (8, 'KO', 'Coca-Cola Company (The)'), (9, 'DIS', 'The
Walt Disney Company'), (10, 'DOW', 'Dow Inc.'), (11, 'XOM', 'Exxon Mobil Corporation'), (12,
'GS', 'Goldman Sachs Group Inc. (The)'), (13, 'HD', 'Home Depot Inc. (The)'), (14, 'IBM',
'International Business Machines Corporation'), (15, 'INTC', 'Intel Corporation'), (16,
'JNJ', 'Johnson & Johnson'), (17, 'JPM', 'JP Morgan Chase & Co.'), (18, 'MCD', "McDonald's
Corporation"), (19, 'MRK', 'Merck & Company Inc. (new)'), (20, 'MSFT', 'Microsoft Corporation'),
(21, 'NKE', 'Nike Inc.'), (22, 'PFE', 'Pfizer Inc.'), (23, 'PG', 'Procter & Gamble Company
(The)'), (24, 'TRV', 'The Travelers Companies Inc.'), (25, 'UTX', 'United Technologies
Corporation'), (26, 'UNH', 'UnitedHealth Group Incorporated (DE)'), (27, 'VZ', 'Verizon
Communications Inc.'), (28, 'V', 'Visa Inc.'), (29, 'WMT', 'Walmart Inc.'), (30, 'WBA',
'Walgreens Boots Alliance Inc.')]
```

Line 4 imports the sqlite3 package, giving it the nickname db. This gives the program access to the Connection and Cursor types to interact with the database.

Line 7 establishes a Connection to the database. Since a SQLite database is stored in a single file, the filename is the only argument needed to be provided. (In the last section of this chapter, we'll look at how to establish a Connection to other types of databases, which will involve more parameters.)

Given the Connection established and stored in the variable `connection`, line 10 uses the `cursor()` method to create a `cursor` variable, which will be used to interact with the database. Line 13 then uses the `execute()` method to submit a query to the database. Note that SQL statements are submitted to the database using the `Cursor` variable as opposed to the Connection variable.

The result of the query is captured on line 16 and is stored in the variable `rows`. The `rows` variable is a Python list where each element is a set of items (stored as a tuple).

Built into Python is the ability to display lists and tuples. Line 19 allows the program to concisely display the results to the screen.

Section Summary

Python can be used with SQL to query relational database data. The approach involves using two types, `Connection` and `Cursor`, to submit a SQL string to a database and get back the results. Results returned by the `fetchall()` method are a matrix (represented as a list of tuples).

Test Your Understanding

1) To work with a relational database and Python, you need to import a specific database package.
   a) True
   b) False
2) In general, the first database-related statement in Python uses:
   a) the `getSQL()` method
   b) the `connect()` function
   c) the `fetchall()` method
   d) any of the above
3) The SQL command is represented in Python as a:
   a) comment beginning with #
   b) a Python string
   c) an optional named parameter for the `connect()` method

14.3 Variations on Processing SQL Results

In the prior section, example code focused on connecting to a database and making queries. The processing of the result table was kept as simple as possible, by displaying the results to the screen with the statement:

```
print(rows)
```

In practice, you will want to process the data returned. In this section, a number of variations will be discussed for accessing the rows and individual data of the SQL result.

Displaying each row

As mentioned, the result of `fetchall()` is a matrix, as represented in a list of tuples. The data can be displayed using a `for` statement. For example:

```
for row in rows:
   print(row)
```

Further, each row will have one or more elements (corresponding to the fields of the SQL result). These can be accessed and displayed in a nested `for` loop:

```
for row in rows:
   for field in row:
      print(field + '|', end='')
   print()
```

Program 14-2 uses the above statements to display the results returned for a SQL query.

Program 14-2

```
1 # query a Database table, and display the results one field at a time
2 import sqlite3 as db
3
4 # establish a Connection to a SQLite file, indicating the filename
5 connection = db.connect('dj.db')
6
7 # the Cursor will be used to submit database requests and get results
8 cursor = connection.cursor()
9
10 # submit a SQL request to the database
11 cursor.execute('SELECT id, ticker, name FROM company')
12
13 # get the full result set as a matrix represented by a list of tuples
14 rows = cursor.fetchall()
15
16 # display the rows returned, accessing each field individually
17 for row in rows:
18    for field in row:
19       print(field, end='|') # display each field with | separator
20    print() # go to next line
```

Program 14-2 Output

```
1|M|Macy's Inc|
2|AXP|American Express Company|
3|AAPL|Apple Inc.|
4|BA|The Boeing Company|
5|CAT|Caterpillar Inc.|
6|CVX|Chevron Corporation|
7|CSCO|Cisco Systems Inc.|
8|KO|Coca-Cola Company (The)|
9|DIS|The Walt Disney Company|
10|DOW|Dow Inc.|
11|XOM|Exxon Mobil Corporation|
12|GS|Goldman Sachs Group Inc. (The)|
13|HD|Home Depot Inc. (The)|
14|IBM|International Business Machines Corporation|
15|INTC|Intel Corporation|
16|JNJ|Johnson & Johnson|
17|JPM|JP Morgan Chase & Co.|
18|MCD|McDonald's Corporation|
19|MRK|Merck & Company Inc. (new)|
20|MSFT|Microsoft Corporation|
21|NKE|Nike Inc.|
22|PFE|Pfizer Inc.|
23|PG|Procter & Gamble Company (The)|
24|TRV|The Travelers Companies Inc.|
25|UTX|United Technologies Corporation|
26|UNH|UnitedHealth Group Incorporated (DE)|
27|VZ|Verizon Communications Inc.|
28|V|Visa Inc.|
29|WMT|Walmart Inc.|
30|WBA|Walgreens Boots Alliance Inc.|
```

In Program 14-2, the nested `for` loops allow the program to visit each element of the matrix. Line 17 iterates through each row of the matrix, then line 18 iterates over each item of the row. Line 19 displays the value of each cell of the table using a vertical bar (`'|'`) as the separator between values. (Note that the `end='|'` argument causes the `print()` function to stay on the same output line as opposed to moving to a new line for the next `print()`.)

There is one other variation in Program 14-2 as compared with Program 14-1. The SQL query used in line 11 is revised from:

```
SELECT * FROM company
```

to

```
SELECT id, ticker, name FROM company
```

It is often preferable to explicitly state the fields of interest in the SQL query, as shown here. The explicit list of field names can help insulate your program from future changes to the structure of the database table.

Accessing and displaying one row at a time

The prior sections have used the method `fetchall()` to get the rows of the SQL result in one statement. Another method, `fetchone()`, allows your program to get the rows of the SQL result one row at a time. For some cases—for example, with certain databases and/or large datasets—this may be more efficient than using `fetchall()`.

In combination with the `while` statement, `fetchone()` can be used to process each row of the SQL result. Program 14-3 shows an example program using `fetchone()` to display each row of a SQL result.

Program 14-3

```
1  # fetchone() gets one result row at a time; used here with while
2  import sqlite3 as db
3
4  # establish a Connection to a SQLite file, indicating the filename
5  connection = db.connect('dj.db')
6
7  # the Cursor will be used to submit database requests and get results
8  cursor = connection.cursor()
9
10 # submit a SQL request to the database
11 cursor.execute('SELECT id, ticker, name FROM company LIMIT 5')
12
13 # Access and display rows one at a time
14 row = cursor.fetchone()
15 while row != None:
16     print(row)
17     row = cursor.fetchone()
```

Program 14-3 Output

```
(1, 'M', "Macy's Inc")
(2, 'AXP', 'American Express Company')
(3, 'AAPL', 'Apple Inc.')
(4, 'BA', 'The Boeing Company')
(5, 'CAT', 'Caterpillar Inc.')
```

In line 14 of Program 14-3, the `fetchone()` method is used to get the first row of the SQL result, if any rows are returned. If the SQL result is an empty table (no rows returned), this statement will return the Python value `None`. Line 16 will print the row (a set of values), and line 17 will again use `fetchone()` to get the next row of the SQL result. After all rows have been processed, the statement on line 17 will return `None`, and the while loop will terminate.

> **Tip**
> **Using `fetchall()` vs. `fetchone()`**
>
> You can use either `fetchall()` or `fetchone()` to process the returned data from a SQL query. In practice, for larger databases, one or the other may be more efficient. You can try both approaches with a particular query to see if there is a significant enough performance difference between these methods for a particular database.

Another way to process one row at a time is using the `for` statement. The `Cursor` variable can be used in combination with `for` to iterate over all the rows of the SQL result. So instead of lines 14–17 in Program 14-3, a more concise alternative would be:

```
for row in cursor:
   print(row)
```

This use of `for` is shown in Program 14-4.

Program 14-4

```
# for can be used with the Cursor to get one result row at a time
import sqlite3 as db

# establish a Connection to a SQLite file, indicating the filename
connection = db.connect('dj.db')

# the Cursor will be used to submit database requests and get results
cursor = connection.cursor()

# submit a SQL request to the database
cursor.execute('SELECT id, ticker, name FROM company LIMIT 5')

# Access and display rows one at a time
for row in cursor:
   print(row)
```

Program 14-4 Output

```
(1, 'M', "Macy's Inc")
(2, 'AXP', 'American Express Company')
(3, 'AAPL', 'Apple Inc.')
(4, 'BA', 'The Boeing Company')
(5, 'CAT', 'Caterpillar Inc.')
```

Getting query results into a pandas DataFrame

SQL results and pandas DataFrames are both tabular, so it is easy to create a DataFrame from a SQL result. In particular, a DataFrame can be created from a matrix, using the list of tuples that is returned from `fetchall()`. Program 14-5 shows an example of this.

Program 14-5

```
1  # query a Database table, and populate the results in a DataFrame
2  import sqlite3 as db
3  import pandas as pd
4
5  # establish a Connection to a SQLite file, indicating the filename
6  connection = db.connect('dj.db')
7
8  # the Cursor will be used to submit database requests and get results
9  cursor = connection.cursor()
10
11 # submit a SQL request to the database
12 cursor.execute('SELECT id, ticker, name FROM company LIMIT 5')
13
14 # get the full result set as a matrix represented by a list of tuples
15 rows = cursor.fetchall()
16
17 # place SQL results in a DataFrame
18 df = pd.DataFrame(rows)
19
20 # Give the DataFrame constructed above meaningful column names
21 df.columns = ['id', 'ticker', 'company']
22
23 # the id column is a good choice for this DataFrame's index
24 df.set_index('id', inplace=True)
25 print(df)
```

Program 14-5 Output

```
   ticker                      name
id
1       M                Macy's Inc
2     AXP  American Express Company
3    AAPL                Apple Inc.
4      BA        The Boeing Company
5     CAT          Caterpillar Inc.
```

In Program 14-5, the DataFrame is created in line 19 by simply providing the rows of the SQL result as the parameter. The default DataFrame column names are 0, 1, 2, . . . , so line 22 renames the DataFrame's columns to more meaningful names. Last, the `id` column is defined as the index of the DataFrame in line 25.

As an alternative to using a Cursor, you can use the pandas `read_sql_query()` function to more simply execute a SQL query and load its result into a DataFrame. Program 14-6 shows an example of this approach.

Program 14-6

```
# use read_sql_query to populate SQL results to a DataFrame
import sqlite3 as db
import pandas as pd

connection = db.connect('dj.db')

# call the read_sql_query() function, and get the resulting DataFrame
df = pd.read_sql_query('SELECT * FROM company LIMIT 5',
                       connection, index_col='id')
print(df)
```

Program 14-6 Output

```
   ticker                      name
id
1       M                Macy's Inc
2     AXP  American Express Company
3    AAPL                Apple Inc.
4      BA        The Boeing Company
5     CAT          Caterpillar Inc.
```

Section Summary

SQL results returned by the `fetchall()` method are a matrix (represented as a list of tuples). Alternatively, results can be retrieved one row at a time using the `fetchone()` method. Another alternative for getting results is to use the `Cursor` with the `for` statement.

Test Your Understanding

1) A `Cursor` method that returns a single row from a query is:
   a) `getNext()`
   b) `fetchone()`
   c) `sqlOne()`
2) Pandas has a function that processes a SQL query and returns a DataFrame result.
   a) True
   b) False
3) To process SQL results, the `for` statement can be used together with the:
   a) `Cursor` variable
   b) `Connection` variable
   c) `Exception` variable

14.4 More about using Python with SQL

Getting the SQL result's field names

In previous examples, the SQL result data values are returned. In addition to the table's data, the SQL result contains the names of the result's columns. The column names are available in a `Cursor`'s description property. This property is a matrix containing one element for each field. Program 14-7 shows a Python program that displays the field names of a SQL query to the screen.

Program 14-7

```
# the Cursor.description property provides field names of the result
import sqlite3 as db
connection = db.connect('dj.db')
cursor = connection.cursor()
cursor.execute('SELECT * FROM company')

# display the field names of the returned result
for fieldinfo in cursor.description:
  print(fieldinfo[0])
```

Program 14-7 Output

```
id
ticker
name
```

Querying multiple tables

As mentioned in Section 14.1, a SQL query can be used to retrieve data stored in multiple tables. This logic for multi-table queries is defined in the SQL string (and the result for multi-table queries is still a single table), so the programs reviewed so far will work with multi-table queries with minimal adjustment—only the string passed to the `execute()` method needs to be changed. An example multi-table query is shown in Program 14-8. In this program, the tables company and daily_measures are both included in the query.

Program 14-8

```
# Use SQL to join 2 tables; statements are similar to 1-table queries
import sqlite3 as db
connection = db.connect('dj.db')
cursor = connection.cursor()

# define a 2-table query
query = """
SELECT ticker, date, closing
FROM company, daily_measures
WHERE company.id = daily_measures.ticker_id
AND company.id = 3
AND date >= '2017-12-26'
ORDER BY date"""

# execute the query and display the result, similar to a 1-table query
cursor.execute(query)
rows = cursor.fetchall()
for row in rows:
    print(row)
```

Program 14-8 Output

```
('AAPL', '2017-12-26', 341.14)
('AAPL', '2017-12-27', 341.2)
('AAPL', '2017-12-28', 342.16)
('AAPL', '2017-12-29', 338.46)
```

Handling database errors with exceptions

As with files, programs that interact with databases can run into errors. Some example database-related errors include:

- The name of the database in the `connect()` statement is incorrect.
- The field or table names used in the SQL statement are incorrect.
- The syntax of the SQL statement is incorrect.

In order to handle different problems that may occur when working with Python and a database, the try/except approach (as discussed in Chapter 7 and Appendix D) should be used. For example, Program 14-9 uses a try block and associated except block to catch any database exception that might occur during program execution.

Program 14-9

```
# As with file IO, Exceptions should be handled with database processing
# This example uses one try/except to cover a set of statements
import sqlite3 as db
try:
    connection = db.connect('dj.db')
    cursor = connection.cursor()
    cursor.execute('SELECT * FROM companies') # correct table name:company
    rows = cursor.fetchall()
    print(rows)
except Exception as e:
    print(e)
```

Program 14-9 Output

```
no such table: companies
```

Program 14-9 uses a try block to cover several statements. In practice, it can be useful to target one or a few statements for each try block in order to better pinpoint where the error occurred. Program 14-10 shows this approach: several try statements are used and more specific errors can be reported using exit codes.

Program 14-10

```
# In contrast with Example 14-10, this uses several try/except statements
# More targeted exception handling can give more control over problems
import sys
import sqlite3 as db
try:
    connection = db.connect('dj.db')
except Exception as e:
    print(1, e)
    sys.exit(1)

try:
    cursor = connection.cursor()
    cursor.execute('SELECT * FROM companies')
except Exception as e:
    print(2, e)
    sys.exit(2)

try:
    rows = cursor.fetchall()
    print(rows)
except Exception as e:
   print(3, e)
   sys.exit(3)
```

Program 14-10 Output

```
2 no such table: companies
```

Updating database data with SQL insert, update, and delete

The primary focus of this chapter has been regarding SQL queries. In addition, data maintenance statements—using the SQL commands INSERT, UPDATE, and DELETE—can be used in Python programs. The same data types (`Connection`, `Cursor`), as well as `execute()`, can be used to submit SQL commands.

However, unlike data queries, methods such as `fetchall()` and `fetchone()` are not applicable for SQL data maintenance statements. Instead, the number of rows impacted by the SQL statement is returned by the `execute()` method. For example:

```
nrows = cursor.execute('DELETE FROM company WHERE id = 10')
```

Care must be taken with data update processing. For example, in some cases, multiple update statements should be grouped together to ensure that all (or none) of the statements are successfully executed. This approach, called transaction processing, is important in maintaining data and can be covered with Python SQL statements. This is described briefly in the last section of this chapter.

Section Summary

In this section, a number of details were covered regarding using SQL with Python. The returned results' field names can be accessed with the Cursor's description property. This section also demonstrated how multiple tables can be queried, which involves using joins in the SQL statement and using similar Python code as the single-table query case.

Database errors follow Python's Exception framework and should be anticipated and handled with try/except blocks. Finally, Python and SQL can be used together to update database data, as well as query databases.

Test Your Understanding

1) A SQL query's result provides the column names in the Cursor's description property.
   a) True
   b) False
2) In order to process one SQL query that references multiple database tables, Python needs to use multiple cursor variables.
   a) True
   b) False
3) Database-related errors should be handled in Python:
   a) with `if` statements
   b) with try/except blocks
   c) by setting the error parameter in the Connection variable

14.5 About SQL Data Types and Python Data Types

Each column in a SQL table has a datatype, and though these are similar to Python types, they are not the same. For example, a database column's type may be TEXT. The data are returned to Python as Python type `str`. As another example, a database field may be of type NUMERIC; these values would be returned to Python as type `float`. In other words, there is a mapping from the SQL data type to the Python data type.

Program 14-11 queries a table and displays the Python types and values for each returned value. For more information on how a specific database's values are mapped to Python types, you can review the database's Python package documentation.

Program 14-11

```
# Query a table and display each value and its Python type using fetchone()
import sqlite3 as db
connection = db.connect('dj.db')
cursor = connection.cursor()
cursor.execute('SELECT id, ticker, name FROM company')
row = cursor.fetchone()
if row != None:
  for field in row:
    print(type(field), '|', field)
```

Program 14-11 Output

```
<class 'int'> | 1
<class 'str'> | M
<class 'str'> | Macy's Inc
```

Section Summary

SQL databases have data types that are different from, but similar to, Python data types. Python SQL packages bridge this difference by providing mappings from each SQL type to specific Python types. In this way, results from SQL queries can be represented as Python types.

Test Your Understanding

1) SQL data types are the same as Python data types.
   a) True
   b) False
2) Data stored as a SQL TEXT type will be returned in the Python result as a string (type `str`).
   a) True
   b) False
3) Data stored as a SQL NUMERIC type will be returned in the Python result as a Python `int`.
   a) True
   b) False

14.6 Working with Database Servers and Python

About database servers

In the prior sections, the programs were using the SQLite database. When starting to work with Python and SQL, SQLite is the most convenient database, since it comes with Python and requires no installation. However, organizations will typically have other types of relational databases that run on database servers. Examples include Oracle, Postgres, SQLServer, and MySQL. These databases are designed to support multiple users accessing or updating data in a relational database.

Fortunately, almost all the code and concepts from the prior sections apply to working with database servers. As we'll see, there are only minor changes needed for a Python program that queries a SQLite database in order for it to query another type of relational database. This is because the types (`Connection`, `Cursor`, and others) defined in the Python relational database standard allow for consistent coding when using Python and various SQL databases.

In particular, the `Connection` to a server database requires some more information in order to locate and log into the server database. The next sections discuss considerations for working with Python and server relational databases.

Using the right package for a particular database server with Python

As described in Section 14.2, a database package is needed to interact with the database. For SQLite, this is the sqlite3 package. Each brand of database will have its own Python package (or you may have a choice of packages). For example, in order to work with Python and the Postgres database, you can use the Python package psycopg2. The import statement could then be:

```
import psycopg2 as db
```

Connecting to database servers with Python

In order to connect to a SQLite database, a Connection can be created by providing the filename of the database. This is all you need because a SQLite database resides in one file, and that file is on the same computer as the Python program that interacts with it. So the statement:

```
connection = db.connect('dj.db')
```

is sufficient to work with the SQLite database named dj.db. However, for server database systems, your program needs to locate the database, since it will often be running on another computer. Also, as a multi-user system, the server database will typically require a username and password before allowing a program to query or update it.

For your program to locate and log into a database server, the `Connection` can be initialized with several named parameters. For example:

```
connection = db.connect(dbname='dj', host='dbhost123',
                        user='jsmith', password='js1')
```

In the above statement, a connection to the `'dj'` database running on the `'dbhost123'` server is attempted by logging in to the database as user `'jsmith'` with database password `'js1'`.

In practice, it is better to keep passwords (as well as the other argument values above) outside of the program. This will be covered shortly in this section.

Closing connections and cursors

When working with database servers, it is good practice to indicate when the program is done at certain points. For example, once the results of a query have been captured by the program, your code can indicate to the server that the related `Cursor` is no longer needed. The `close()` method indicates this. For example, if the `Cursor` variable is `cursor`, the close method is called using:

```
cursor.close()
```

Similarly, when the program is done interacting with the database, the program should close the `Connection` with a close method like:

```
connection.close()
```

To highlight the database aspects for servers covered so far, here is a simple program that connects to a Postgres database and queries one of its tables.

Program 14-12

```
1  # a minimal Python + SQL example, using Postgres
2
3  import psycopg2 as db
4
5  # establish a Connection to the Postgres server
6  connection = db.connect(dbname='dj', host='dbhost123',
7                          username='jsmith', password='js1')
8
9  # the Cursor will be used to submit database requests and get results
10 cursor = connection.cursor()
11
12 # submit a SQL request to the database
13 cursor.execute('SELECT * FROM company')
14
15 # get the full result set as a matrix represented by a list of tuples
16 rows = cursor.fetchall()
17
18 # display the rows returned from the database
19 print(rows)
20
21 # close the cursor and connection
22 cursor.close()
23 connection.close()
```

Note that the above program is very similar to Program 14-1. Differences are limited to the package imported (line 3), the Connection creation (lines 6–7), and the addition of `close()` methods (lines 22–23). All other statements in Program 14-12 are the same as those in Program 14-1.

Working with transactions

Programs that update server databases may need to take care that a set of related updates happen together. Otherwise, if only some of the updates happen and others do not, the resulting database data may be invalid. The approach used to cover this is called *transaction processing*.

Python database packages provide for transaction processing by including a number of Connections methods, including `commit()` and `rollback()`. Used in combination with Exception handling, a Python program can include logic to group updates through transaction processing. As this chapter is focused on queries (as opposed to updates), examples and further details are not included here.

Connection parameters via the dotenv package

As noted earlier in this chapter, passwords (and other connection parameters) are better placed outside of the program. In particular, connection parameters (dbname, host, username, and password) can be placed in a file and read by the program. In this way, password access can be limited. Also, changes to these parameters can be made without changing and releasing code but instead by changing a file.

A package can be used to simplify the reading and use of file-based parameters. One such package is python-dotenv. This package can read a file of parameter settings and make them available to the program using environment variables.

For example, if we create a file called dbparams.env with the following entries:

```
DBNAME='dj'
HOST='dbhost123'
USERNAME='jsmith'
PASSWORD='js1'
```

then the following program can be used to read these settings from the dbparams.env file, and create the connection to the database:

Program 14-13

```
# Using a properties file, to connect to a Postgres database
import psycopg2 as db
from os import getenv
from dotenv import load_dotenv
load_dotenv('dbparams.env')
try:
   connection = db.connect(dbname=getenv('DBNAME'), host=getenv('HOST'),
                     user=getenv('USERNAME'),password=getenv('PASSWORD'))
except (Exception, db.Error) as error:
   print('Error while connecting to Postgres', error)

connection.close()
```

Though Program 14-13 does not query the database, it can be useful to check whether all the parameters needed to connect to a SQL database are set up to make a connection.

Section Summary

Python can interact with SQL server databases. The choice of Python database package must be appropriate for a given brand of database. By specifying appropriate properties, a Connection can be established to a particular database server. The `close()` method is appropriate to use when database interaction is complete in a program. Further, if multiple SQL statements must be executed together, the Python database standard provides for methods such as `commit()` and `rollback()`, to ensure groups of SQL statements are applied together. Lastly, it is good practice to keep settings such as username and password in a file separate from a program, and these settings can be loaded and used by a program in order to connect to a particular database server.

Test Your Understanding

1) Connecting to a database server from a Python program involves setting named parameters in the:
   a) `execute()` function
   b) `fetchall()` function
   c) `connect()` function

2) Values for Connection properties can be stored in a text file and accessed by a Python program using the os and dotenv packages.
   a) True
   b) False

Chapter Review

Python programs can access relational databases using a database package. This chapter covered how SQL can be sent to a database by a Python program, as well as how the results can be processed by Python. The Python relational database standard defines the `Connection` type, which can be used to identify a database to interact with. The other primary database type, `Cursor`, allows your program to issue queries to the database and access the result. `Connection` and `Cursor` types are available in the sqlite3 package for working with SQLite databases. Since SQLite is included with Python distributions, and it is a self-contained database system, SQLite is a good option for getting started with database programming in Python.

The steps for querying a database from Python are: (1) importing a database package, (2) establishing a `Connection` to the database with the `connect()` method, (3) getting a `Cursor` with the `cursor()` method, (4) submitting a SQL query using the `execute()` method, and (5) processing the results. Alternatives for processing the results are the `fetchall()` method, the `fetchone()` method, or using a `for` statement with the `Cursor` variable. Another approach to querying a database is to use the pandas `read_sql_query()` function, which issues a query using a `Connection` and populates a DataFrame with the results.

The chapter also covered some additional aspects when working with databases and Python. A SQL query involving multiple tables using a join can be performed in Python similarly to single-table queries. For all Python database statements, potential database errors should be anticipated and handled with Python's try/except statements.

SQL data types are different from Python data types. Database packages bridge this difference by mapping SQL result data to Python data types. For example, a SQL NUMERIC type of data will be returned as a Python float value.

A variety of database packages, which implement Connection and Cursor, are available to work with database servers. This standardization allows you to implement Python programs that work with one brand of database, and with minimal changes, adjust your program to work with another database brand. The primary changes needed to work with a different database server involve adjusting the values for the named parameters of the `connect()` method, to indicate the database location and login information for a particular database server.

Key Terms

Connection	A variable type that represents a particular database in a Python program.
Cursor	A variable type that allows database queries to be submitted to the database, as well as provides methods, `fetchall()` and `fetchone()`, for processing the results returned from the database.
database package	In order for a Python program to access or update database data, a specific database package must be used by the program. An example Python database package is sqlite3.
database system	A system for storing, maintaining, and allowing access to a collection of data.
database transaction	Database updating may require sets of statements to be processed together. Database transactions allow for a group of statements to be processed together.
join	A SQL query involving data from multiple tables can match records between tables, using join logic in order to form the result.
query	The SQL SELECT command provides a variety of ways to request (query) data from a database.
relational database	A type of database that stores its data in tables.
SELECT statement	The SELECT statement is a type of SQL command that is used to issue queries to a database.
server database	A type of database that runs on a separate computer.
SQL	A language for querying and managing data in relational databases.
SQLite	A library that implements a self-contained SQL engine that does not require any setup.
sqlite3	A Python package that can be used to work with SQLite databases.
tuple	A read-only ordered set of data, similar to a list.
WHERE clause	In SQL, a WHERE clause provides a variety of ways to specify the condition for the data of interest.

Programming Problems

For all programming problems, use the sqlite3 package and the dj.db SQLite database provided for use with this chapter. The database tables used are daily_measures (problems 1 through 6) and company (problem 7).

1) Querying a Table—`fetchall()`

Using Python (but not pandas), determine and display the first five dates and closing prices for the stock with id=1. Use the SQL statement:

```
SELECT date, closing FROM daily_measures WHERE ticker_id = 1 LIMIT 5
```

Display the results using csv format, including a header row.

2) Querying a Table—pandas

Using Python and pandas, determine and display the first five dates and closing prices for the stock with id=1. Use the SQL statement:

```
SELECT date, closing FROM daily_measures WHERE ticker_id = 1 LIMIT 5
```

Display the result to the screen.

3) Finding Date Range of Data—Python

Using Python (and not pandas), determine the span of the data in the daily_measures table. Use the SQL statement:

```
SELECT min(date), max(date) FROM daily_measures
```

Note that this SQL statement should return one row for the result. As part of your solution, use the `fetchone()` method.

4) Finding Date Range of Data—pandas

Using Python and pandas, determine the span of the data in the daily_measures table. Use the SQL statement:

```
SELECT ticker_id, date FROM daily_measures
```

5) Finding Minimum and Maximum—Python and for

Using Python (but not pandas), for the stock with ticker_id=1, find the minimum and maximum closing prices and their corresponding dates. As part of your solution, use the `for` statement with a Cursor variable.

6) Finding Minimum and Maximum—pandas

Using Python and pandas, for the stock with ticker_id=1, find the minimum and maximum closing prices and their corresponding dates.

7) Querying a Table and Saving Results to a .csv

Using pandas, for the data in the companies table, query the data and save the results to a csv format file. Use the SQL statement:

```
SELECT id, ticker, name FROM company
```

Chapter 15

Accessing Spreadsheets with pandas and openpyxl

Chapter Objectives

In this chapter, you'll learn:

- how you can concisely access tabular spreadsheet data with pandas
- how pandas maps spreadsheet data types to Python, pandas, and NumPy types
- how you can access spreadsheet data with openpyxl, when data are arranged in a non-tabular format
- how openpyxl maps spreadsheet data types to Python types

Chapter Outline

15.1 About Spreadsheets and Python

As spreadsheets are widely used, there may be cases where a spreadsheet is the only place you can get data for an analysis you are doing or an application you are writing. In this chapter, we look at two cases:

- spreadsheet data of interest in a table
- spreadsheet data of interest in a less structured arrangement than a table

For the first case, pandas provides a concise way to read Excel data into a DataFrame. Once the data are in a DataFrame, you can use any of the features that pandas provides to do concise data manipulation and analysis.

For the second case, where data are more flexibly arranged in a spreadsheet, a different Python package can be used. In this chapter, we'll look at how the openpyxl package can read an .xlsx file and access any of its cells, regardless of layout. For .xls files, xlrd and xlwt are packages that can be used to read and write the older .xls file format, although we won't cover these packages.

Table 15-1 summarizes approaches for reading and writing Excel files with selected Python packages. In this chapter, we'll focus on reading spreadsheets, but as noted in the table, the packages can write spreadsheet files as well.

Layout	Excel file type	Reading	Writing
tabular only	.xls or .xlsx	pandas `read_excel()`	pandas `to_excel()`
tabular or non-tabular	.xlsx	openpyxl package	openpyxl package
	.xls	xlrd package	xlwt package

Table 15-1: Some options for reading and writing Excel files

15.2 Reading Excel Tables with pandas

If you need to read Excel data into a Python program, there are two main cases—the data layout is a table or it is non-tabular. For tabular data, pandas provides the `read_excel()` function, which lets us in one line of code read an Excel sheet's data and returns a DataFrame. The `read_excel()` function is similar to `read_csv()` except instead of providing the name of a text file, we provide the name of an Excel file (.xls or .xlsx). And like `read_csv()`, `read_excel()` has several named parameters to guide the reading and translation to a DataFrame.

In this section, we'll read a small spreadsheet whose entries include all types of spreadsheet cells. By covering the variety of cell entries, we can see how pandas will map these to Python and NumPy types when constructing a DataFrame. The file we'll use is PythonExcelTable.xlsx, as shown in Figure 15-1. It has a header row, as well as two data rows.

	A	B	C	D	E	F	G	H	I
1	integer	float	date	datetime	text	formula	boolean	error	empty
2	1	1.5	5/31/19	5/31/19 12:00	hello	2.5	FALSE	#NAME?	
3	2	2.5	6/30/19	6/30/19 12:00	goodbye	4.5	TRUE	#NAME?	

Figure 15-1: PythonExcelTable.xlsx

Table 15-2 provides information about the columns in this spreadsheet.

Column	Column heading	Example cell value
A	integer	1
B	float	1.5
C	date	5/31/19
D	datetime	6/30/19 12:00
E	text	hello
F	formula	=A2+B2
G	boolean	B2>2
H	error	=@unknown()
I	empty	

Table 15-2: Columns in PythonExcelTable.xlsx

In its simplest use, we can call the `read_excel()` function with just the name of the spreadsheet to read, like:

```
[1]: import pandas as pd
     df = pd.read_excel('PythonExcelTable.xlsx')
     df
```

	integer	float	date	datetime	text	formula	boolean	error	empty
0	1	1.5	2019-05-31	2019-05-31 12:00:00	hello	2.5	False	NaN	NaN
1	2	2.5	2019-06-30	2019-06-30 12:00:00	goodbye	4.5	True	NaN	NaN

The resulting DataFrame has column headings that match the spreadsheet headings. Also, the DataFrame's column types are mapped to the data types in the spreadsheet. We can see this from the output of the `info()` method:

```
[2]: df.info()
```

```
<class 'pandas.core.frame.DataFrame'>
RangeIndex: 2 entries, 0 to 1
Data columns (total 9 columns):
integer     2 non-null int64
float       2 non-null float64
date        2 non-null datetime64[ns]
datetime    2 non-null datetime64[ns]
text        2 non-null object
formula     2 non-null float64
boolean     2 non-null bool
error       0 non-null float64
empty       0 non-null float64
dtypes: bool(1), datetime64[ns](2), float64(4), int64(1), object(1)
memory usage: 210.0+ bytes
```

From the `info()` results above, we can see the mappings are generally what's expected. A few notes:

- Excel dates and datetimes are mapped to the pandas Timestamp type.
- The result of a numeric formula is mapped to a NumPy `float64`.
- The result of an unknown formula is mapped to `float64`.
- An empty cell is mapped to a numeric missing value (NumPy `NaN`).

Although not demonstrated in this section, `read_excel()` has several named parameters allowing you to refine how the data are loaded and mapped.

Section Summary

For tabular spreadsheet data, the pandas `read_excel()` function can readily read data and construct a DataFrame. This function maps the Excel data types to equivalent Python, pandas, or NumPy types in the returned DataFrame. Like `read_csv()`, `read_excel()` also has several named parameters that allow you to tailor its functionality.

Test Your Understanding

1) The pandas `read_excel()` function can be used to read both tabular and non-tabular data.
   a) True
   b) False

2) When using the pandas `read_excel()` function to read a spreadsheet, the spreadsheet cells containing dates will be mapped to which type in the resulting DataFrame?
   a) string
   b) integer
   c) Timestamp
   d) none of the above

15.3 Combining Excel Tables with pandas

In the previous section, we saw how you can use the pandas `read_excel()` function to read tabular data from an Excel file. The prior section had an elemental spreadsheet; in this section, we'd like to work with a more realistic set of spreadsheets. In particular, in this section, we'll see how to combine spreadsheet data from multiple Excel files that each have the same tabular layout. For example, if you received daily sales data, with each day's sales dataset in its own spreadsheet, and wanted to combine these data into one file, you could use `read_excel()` for this.

Further, let's assume we'll place the set of spreadsheets to be consolidated in a common directory. In addition to using `read_excel()`, we can programmatically determine all the .xlsx files in a certain directory. In this way, we don't need to hardcode a list of filenames in our program. To do this, we'll use the Python os package's `listdir()` function. This function returns the files and subdirectories for a given directory. By default, it checks the current directory, but you can provide an argument indicating the directory of interest.

Program 15-1 uses `listdir()` to display the list of all .xlsx files in the directory named `'data'`.

Program 15-1

```
# display the list of .xlsx files in a given directory
import os

def getFileList(ext, dir='.'):
    filenames = os.listdir(dir)
    filenameList = []
    for filename in filenames:
        if filename.endswith(ext):
            filenameList.append(filename)
    return filenameList

filenames = getFileList('.xlsx', dir='data')
print(filenames)
```

Program 15-1 Output

```
['sales1.xlsx', 'sales2.xlsx', 'sales3.xlsx']
```

Using `read_excel()` and `listdir()`, we can read all .xlsx files and consolidate them into a DataFrame. Further, we can output the consolidated data, either to a text file using `to_csv()` or to a new .xlsx using `to_excel()`. Program 15-2 reads all .xlsx files from the current directory and writes the data to a text file.

Program 15-2

```
# Read all .xlsx files in the 'data' directory and output data to a text file
import os
import pandas as pd

def getFileList(ext, dir='.'):
   filenames = os.listdir(dir)
   filenameList = []
   for filename in filenames:
      if filename.endswith(ext):
         filenameList.append(filename)
   return filenameList

filenames = getFileList('.xlsx', dir='data')
if len(filenames) > 0:
   os.chdir('data') # change the working directory to data 'data' sub-folder
   dfall = pd.read_excel(filenames[0])
   for i in range(1, len(filenames)):
      df = pd.read_excel(filenames[i])
      dfall = dfall.append(df , ignore_index=True, sort=False)

   dfall.to_csv('all.txt')
```

Note that in the above program, the final line writes the DataFrame to a text file using `to_csv()`. Alternatively, you can save a DataFrame's data as an Excel workbook file using the `to_excel()` method.

As of pandas version 1.4.1, there are 26 named parameters available in `read_excel()`. Many of these are similar to those available in `read_csv()`. For example, `index_col` indicates which column to set as the index of the DataFrame. The `sheetname` parameter controls which spreadsheet tabs are loaded. By default, the first sheet is loaded. Alternatively, a specific sheet can be loaded by its name or position number. Further, multiple sheets can be loaded, in which case a dictionary is returned, with `sheetname` as the key, and a DataFrame as the corresponding value.

Section Summary

In cases where you need to read and process a set of spreadsheets, the os function `listdir()` can be used to discover spreadsheets in a directory. Assuming these spreadsheets have a tabular data arrangement, they can be processed in a loop using pandas `read_excel()`.

Test Your Understanding

1) To get the list of all filenames in a directory, what function within the os package could you use?
   a) `getFilenames()`
   b) `dir()`
   c) `listdir()`
   d) none of the above

2) Which pandas DataFrame method will save data to a csv file?
   a) `to_csv()`
   b) `write_csv()`
   c) `print()`
   d) none of the above

15.4 Reading Excel Data with openpyxl

In the previous sections, we've seen how to read Excel tabular data using pandas. Although data in table form is common, spreadsheets allow any data to be placed anywhere in a sheet. Given this flexibility, you may need to read a spreadsheet with data in a layout other than a table. The openpyxl package allows your program to read any data located in any cell, regardless of layout, in order to collect the data you need.

As spreadsheets may contain a variety of data types—numbers, strings, dates, times, Booleans—we will need to see how openpyxl handles these. In this section, we'll use openpyxl to read a simple spreadsheet that has a variety of cell types, to see what Python type is returned. This is similar to Section 15.2, which uses a table with a variety of types. The spreadsheet we'll use is PythonExcel.xlsx, pictured in Figure 15-2.

	A	B
1	integer	1
2	float	1.5
3	date	5/31/19
4	datetime	5/31/19 12:00
5	text	hello
6	formula	2.5
7	boolean	TRUE
8	error	#NAME?
9	empty	

Figure 15-2: PythonExcel.xlsx

To start working with a workbook in openpyxl, we use its `load_workbook()` function, which returns a `Workbook` object. This function takes the name of the file as its first argument. We can also indicate that we'd like to access the results of formulas, as opposed to the formula logic; this is done by specifying `data_only=True`. For example:

```
[1]: from openpyxl import load_workbook
     workbook = load_workbook('PythonExcel.xlsx', data_only=True)
```

We can see the type for `workbook` using:

```
[2]: workbook
     <openpyxl.workbook.workbook.Workbook at 0x7fcd504b64f0>
```

Note that the displayed result above from a `Workbook` indicates only its type and address in memory, as opposed to showing its contents—for example, as we see when we display a DataFrame.

As workbooks are made up of one or more worksheets, we next need to indicate which worksheet we want to work with. This can be done with the square bracket accessors `[]`. For example:

```
[3]: worksheet = workbook['Sheet1']
     worksheet
     <Worksheet "Sheet1">
```

As shown above, the variable worksheet is of type `Worksheet`.

A `Workbook` can also provide the list of its `Worksheet` names using the `sheetnames` property:

```
[4]: workbook.sheetnames
     ['Sheet1', 'Sheet2']
```

With a `Worksheet` object, we can now navigate to specific cells. Using the `Worksheet` method `cell()`, we can access a specific cell, including several of the cell's properties. For example, to access cell B2, we can use:

```
[5]: cell = worksheet.cell(row=2, column=2) # access cell B2
     cell
     <Cell 'Sheet1'.B2>
```

Note that openpyxl uses 1 as the first item's row and column indexes as opposed to 0. This is for consistency with Excel's numbering—for example, cell B2 is column 2, row 2.

The `cell()` method returns a type of `Cell`. Note that the types used in openpyxl mirror the building blocks of an Excel file: `Workbook`, `Worksheet`, and `Cell`. A `Cell` has several properties; the primary one is `value`, which provides the value entered in the spreadsheet cell as a Python type such as `str`, `int`, `float`, or others.

```
[6]: cell.value
```

```
1.5
```

Another key property of Cell is `data_type`. The `data_type` is a single-letter code used by openpyxl to indicate the type of an Excel cell. For example, for cell B2, we can see the `data_type` is `'n'`:

```
[7]: cell.data_type
```

```
'n'
```

The openpyxl data_types and corresponding Python types returned by the cell.value property are shown in Table 15-3.

Type Code of .data_type	Meaning of .data_type	Python Type of .value
b	boolean	bool
d	date or time	datetime
e	error	`str`
f	formula	`int`, float, `str`, bool, or datetime
n	number or none	`int`, float, NoneType
s	string	`str`

Table 15-3: openpyxl cell types

Figure 15-3 summarizes the steps for loading a workbook and accessing cell values and types.

To recap the sequence of steps described above, Program 15-3 reads the PythonExcel.xlsx spreadsheet and displays the cells' values and their types.

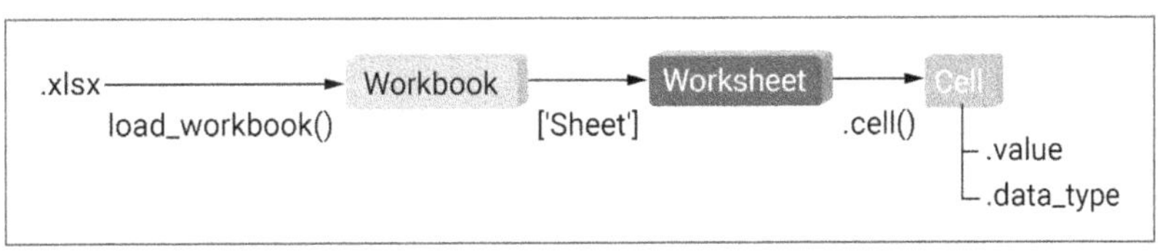

Figure 15-3: openpyxl flow sequence

Program 15-3

```
# access and display cells of different types from a workbook
from openpyxl import load_workbook

# Read the .xlsx file
workbook = load_workbook('PythonExcel.xlsx', data_only=True)

# access the 'Sheet1' worksheet tab
worksheet = workbook['Sheet1']

# for cells in the worksheet, display their values and types
for row in range(1, 10):
    cell1 = worksheet.cell(row=row, column=1)
    cell2 = worksheet.cell(row=row, column=2)
    print(cell1.value, cell2.value, cell2.data_type,type(cell2.value), sep='|')
```

Program 15-3 Output

```
integer|1|n|<class 'int'>
float|1.5|n|<class 'float'>
date|2019-05-31 00:00:00|d|<class 'datetime.datetime'>
datetime|2019-05-31 12:00:00|d|<class 'datetime.datetime'>
text|hello|s|<class 'str'>
formula|2.5|n|<class 'float'>
boolean|True|b|<class 'bool'>
error|#NAME?|e|<class 'str'>
empty|None|n|<class 'NoneType'>
```

Program 15-3 is written for a specific spreadsheet layout. We can generalize this by navigating to all the cells within a workbook. In order to do this, we need to know all the tabs in the workbook. As mentioned above, this is found using the `Workbook sheetnames` property, which returns a list. Further, we can use the `Worksheet rows` property, which returns a tuple (read-only list) of the rows of the spreadsheet. Program 15-4 is a general program to access and display all the cells of a workbook.

Program 15-4

```
# access and display all cells from a workbook
from openpyxl import load_workbook

# Read an .xlsx file
filename = 'PythonExcel.xlsx'
workbook = load_workbook(filename, data_only=True)

# access the list of sheets
sheetnames = workbook.sheetnames

for sheetname in sheetnames:
    print('=== Worksheet:', sheetname, '===')
    worksheet = workbook[sheetname]
    for row in worksheet.rows:
       for cell in row:
          if cell.value != None:
             print(cell.coordinate, cell.row, cell.column,
                   cell.data_type, cell.value, sep='|')
```

Section Summary

For both tabular and non-tabular spreadsheet data, the openpyxl package can be used to read a spreadsheet and access its cells' data. The primary types provided by the package are `Workbook`, `Worksheet`, and `Cell`. The `Cell` value types are mapped to appropriate Python types such as `int`, `float`, `str`, and `datetime`.

Test Your Understanding

1) The openpyxl package is limited to reading Excel data arranged in a tabular layout.
   a) True
   b) False

2) Which is a type that is defined in the openpyxl package?
   a) `Workbook`
   b) `Worksheet`
   c) `Cell`
   d) all of the above

15.5 Navigating Excel with openpyxl

In the previous section, we looked at the key functions, methods, and properties in openpyxl to access spreadsheet data. In this section, we'll look at an applied example using openpyxl. In particular, assume we are provided income statement information from several divisions that we need to combine into a file. Further, while there is some consistency expected in these spreadsheets (row and column names), each division's spreadsheet may include blank columns and/or rows. Figure 15-4 shows an example spreadsheet.

Given the presence of blank rows and columns, let's consider using openpyxl to navigate the spreadsheets and locate and extract the data of interest. (In fact, we might be able to use pandas `read_excel()` to read the data by providing this function additional arguments.)

For this section, we'll assume submitted income statements have standard row names, indicated in Table 15-4. The columns are expected to have two years of history (`'Act2019'`, `'Act2020'`) as well as one column of forecasted results (`'Proj2021'`).

	A	B	C	D	E	F
1	# Division A - Income Statement - actual and year-ahead plan					
2						
3	CODE	Description	Act2019	Act2020		Proj2021
4						
5	SALE	Sales Revenue	43.9	40.3		50.2
6	CGS	Cost of Goods Sold	17.5	18		18.2
7	PROF	GROSS PROFIT	26.4	22.3		32
8						
9	SGA	Selling, General, and Admin	2.1	1.8		2.1
10	ADV	Advertising	1.3	1.2		1.5
11	DEP	Depreciation	2.6	2.5		3.2
12	RENT	Rental Expense	2.2	2.2		2.2
13	OTHX	Other Expense	3.1	3.2		3.4
14	EXP	TOTAL EXPENSES	11.3	10.9		12.4
15						
16	EBT	EARNINGS BEFORE TAXES	15.1	11.4		19.6

Figure 15-4: Example income statement spreadsheet

Row code	Meaning
SALE	Sales
CGS	Cost of Goods Sold
PROF	Profit (SALE-CGS)
SGA	Selling, General, and Admin Expenses
ADV	Advertising Expense
DEP	Depreciation
RENT	Rent Expense
OTHX	Other Expenses
EXP	Total Expenses (SGA+ADV+DEP+RENT+OTHX)
EBT	Earnings Before Taxes (PROF-EXP)

Table 15-4: Row codes in income statement spreadsheets

For each submitted income statement spreadsheet, how can we discover its layout? One approach is to use openpyxl to find all the row codes and column headings. We can iterate over all the cells, and when there is a match with a row code or column header, we can record that cell's coordinates (column and row indexes) in a dictionary. Note that we use "column, row" pairs as opposed to "row, column" to match Excel naming convention—for example, cell B3 is column 2, row 3 and represented by (2, 3). Program 15-5 implements this label search.

Program 15-5

```
# find row labels and column headings in a worksheet
from openpyxl import load_workbook

# Read an .xlsx file, access a specific worksheet
divname = 'diva'
filename = 'incstmt-' + divname + '.xlsx'
workbook = load_workbook(filename, data_only=True)
worksheet = workbook['IncomeStatement']

rowcodes = ('SALE', 'CGS', 'SGA', 'ADV', 'DEP', 'RENT', 'OTHX')
colcodes = ('Act2019', 'Act2020', 'Proj2021')
rowdict = {}
coldict = {}
for row in worksheet.rows:
    for cell in row:
        #print(cell.row, cell.column, cell.coordinate, cell.value)
        if cell.value in rowcodes:
            rowdict[cell.value] = (cell.column, cell.row)
        elif cell.value in colcodes:
            coldict[cell.value] = (cell.column, cell.row)
print(rowdict)
print(coldict)
```

Program 15-5 Output

```
{'SALE': (1, 3), 'CGS': (1, 4), 'SGA': (1, 6), 'ADV': (1, 7), 'DEP': (1, 8), 'RENT': (1, 9),
'OTHX': (1, 10)}
{'Act2019': (3, 1), 'Act2020': (4, 1), 'Proj2021': (5, 1)}
```

Given the above logic, we can now use the row and column label coordinates to locate and save the data of interest. Program 15-6 accesses the income statement values for non-calculated rows and the three columns of interest and writes these results to a file.

Program 15-6

```
1  # export spreadsheet data by locating row labels and column headings
2  from openpyxl import load_workbook
3
4  # for an .xlsx income statement, locate row and column labels
5  def getRowColCoords(worksheet, rowcodes, colcodes):
6      rowdict = {}
7      coldict = {}
8      for row in worksheet.rows:
9          for cell in row:
10             if cell.value in rowcodes:
11                 rowdict[cell.value] = (cell.column, cell.row)
12             elif cell.value in colcodes:
13                 coldict[cell.value] = (cell.column, cell.row)
14     return rowdict, coldict
15
16 # given the row/col name locations, locate data and save to a file
17 def saveData(divname, worksheet, rowcodes, colcodes, rowdict, coldict, f):
18     for rowname in rowcodes:
19         rowix = rowdict[rowname][1]
20         print(divname,rowname, sep='|', end='|', file=f)
21         for colname in colcodes:
22             colix = coldict[colname][0]
23             cell = worksheet.cell(row=rowix, column=colix)
24             print(cell.value, end='|', file=f)
25         print(file=f)
26     f.close()
27
28 divname = 'diva'
29 filename = 'incstmt-' + divname + '.xlsx'
30 workbook = load_workbook(filename, data_only=True)
31 worksheet = workbook['IncomeStatement']
32
33 rowcodes = ('SALE', 'CGS', 'SGA', 'ADV', 'DEP', 'RENT', 'OTHX')
34 colcodes = ('Act2019', 'Act2020', 'Proj2021')
35 rowdict, coldict = getRowColCoords(worksheet, rowcodes, colcodes)
36
37 f = open('incstmt-' + divname + '.txt', 'w')
38 saveData(divname, worksheet, rowcodes, colcodes, rowdict, coldict, f)
```

In Program 15-6, we find the row and column coordinates for specific labels using the `getRowColCoords()` function, which saves this information to two dictionaries. The `getRowColCoords()` function is based on logic developed in Program 15-5. The `saveData()` function then uses these coordinates to locate the data values and we save these to a file using pipe | separated lines, as specified by `end='|'` on line 24. For example, an example output line as written by `saveData()` might be:

```
diva|SALE|43.9|40.3|50.2|
```

Program 15-6 separates the statements for locating labels from the statements for writing to a file. In this way, if we later decide to change the layout of the output file, we can focus on changes to the `saveData()` function without changing the `getRowColCoords()` function. For example, if we decide to change our output file so each row contains one value, like:

```
diva|SALE|2019Act|43.9|
```

we could focus our changes on lines 24 and 25 of Program 15-6.

The example program extracts data from one spreadsheet. Using an approach similar to Section 15.3, this program could be enhanced to loop over all spreadsheets in a directory and create a composite file with historical and project values for all divisions.

Section Summary

In this section, we looked at an applied example for locating a spreadsheet's row and column labels in order to navigate to and extract data of interest. In general, openpyxl provides functionality to read spreadsheets with any layout and to access data as specific locations or using rules to locate the data of interest. Once read into Python variables, the data can be further processed as needed.

Chapter Review

pandas and openpyxl provide alternative ways to read spreadsheet data into a Python program. For tabular data, pandas can create a DataFrame from Excel data. For access to tabular or non-tabular data, openpyxl allows you to read a workbook and visit and process its cells. Both packages convert Excel cell data into appropriately mapped data types.

Key Terms

Cell	An openpyxl type corresponding to one cell of a worksheet, with properties including a cell's value, type, and others.
list_dir()	An os package function that returns all file and subdirectory names at a given directory location.
load_workbook()	An openpyxl function that reads an Excel .xlsx file into memory as a Workbook object.
read_excel()	A pandas function that reads an Excel file and returns a DataFrame or dictionary of DataFrames.
Workbook	An openpyxl type corresponding to an Excel workbook.
Worksheet	An openpyxl type corresponding to an Excel worksheet (tab).

Programming Problems

1) Data Quality Checks with a Spreadsheet and pandas

Assume sales data are stored in a spreadsheet. Each row represents an item bought by a customer for a particular product on particular data. The data in the spreadsheet are in a table format with the following columns:

sale_id (integer)
customer_id (integer)
date (Excel date)
product_id (string)
count (integer)
unit_price (number)

a) Write a program that uses pandas `read_excel()` to read this spreadsheet into a DataFrame and locates rows where the count field is zero or negative.
b) Enhance your program to find all rows where any value is missing.

2) Data Quality Checks with a Spreadsheet and openpyxl

For the spreadsheet and questions in part 1 above, write a program to solve parts (a) and (b) using openpyxl instead of pandas.

3) Income Statement and pandas

For the spreadsheet used in Section 15.5, use pandas `read_excel()` to read the spreadsheet into a DataFrame. For the resulting DataFrame, use the CODE column as the index. Also remove rows that are all blank. Named parameters in `read_excel()` that are useful for this are:

skiprows
usecols
index_col

4) Income Statement Assumptions with openpyxl—One File

For the spreadsheet used in Section 15.5, assume that divisions are instructed to include spreadsheet cells with specific assumptions. In particular:

Cell C3 will contain the sales growth assumption from 2020 to 2021.
Cell C4 will contain the cost of goods as a percentage of sales.

Write a program using openpyxl that reads a spreadsheet in this form and displays to the screen a row in the form:

division,sales_growth,cgs_percent

5) Income Statement Assumptions with openpyxl—Multiple Files

For the spreadsheet described in problem 4, assume each division submits a spreadsheet in this form. Write a program that reads each spreadsheet (assume spreadsheets are in a particular directory) and locates and displays each spreadsheet's sales growth and assumptions for the percentage cost of goods sold. For example, for two spreadsheets, output might look like:

Division,SalesGrowth,CGS%
diva,.25,.45
divb,.12,.12

PART III

Python for Analytics

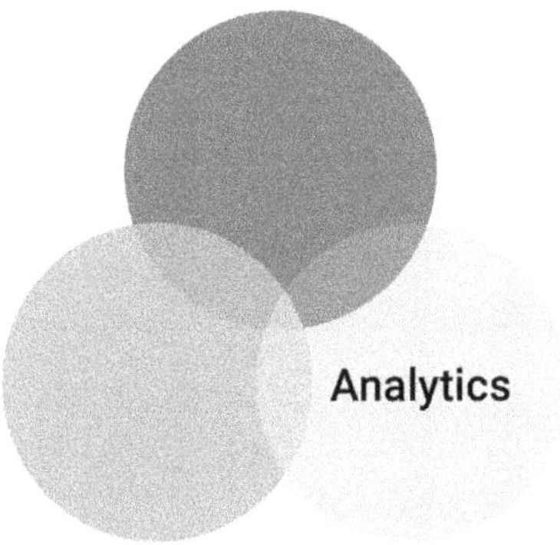

In Part III, we cover three types of analytics: visualization, statistics, and text analysis. In addition, the final chapter is a case study using a statistical approach to improving customer retention for a fictional telecom company.

Chapter 16 covers a variety of business charts using the pandas, Matplotlib, and seaborn packages. Example datasets included are electric car adoption over time by countries as well as membership data from a fictional health club.

Chapter 17 covers several statistical topics and how Python and packages can be used for these calculations. Topics include descriptive statistics, pivot tables, time series, comparing populations, confidence intervals, distribution and outlier review, and regression analysis.

Chapter 18 provides an introduction to text analysis using the nltk and scikit-learn packages. Topics include word frequency analysis, sentiment analysis, and text classification. Data includes fictional product reviews as well as a set of emails with both legitimate and spam emails.

Chapter 19 is a case study of a fictitious telecom company seeking to understand what types of customers are more likely to leave their service. The case uses a statistical model and techniques presented in prior chapters to identify at-risk customers and consider a strategy for customer retention.

Chapter 16

Visualization

Chapter Objectives

In this chapter, you'll learn:

- how you can use pandas to generate a chart from a Series or DataFrame
- about some of the different types of charts that pandas can generate, including line, bar, pie, box, histogram, and scatter charts
- some statements for preparing your data to support a desired chart
- a few ways to use Matplotlib to refine your chart to improve its impact
- how you can use the seaborn package for concise exploratory charting

Chapter Outline

16.1 About pandas, Matplotlib, and seaborn Charting

Tables of data can often be visualized in one or more types of charts, to provide insights about measures, comparisons, trends, and relationships. pandas provides a concise way to turn its DataFrame data into a variety of charts. It does this by using the Matplotlib package. For simple charts, you can use pandas alone, without knowing Matplotlib. But to refine your charts. you can add Matplotlib statements.

You can also use a visualization package such as Matplotlib or seaborn, and have your statements reference Python or pandas data. Given the wide variety of chart types available in visualization packages, our focus will be targeted. Once you're familiar with the basics of Python charting, you can explore visualization packages further to learn about the many other useful and creative visualizations available.

In this chapter, we'll visualize three datasets. Initially we'll look at how electric cars have been adopted over time in different countries. For example, if you were working in the marketing area of an electric car company, you may want to find opportunities for expansion. For this analysis, there are two data tables: the percentage of sales that are electric, and the number of electric cars sold. Using these datasets, we'll see how to generate different charts using pandas—including line, bar, and pie charts—in order to gain insight into the global electric car market. In some cases, in order to better visualize your data, you may need to rearrange them, and we'll cover a few approaches for doing this as well.

In the middle part of the chapter, we'll look at daily stock and index prices, and chart prices and returns. These can be viewed as time series in a line chart, as well as looked at statistically in terms of their distributions using a histogram chart and a box plot. Lastly, we'll look at how to use a scatter chart to check if there is a relationship between two variables.

In the last part of the chapter, we'll examine the membership data from a fictional health club. We'll use these data to look for patterns or trends that may lead to recommendations to increase membership. We'll see how we can do this with some basic seaborn charts.

There are many other types of charts that Matplotlib and seaborn can generate. For your specific data, those covered in this chapter may suffice, or you may find other types of charts will better help uncover a pattern or convey a message. The material in this chapter will give you a strong foundation on how to create charts, allowing you to quickly become acquainted with other chart types available in these powerful visualization libraries.

16.2 Electric Car Data

For the early part of this chapter, we'll look at electric car sales and how they have evolved over time in different countries. In particular, we have previously downloaded two tables. The file cars_pct.csv includes what percentage of new car sales were electric for 22 countries for the years from 2013 to 2020. Another file, cars_sales.csv, has the number of new electric cars sold from 2017 to 2020 for 10 countries. We can use different types of charts to compare the trends of electric car adoption in these different countries and decide which best conveys the message we want to show.

To get acquainted with the data, we'll load the files into DataFrames and browse the values. First we can use the pandas `read_csv()` function to load the percentage data into the `dfpct` DataFrame:

```
[1]: import pandas as pd
     dfpct = pd.read_csv('cars_pct.csv', index_col='Country')
```

In the above, we chose Country as the `index_col`, since the values in this column are distinct for each row. To confirm our data have loaded correctly, we can browse the first few rows using the `head()` method:

```
[2]: dfpct.head()
```

	2020	2019	2018	2017	2016	2015	2014	2013
Country								
Norway	74.7	55.9	49.1	39.20	29.1	22.40	13.80	6.10
Iceland	45.0	22.6	19.0	14.05	4.6	2.93	2.71	0.94
Sweden	32.2	11.4	8.2	5.20	3.5	2.62	1.53	0.71
Netherlands	24.6	14.9	6.2	2.20	6.7	9.90	3.87	5.55
Finland	18.1	6.9	4.7	2.57	1.2	NaN	NaN	NaN

In the above table, note the columns are years, and further, these are displayed in reverse chronological order (with more recent years shown on the left). The data layout of the table will influence how the charts will be arranged, as we'll see later in the chapter. This is OK, as we can easily rearrange the table layout with pandas in order to create the charts we'd like.

Similarly, we can load the sales data into the `dfsales` DataFrame using:

```
[3]: dfsales = pd.read_csv('cars_sales.csv', index_col='Country')
```

As this dataset has fewer countries than the percentage data, we'll review the full table here:

```
[4]: dfsales
```

Country	2020	2019	2018	2017
China	1246000	1204000	1016002	579000
United States	297939	329528	361307	199818
Germany	394632	108839	67504	54492
Norway	113588	88443	86290	71737
France	194881	69466	53745	42799
United Kingdom	175082	72584	59911	49182
Netherlands	89362	67520	29187	11085
Japan	29000	43867	52013	54100
Sweden	96054	41784	29909	19793
Canada	47000	50960	44150	18746

As with the percentage data, the columns are years, in reverse chronological order. Although the table is smaller, with 40 points, there are enough data here that visualizations through charting may provide more insight than reviewing the numbers directly.

16.3 Electric Car Adoption

Line chart

Our data have two dimensions: time and country. If we'd like to review how electric car adoption has evolved over time, we can use a line chart. The pandas `plot()` method provides several different types of charts, as we'll see in this chapter. The line chart is the default chart type. pandas makes charting easy—you can call the plot method for a Series or DataFrame. Let's do this for our percentage data. As in the prior section, we'll load the data into a DataFrame using:

```
[1]: import pandas as pd
     dfpct = pd.read_csv('cars_pct.csv', index_col='Country')
```

To visualize the data in a line chart, we can use:

```
[2]: dfpct.plot()
```

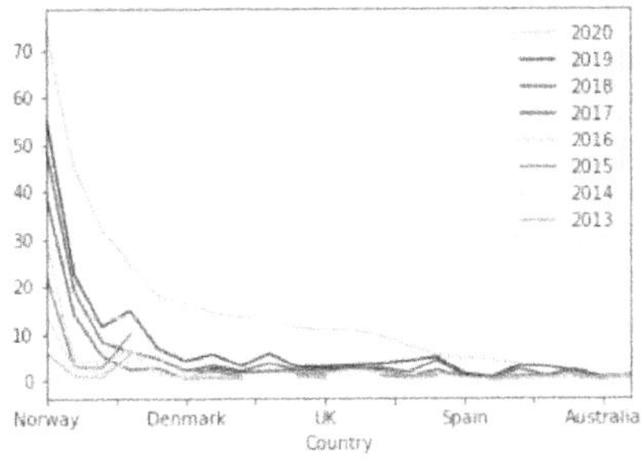

Although we do get a chart, this is not meaningful yet, as the different years are represented by different lines. To get a useful time series chart, we would instead want years to be on the *x* axis, with different countries represented by different lines. In order to do this, we need to transpose the DataFrame so that years are the rows and countries are the columns. We can do this using:

```
[3]: dfpct = dfpct.transpose()
```

To check the result of the transpose, we can use the `iloc[]` accessor to see the first five rows and columns of the DataFrame:

```
[4]: dfpct.iloc[:5, :5]
```

Country	Norway	Iceland	Sweden	Netherlands	Finland
2020	74.7	45.00	32.2	24.6	18.10
2019	55.9	22.60	11.4	14.9	6.90
2018	49.1	19.00	8.2	6.2	4.70
2017	39.2	14.05	5.2	2.2	2.57
2016	29.1	4.60	3.5	6.7	1.20

When reviewing the table, it would be helpful to give the row index a name:

```
[5]: dfpct.index.name = 'Year'
     dfpct.iloc[:5, :5]
```

Country	Norway	Iceland	Sweden	Netherlands	Finland
Year					
2020	74.7	45.00	32.2	24.6	18.10
2019	55.9	22.60	11.4	14.9	6.90
2018	49.1	19.00	8.2	6.2	4.70
2017	39.2	14.05	5.2	2.2	2.57
2016	29.1	4.60	3.5	6.7	1.20

With the DataFrame rearranged, let's try again to chart it using `plot()`:

```
[6]: dfpct.plot()
```

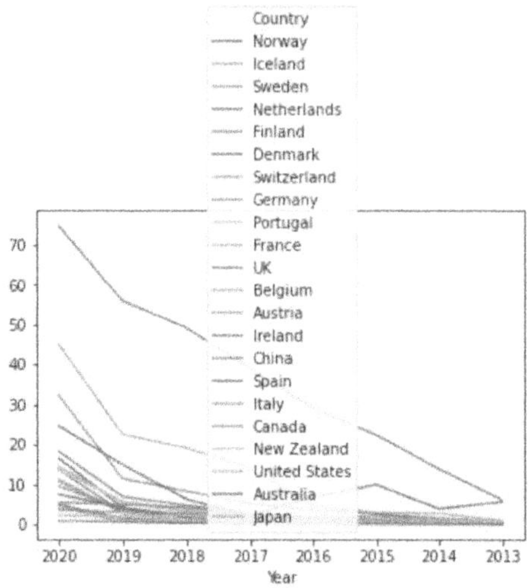

We now have a time series chart—however, time is reversed from the typical order of earliest to latest from left to right. This is because our original data order is latest to earliest. To address this, we can sort the rows of the DataFrame using:

```
[7]: dfpct = dfpct.sort_index()
     dfpct.iloc[:5, :5]
```

Country	Norway	Iceland	Sweden	Netherlands	Finland
Year					
2013	6.1	0.94	0.71	5.55	NaN
2014	13.8	2.71	1.53	3.87	NaN
2015	22.4	2.93	2.62	9.90	NaN
2016	29.1	4.60	3.50	6.70	1.20
2017	39.2	14.05	5.20	2.20	2.57

Now using `plot()`, we get the following improved time series chart:

```
[8]: dfpct.plot()
```

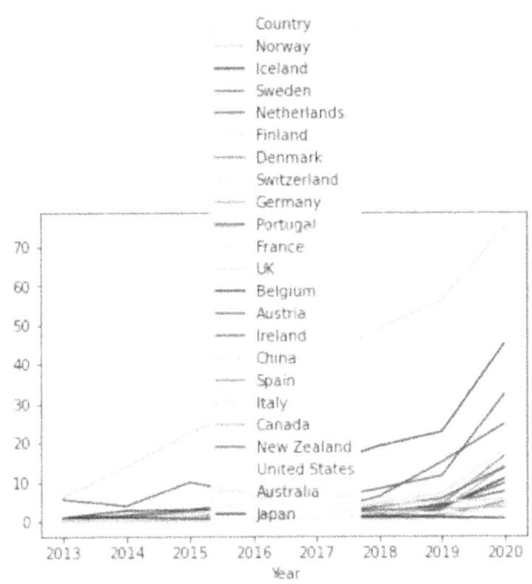

The chart is a little dense, but we can see some outliers. In particular, Norway's high electric car adoption rate is clear, and several other countries follow with high rates (Iceland, Sweden, the Netherlands). Perhaps we would like to see a chart featuring these three countries. Recall that we can select a subset of a DataFrame's columns using the double-bracket notation and again use `plot()`:

```
[9]: dfpct[['Iceland', 'Sweden', 'Netherlands']].plot()
```

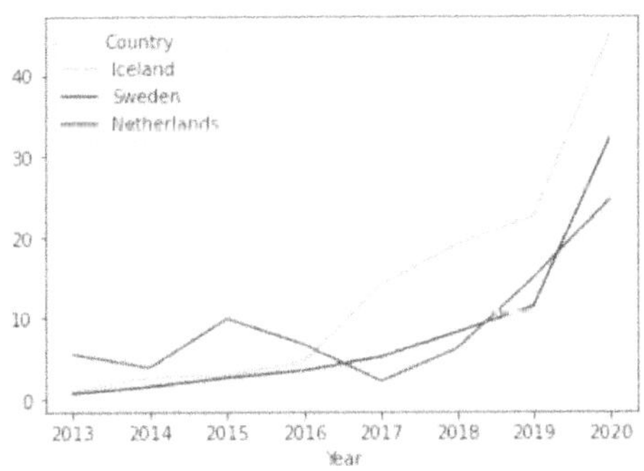

Any subset of countries can be charted in this way. For example, if we were interested in a regional chart, say for Asian/Pacific countries, we could use:

```
[10]: dfpct[['China', 'Japan', 'Australia', 'New Zealand']].plot()
```

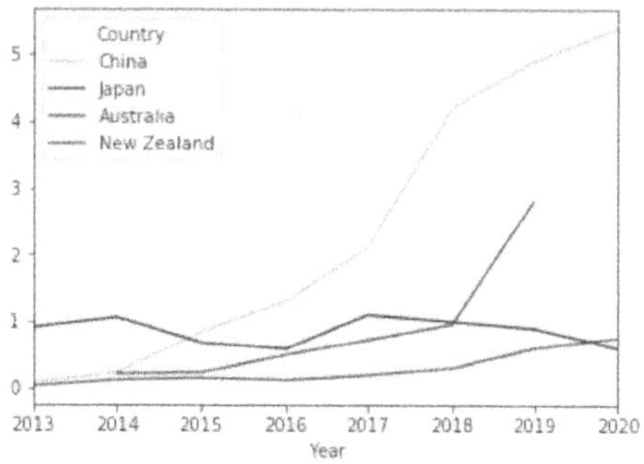

Some insights from the above chart are:

- China has had significant growth in electric car sales and, as of 2020, has the highest percentage of electric car sales among these countries.
- New Zealand has had a recent jump in electric car adoption, although we are missing the latest data for 2020.
- Japan's electric car adoption has been relatively flat over time.

Clustered bar chart

Since we have relatively few years of data, we may also want to visualize the data using a cluster bar chart, which may be clearer since the line chart interpolates between years. The `plot()` method has a `kind` parameter that allows us to choose between different types of charts. For example, we can use:

```
[11]: dfpct[['China', 'Japan', 'Australia', 'New Zealand']].plot(kind='bar')
```

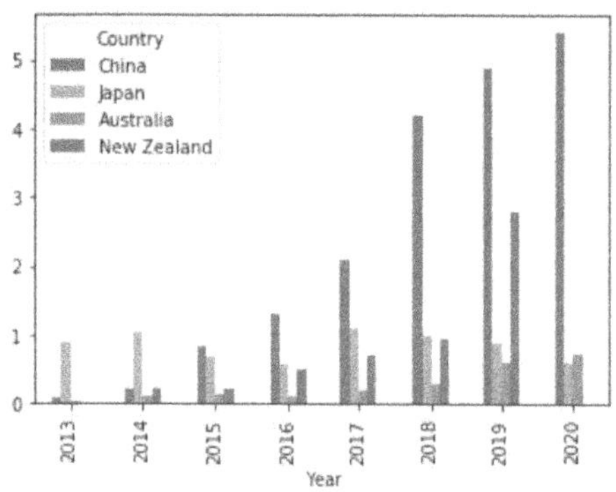

Horizontal bar chart

The previous charts have looked at multiple countries over time. If we are interested in one slice of time—for example, just the latest year—we can visualize this using a horizontal bar chart. If time is not a dimension, horizontal bar charts may be preferred, since, for example, the labels can be easier to read. The `plot()` method's `kind` parameter can be set to `'barh'` for this chart type. Further, we can select the latest year's data by using the `.iloc[]` accessor. Using these in combination, we have:

```
[12]: dfpct.iloc[-1].plot(kind='barh')
```

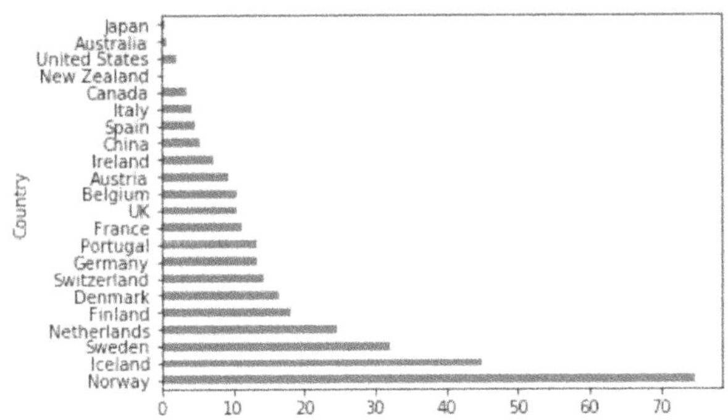

As we see above, the larger values are shown at the bottom of the chart, since the order of the bars reflects the order of the data within the DataFrame. As an alternative, we may prefer to have the largest bars at the top of the chart to show first the countries with the highest rates of electric car adoption. To do this, we can reorder the data using `sort_values()`:

```
[13]: dfpct.iloc[-1].sort_values(ascending=True).plot(kind='barh')
```

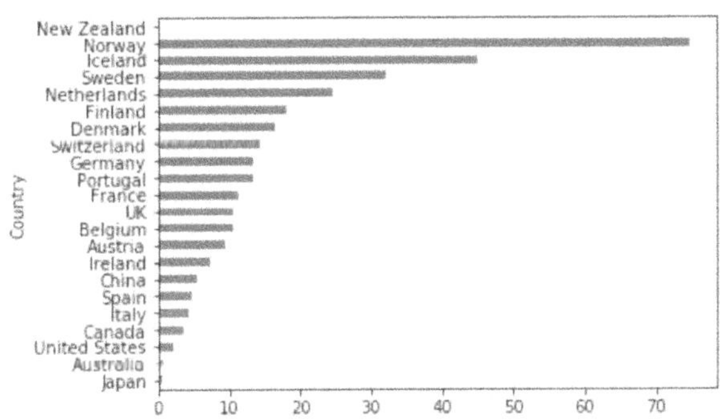

One other refinement relates to missing data. In particular, New Zealand has no data for 2020, though it shows at the top of our chart. We can filter out the New Zealand data using `dropna()`:

```
[14]: dfpct.iloc[-1].dropna().sort_values(ascending=True).plot(kind='barh')
```

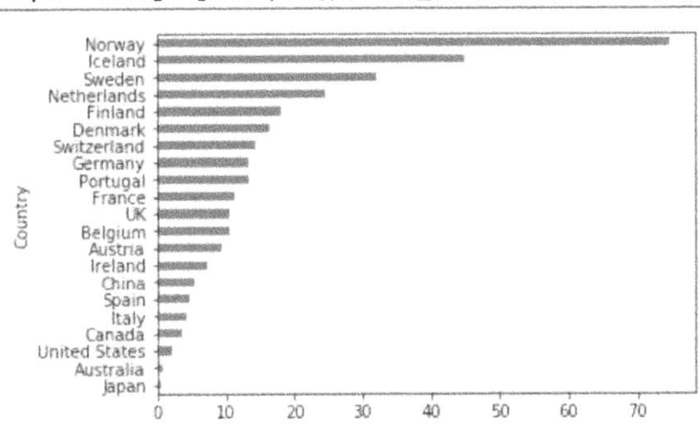

Section Summary

The pandas DataFrame `plot()` method can be used to visualize tables of data. By default, the chart type is a line chart. For line charts, each DataFrame row should represent a point or period of time. You can use the `kind` parameters of `plot()` to create other types of plots, including certain bar charts shown in this section as well as many other types available. In general, you may find it helpful to successively refine your data orientation and selection, as well as chart type, in order to find one or more good visualizations of your data.

Test Your Understanding

1) In order to visualize data in a pandas DataFrame, you can use its:
   a) `chart()` method
   b) `plot()` method
   c) `visualize()` method
   d) all of the above

2) The default chart provided by a pandas DataFrame is a:
   a) line chart
   b) bar chart
   c) pie chart
   d) none of the above

16.4 Electric Car Sales Charts

In this section, we turn our attention to unit sales data for electric cars in the file cars_sales.csv. These data provide us with a different perspective on electric cars:

- Unit sales reflect population size differences, since although Norway has high adoption of electric cars, its smaller population will likely translate to fewer numbers of cars than other countries.
- Sales units can be aggregated so we can see totals visually across countries.

Data preparation

As shown in Section 16.2, we can load the unit sales data with:

```
[1]: import pandas as pd
     dfsales = pd.read_csv('cars_sales.csv', index_col='Country')
     dfsales
```

Country	2020	2019	2018	2017
China	1246000	1204000	1016002	579000
United States	297939	329528	361307	199818
Germany	394632	108839	67504	54492
Norway	113588	88443	86290	71737
France	194881	69466	53745	42799
United Kingdom	175082	72584	59911	49182
Netherlands	89362	67520	29187	11085
Japan	29000	43867	52013	54100
Sweden	96054	41784	29909	19793
Canada	47000	50960	44150	18746

Before we look at charts, let's consider the data as they are displayed in the above table. The numbers are large, so we may want to scale them—for example, in thousands—so they are easier to read. In pandas, this is easy to do with:

```
[2]: dfsales /= 1000
     dfsales
```

	2020	2019	2018	2017
Country				
China	1246.000	1204.000	1016.002	579.000
United States	297.939	329.528	361.307	199.818
Germany	394.632	108.839	67.504	54.492
Norway	113.588	88.443	86.290	71.737
France	194.881	69.466	53.745	42.799
United Kingdom	175.082	72.584	59.911	49.182
Netherlands	89.362	67.520	29.187	11.085
Japan	29.000	43.867	52.013	54.100
Sweden	96.054	41.784	29.909	19.793
Canada	47.000	50.960	44.150	18.746

While the magnitude of the numbers is improved, the displayed DataFrame now displays with three decimal places. Further, we may want to use commas within the values. We can control the number format for easier readability using the pandas `float_format` property:

```
[3]: pd.options.display.float_format = '{:,.0f}'.format
     dfsales
```

	2020	2019	2018	2017
Country				
China	1,246	1,204	1,016	579
United States	298	330	361	200
Germany	395	109	68	54
Norway	114	88	86	72
France	195	69	54	43
United Kingdom	175	73	60	49
Netherlands	89	68	29	11
Japan	29	44	52	54
Sweden	96	42	30	20
Canada	47	51	44	19

The above statement indicates to use commas and no decimal places, giving us an easier table to review as compared with the original data and format.

To visualize these data in a chart, as with the percentage data in the prior section, we next transpose and sort the DataFrame:

```
[4]: dfsales = dfsales.transpose()
     dfsales.rename_axis(index='Year')
     dfsales = dfsales.sort_index()
     dfsales.iloc[:5, :5]
```

Country	China	United States	Germany	Norway	France
2017	579	200	54	72	43
2018	1,016	361	68	86	54
2019	1,204	330	109	88	69
2020	1,246	298	395	114	195

Stacked bar chart

We are now ready to visualize our unit sales data in charts. Since we can aggregate the data across countries, a stacked bar chart is appropriate, as the sum of the bars equate to an overall total. We can set the kind to `'bar'`, and provide an additional parameter `stacked`:

```
[5]: dfsales.plot(kind='bar', stacked=True)
```

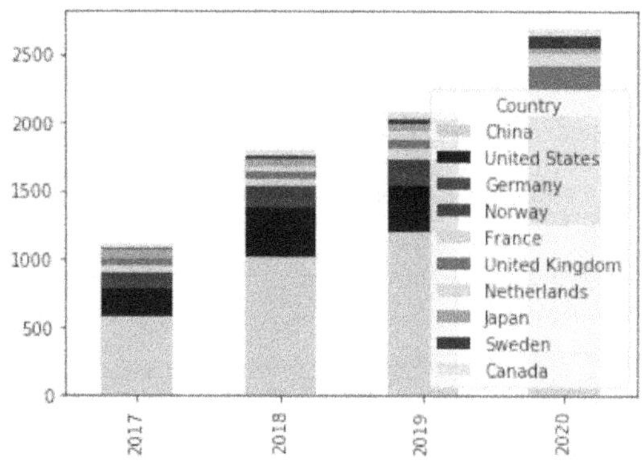

Although the chart provides a good visual, the default position of the legend is not ideal, since it covers part of the chart. As pandas uses the Matplotlib visualization package to generate charts, we can refine the chart using Matplotlib directly. In order to do this, we import Matplotlib using:

```
[6]: import matplotlib.pyplot as plt
```

For the above chart, the legend can be located using these statements:

[7]:
```
dfsales.plot(kind='bar', stacked=True)
plt.legend(loc='upper left', fontsize='x-small')
plt.show()
```

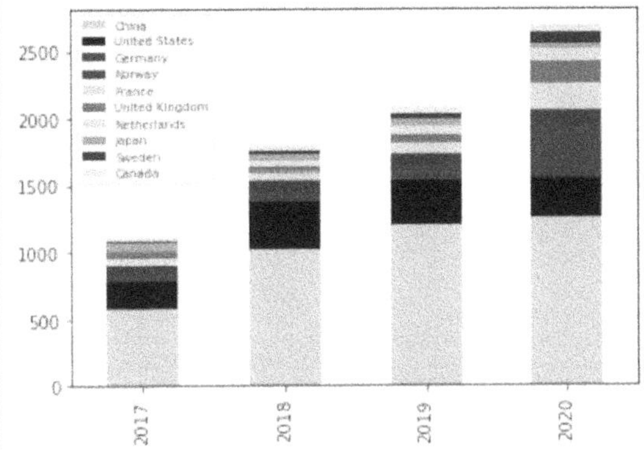

Alternatively, the legend can be placed next to the chart. For example:

[8]:
```
dfsales.plot(kind='bar', stacked=True)
plt.legend(bbox_to_anchor=(1.05, 1))
plt.show()
```

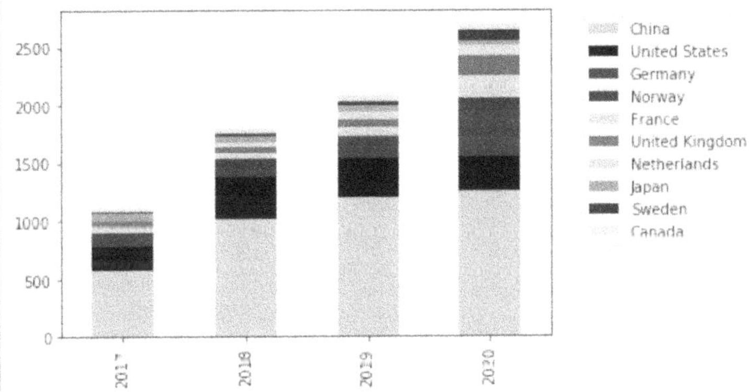

Pie chart

For these data, we can use a pie chart to compare countries' sales for a given year. This can be done with the `kind` parameter set to `'pie'`. For example:

[9]:
```
dfsales.loc['2020'].plot(kind='pie')
```

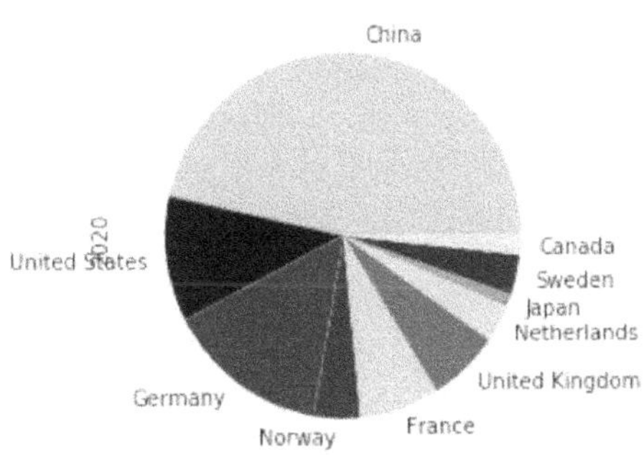

```
[10]: dfsales.loc['2017'].plot(kind='pie')
```

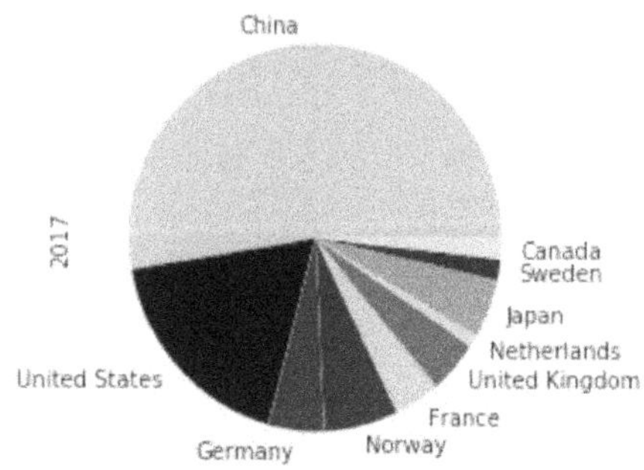

Section Summary

In this section, we looked at a dataset for electric car sales by country over time in order to uncover trends in the data. As in the prior section, we used the pandas `plot()` method to visualize the data using stacked bar and pie charts. To refine your charts, you can use Matplotlib directly—for example, to adjust the placement or size of the legend.

Test Your Understanding

1) The named parameter of the DataFrame `plot()` method to select the type of chart is called:
   a) type
   b) kind
   c) chart
   d) none of the above

2) In order to refine a pandas chart, you can use which package?
   a) pychart
   b) Matplotlib
   c) numpy
   d) none of the above

16.5 Stock Price Data

In the prior sections, we looked at small datasets for electric car sales over time by country. In this and the next section of the chapter, we'll look at a larger dataset of stock and index prices to review some different chart types that are more related to statistics. In particular, we have daily stock prices of a fictional company, Orange Technologies (symbol ORNG), as well as daily levels of a hypothetical stock market index (symbol USIX) over a five-year period. The index is one summary measure of the stock market, whose value is based on the market price of many stocks. From these daily time series we can calculate the daily returns of Orange stock and the index, and use charts to visualize any relationship between these two datasets.

The price data for both the stock and index have several columns of information. We are interested in analyzing the ***adjusted closing price***, as this data series is restated for events such as stock splits and dividends paid, which provides a more consistent history than closing stock prices. We can load Orange's stock price data using:

```
[1]: import pandas as pd
     stockSymbol = 'ORNG'
     stock = pd.read_csv(stockSymbol +'.csv', index_col=0)
     stock.head()
```

Date	Open	High	Low	Close	AdjustedClose	Volume
2017-01-03	57.900	58.165	57.380	58.075	54.595	230255200
2017-01-04	57.925	58.255	57.875	58.010	54.534	168944800
2017-01-05	57.960	58.430	57.905	58.305	54.812	177548800
2017-01-06	58.390	59.080	58.235	58.955	55.423	254015200
2017-01-09	58.975	59.715	58.970	59.495	55.930	268495200

To check the time span of the data, we can see the last row using `tail()`:

```
[2]: stock.tail(1)
```

Date	Open	High	Low	Close	AdjustedClose	Volume
2021-12-30	358.94	361.14	356.18	356.4	355.947	119546000

Since we are interested only in the adjusted closing price, we can narrow and rename the data using:

```
[3]: stock = stock[['AdjustedClose']]
     stock.rename(columns={'AdjustedClose':stockSymbol}, inplace=True)
     stock.head()
```

Date	ORNG
2017-01-03	54.595
2017-01-04	54.534
2017-01-05	54.812
2017-01-06	55.423
2017-01-09	55.930

For the analysis, rather than analyzing price changes, we'd like to focus on the daily price returns, since these are more consistent if our data span a wide range of prices. To calculate daily returns, we can use the pandas method `pct_change()`:

```
[4]: stock[stockSymbol + '_Return'] = stock.pct_change()
     stock.head()
```

	ORNG	ORNG_Return
Date		
2017-01-03	54.595	NaN
2017-01-04	54.534	-0.001117
2017-01-05	54.812	0.005098
2017-01-06	55.423	0.011147
2017-01-09	55.930	0.009148

In the above result, the missing value for the first Return is expected since there is no price data available in the DataFrame for the prior date. Further, the ORNG_Return column is stated in decimal (even though the pandas method is called `pct_change()`), so -0.00117 indicates a daily change of -0.117% from 1/3/2017 to 1/4/2017.

Similarly, we can read and augment the USIX index data with the statements:

```
[5]: indexSymbol = 'USIX'
     index = pd.read_csv(indexSymbol +'.csv', index_col=0)
     index = index[['AdjustedClose']]
     index.rename(columns={'AdjustedClose':indexSymbol}, inplace=True)
     index[indexSymbol +'_Return'] = index.pct_change()
     index.head()
```

	USIX	USIX_Return
Date		
2017-01-03	412.295	NaN
2017-01-04	414.748	0.005950
2017-01-05	414.418	-0.000796
2017-01-06	415.901	0.003579
2017-01-09	414.528	-0.003301

Rather than working with two DataFrames, we can combine them into one with the `join()` method using:

```
[6]: df = index.join(stock)
     df.head()
```

	USIX	USIX_Return	ORNG	ORNG_Return
Date				
2017-01-03	412.295	NaN	54.595	NaN
2017-01-04	414.748	0.005950	54.534	-0.001117
2017-01-05	414.418	-0.000796	54.812	0.005098
2017-01-06	415.901	0.003579	55.423	0.011147
2017-01-09	414.528	-0.003301	55.930	0.009148

Above we use `join()` since it will match the dates between our stock and index DataFrames. This matching will handle the case where our ORNG stock data have dates that aren't present in the USIX (or vice versa). With the prices loaded, returns calculated, and data joined into one DataFrame, we'll look in the next section at visualizing the data.

16.6 Stock Price and Return Charts

Time series

As with electric car data, we can visualize our stock data with the `plot()` method. For example, if we'd like to see the time series of the stock and index prices, we can use:

```
[7]:  df[[stockSymbol,indexSymbol]].plot()
```

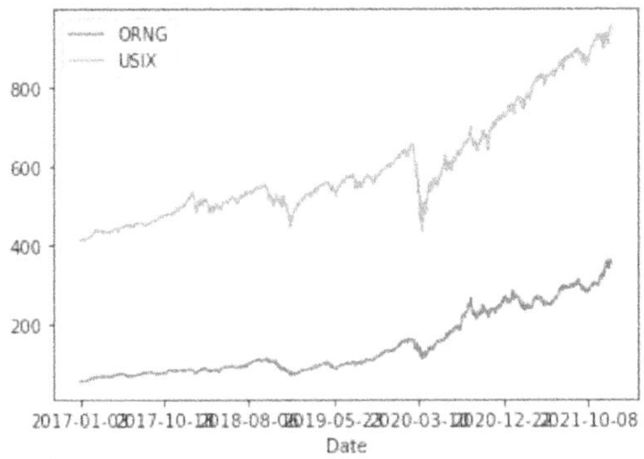

There are some refinements we can make to this chart. As the stock and index prices are at different levels, we can use a dual *y* axis chart to visualize the time series closer together. Also, since the *x* axis labels are overlapping; to resolve this, we can create a larger chart than the default size. To set a different chart size, we use the `figsize` parameter to specify the chart's width and height in inches. We can also add a chart title and a *y* axis title. In order to make these refinements, we can use Matplotlib and the following statements:

```
[8]: import matplotlib.pyplot as plt
     df[[stockSymbol, indexSymbol]].plot(secondary_y=indexSymbol, figsize=(9,7))
     plt.title(stockSymbol+' vs '+indexSymbol)
     plt.ylabel('Adj Closing Price')
     plt.show()
```

Histogram

Having reviewed the price data, let's focus on the calculated daily returns. For a variable with a large number of observations, we may be interested in its distribution, to see how the data are spread out. For this, the `plot()` `type` of `'hist'` can be used to display a histogram of the data. For this type of chart, we can control the number of bins (also known as buckets) that the histogram will use to tailor our histogram. The height of each bar in the histogram is the number of observations falling within each bin. For example, here is a histogram of Orange's daily stock returns with 30 bins:

```
[9]: df[stockSymbol+'_Return'].plot(kind='hist', bins=30)
```

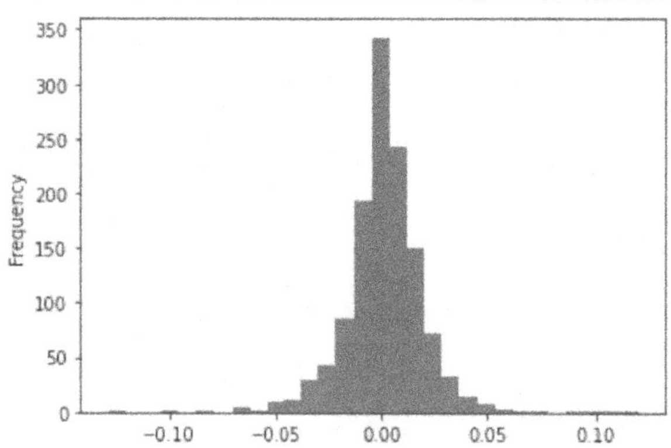

In the above histogram chart, the x axis is displayed in decimal, since the return data are in decimal. Therefore, the maximum label of 0.10 corresponds to a 10% daily return. Here we see almost all daily returns are between -5% and 5%.

The histogram provides the count of observations within each bin. We can also visualize a smooth histogram-like curve using the chart type of `'kde'` (short for kernel density estimate). For the stock returns, this is:

```
[10]: df[stockSymbol+'_Return'].plot(kind='kde')
```

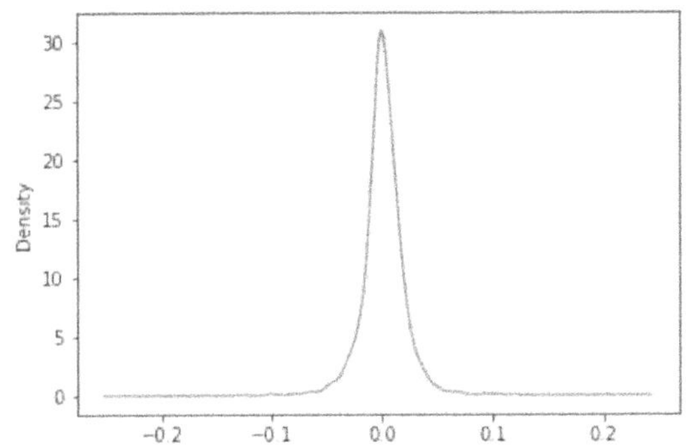

Box plot

Another type of distribution visualization is the box plot, which shows the median, second and third quartiles, and other points. The `plot()` kind for this is `'box'`; here is this chart showing both the stock and index returns:

```
[11]: df[[stockSymbol+'_Return',indexSymbol+'_Return']].plot(kind='box')
```

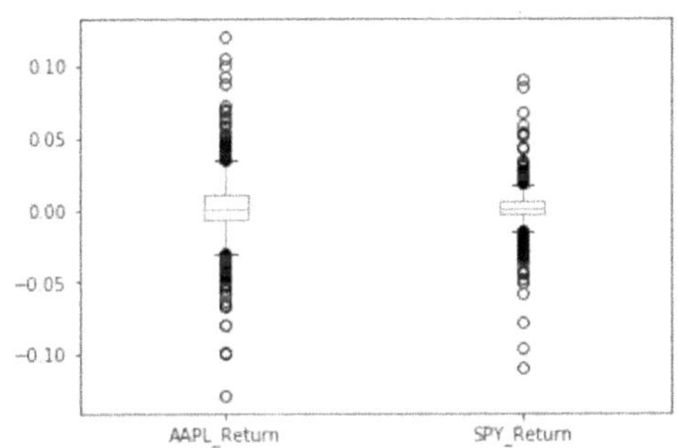

Here we see that the stock and index median returns are close to 0 and the spread of Orange's stock returns is greater than that of the spread of the USIX returns.

Scatter chart

Some of the charts in this section provide some comparisons between the stock and the index prices and returns. For example, the time series charts give an indication that the prices of the stock and index tend to move together. We can also visualize this relationship by looking at a scatter chart of the returns. The `plot()` kind for this is `'scatter'`, and we can set the x axis and y axis values using named parameters. For example:

```
[12]: df.plot(kind='scatter', x=indexSymbol+'_Return', y=stockSymbol+'_Return')
```

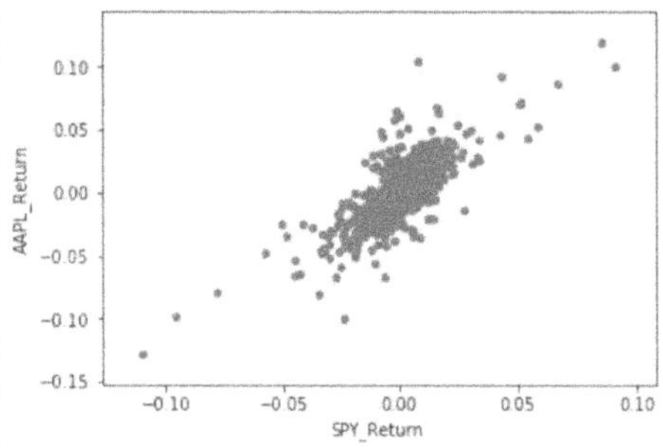

Here we confirm the relationship between the stock and index movements. Note that there is some variability to this, as the chart does not show a tightly defined line.

Section Summary

In this section, we looked at a different dataset, covering time series of a stock's price, as well as a stock index's prices. Line charts with one or two *y* axes are suited for reviewing the time series data and can be easily generated using the pandas DataFrame `plot()` method. To review the distribution of data, we used a histogram as well as a KDE (kernel density estimate) chart. A box plot also provides a view of the distribution of data, through which we compared the stock and index price data. Last we looked at the relationship between the stock and index data using a scatter plot.

Test Your Understanding

1) To review the distribution of data, the DataFrame `plot()` method's kind parameter can be set to:
   a) `'dist'`
   b) `'hist'`
   c) `'bin'`
   d) none of the above

2) A chart type for reviewing a potential relationship between two variables is:
   a) scatter
   b) KDE
   c) box
   d) all of the above

16.7 Health Club Membership Data

In the last two sections of this chapter, we'll look at a membership dataset from a fictional health club. This club has seen declining membership and has asked us to analyze their records to uncover any patterns that may lead to recommendations. In contrast with the prior sections of the chapter, where we had an idea of the specific charts we wanted to generate, our analysis will now be more exploratory—we're initially not sure what we'll find. In order to analyze the data, we will continue to use pandas and also leverage the seaborn visualization package. Seaborn offers a variety of charts that can be generated using very few lines of code and is often a good tool for data exploration.

The health club has been maintaining records of every potential member who has visited the club to have a tour. Some of these visitors decided to join and become members. Data about visitors and members are stored in the files visitors.csv and members.csv. We decide to analyze the membership data, members.csv, which has the following fields:

field	description
guestID	An integer assigned to each visitor
visitDate	The date the member first visited the club, before they joined
gender	M=Male; F=Female
age	Age of member when they visited the club
married	M=Married; S=Single
pool	y=is a pool member n=not a pool member
gym	y=is a gym member n=not a gym member
status	active: is still a member inactive: no longer a member
endDate	If inactive, date when membership ended

Since members.csv is a tabular comma-separated file, we can load it into pandas as:

```
[1]: import pandas as pd
     dfmembers = pd.read_csv('members.csv', parse_dates=['visitDate', 'endDate'])
```

Note that we use the `parse_dates` named parameter to indicate which fields are dates so that pandas will store these values as the `datetime64` datatype rather than strings. The `datetime64` datatype will be useful later to analyze trends over time.

We can confirm the data have loaded correctly using the DataFrame `head()` method:

```
[2]: dfmembers.head()
```

	guestID	visitDate	gender	age	married	pool	gym	status	endDate
0	6	2015-01-05	M	47	M	n	y	inactive	2017-08-01
1	7	2015-01-05	F	46	S	n	y	inactive	2016-11-01
2	10	2015-01-06	M	61	S	y	n	inactive	2016-12-01
3	17	2015-01-09	F	36	S	y	n	inactive	2017-01-01
4	24	2015-01-13	M	45	M	y	n	inactive	2016-09-01

We can check the field types and number of rows in the data with `info()`:

```
[3]: dfmembers.info()
```

```
<class 'pandas.core.frame.DataFrame'>
RangeIndex: 871 entries, 0 to 870
Data columns (total 9 columns):
guestID      871 non-null int64
visitDate    871 non-null datetime64[ns]
gender       871 non-null object
age          871 non-null int64
married      871 non-null object
pool         871 non-null object
gym          871 non-null object
status       871 non-null object
endDate      723 non-null datetime64[ns]
dtypes: datetime64[ns](2), int64(2), object(5)
memory usage: 61.3+ KB
```

Above we see, there are 871 members (past and present) reflected in the data. As there are 723 rows with a non-null `endDate`, there are 871 – 723 = 148 active members. We can also confirm this with the statement:

```
[4]: len(dfmembers[dfmembers.status=='active'])
```

```
148
```

Also from the above, we see the date fields were loaded as `datetime64`. Further note that the data are fully populated for all the other fields, as all columns (except endDate) have 871 non-null values.

16.8 Health Club Membership Charts

Having loaded the membership data into a pandas DataFrame, we can now visualize them through some charts by using seaborn. To import seaborn, we use the statement:

```
[5]: import seaborn as sns
```

To start, let's review categorical data: married and gender. To see the count of observations, we can use the seaborn `countplot()` function:

```
[6]: sns.countplot(data=dfmembers, x='married')
```

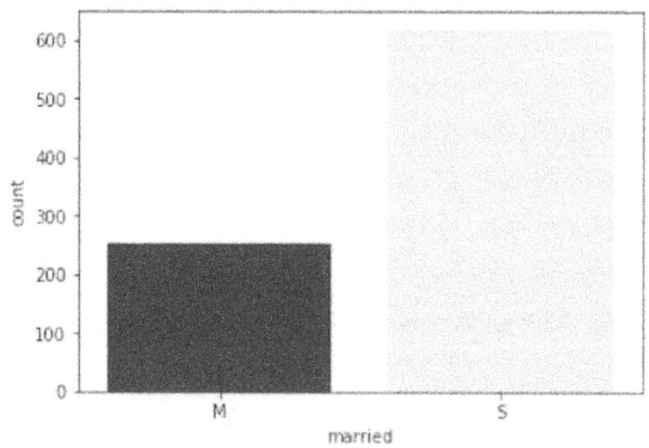

From the above chart, we see our member data show over twice as many single as married members.
We can also review the counts of male and female members:

```
[7]:  sns.countplot(data=dfmembers, y='gender')
```

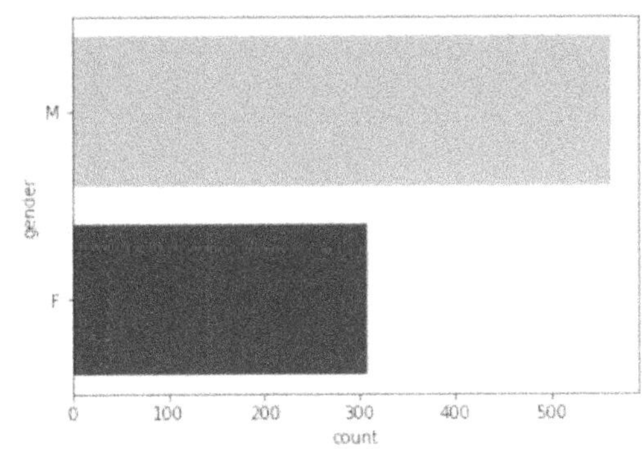

In contrast to the married chart, above we use the `y` named parameter in order to generate a horizontal bar chart. Either vertical or horizontal bar charts are suitable, although in charts without time as a dimension, a vertical bar may be preferred. Note also the above two charts span all the member data—both active and inactive—so these charts do not show potential changes in these demographics over time.

Another demographic dimension we can look at is age. Although age has many more values than gender or marital status, we can still initially use `countplot()`:

```
[8]:  sns.countplot(data=dfmembers, x='age')
```

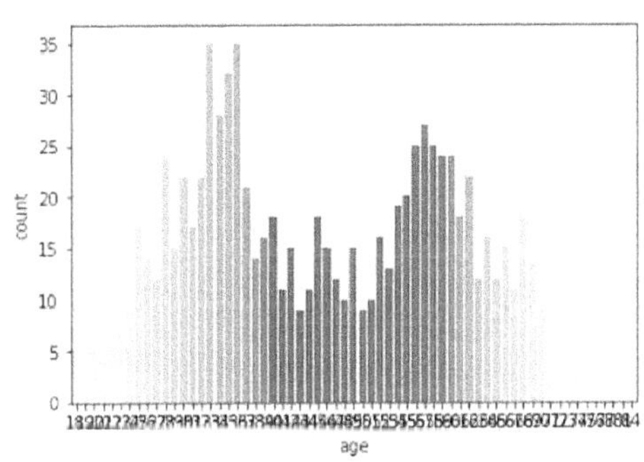

The above chart gives us an interesting finding—the members' ages are clustered around two peaks. Unfortunately, the default chart shows all the *x*-axis values, making them unreadable. As seaborn is built on Matplotlib, we can use the latter to refine our plot as follows:

```
[9]:  import matplotlib.pyplot as plt
      g = sns.countplot(data=dfmembers, x='age')
      for label in g.get_xticklabels():
          visible = int(label.get_text()) % 10 == 0
          label.set_visible(visible)
      g
```

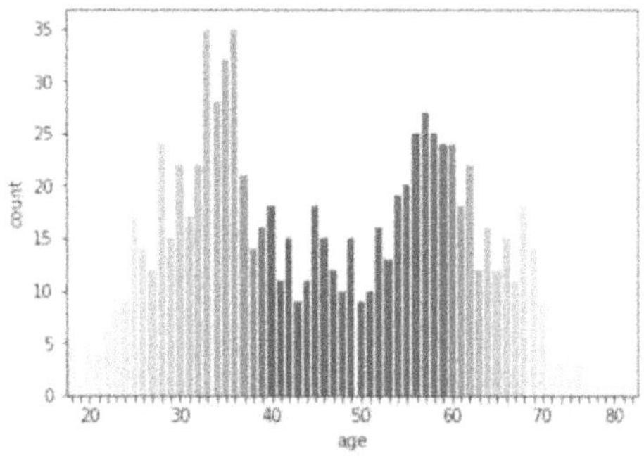

We can also smooth out the age distribution by using a histogram with the seaborn `displot()` function:

```
[10]:  sns.displot(dfmembers.age, bins=15)
```

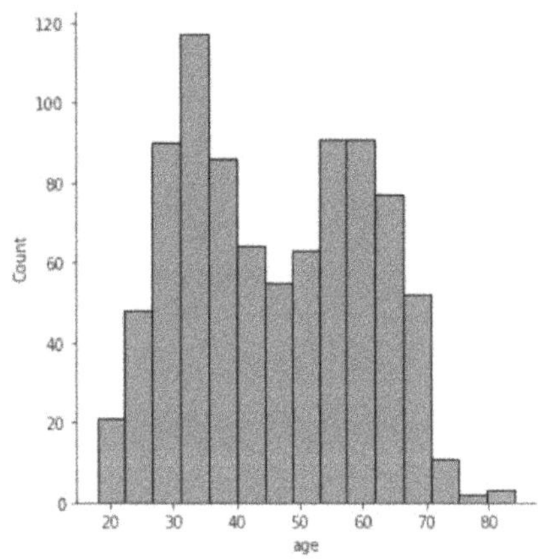

As the age profiles above show two age peaks and a corresponding dip in the middle, this may lead to a recommendation to focus on marketing to potential members in the 40–55 age range.

We may also be interested to see if there is a relationship between membership type (pool vs. gym) and age. For this we can use the FacetGrid type:

```
[11]: g = sns.FacetGrid(dfmembers, col='pool', height=5)
      g = g.map(sns.histplot, 'age')
```

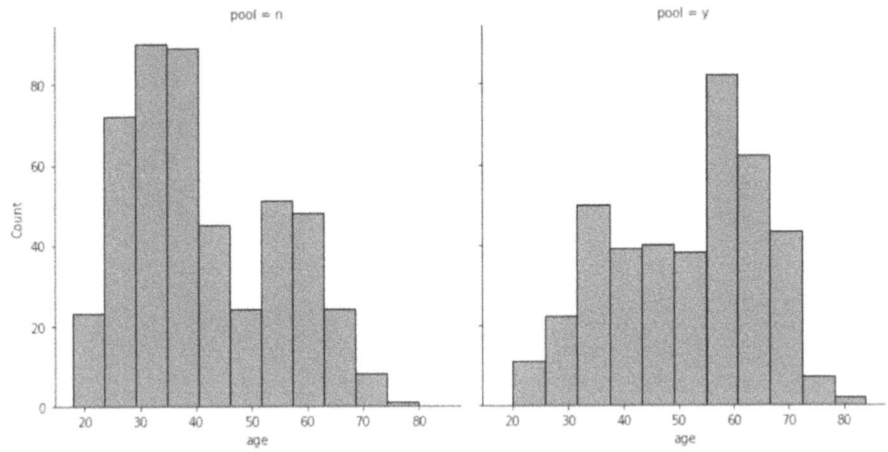

From the above charts, we see that pool members skew toward older members.

Another question we can check is whether any members join for both the gym and the pool. This can be done with a filter condition:

```
[12]: len(dfmembers[(dfmembers.gym == 'y') & (dfmembers.pool == 'y')])
      0
```

The result is 0, which tells us the club has been signing up members for either the gym or the pool. Although other clubs may bundle gym and pool memberships, this is not the case for our health club. This finding indicates an opportunity to provide a combination membership for both facilities.

The above charts give us a sense of the full dataset, including both inactive and currently active members. We may want to delve in further to see if there are any trends of interest. For example, has the age of members been changing over time? One way to view this is with a scatter chart of individual members showing their age when they joined:

```
[13]: sns.scatterplot(x=dfmembers.visitDate, y=dfmembers.age)
```

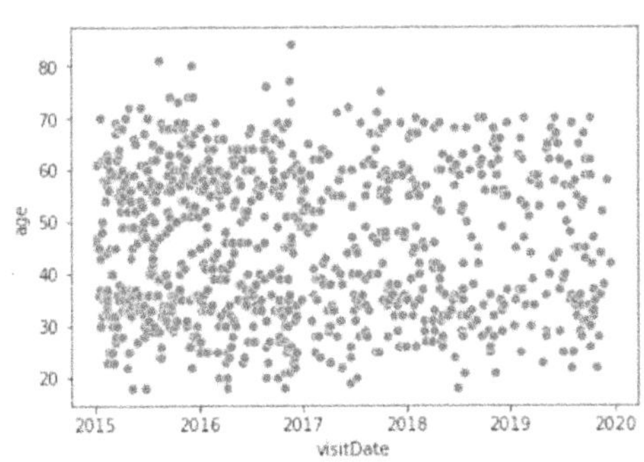

The above chart doesn't show an apparent change in age distribution of joining members over time. However, we see the chart become less dense over time, indicating that fewer people are joining over time.

To review the join rate of members over time, we'll use a few steps to transform and visualize the data. First, we'll introduce a new column for the year when each member joined using:

```
[14]: dfmembers['year'] = dfmembers.visitDate.dt.year
      dfmembers.head()
```

	guestID	visitDate	gender	age	married	pool	gym	status	endDate	year
0	6	2015-01-05	M	47	M	n	y	inactive	2017-08-01	2015
1	7	2015-01-05	F	46	S	n	y	inactive	2016-11-01	2015
2	10	2015-01-06	M	61	S	y	n	inactive	2016-12-01	2015
3	17	2015-01-09	F	36	S	y	n	inactive	2017-01-01	2015
4	24	2015-01-13	M	45	M	y	n	inactive	2016-09-01	2015

Next, we'll count the number of members who joined each year. One way to do this is using the pandas `pivot_table()` function, which provides a summary of our data. Pivot tables are covered in the next chapter, on "Statistics," in more detail. Here we use:

```
[15]: dfmembers.pivot_table(values='guestID', index='year', aggfunc='count')
```

	guestID
year	
2015	276
2016	225
2017	142
2018	128
2019	100

The decline in people joining over time is clear from the above table. We can further check if there is an additional factor—for example, if this trend differs between men and women. The `pivot_table()` function can break this down:

```
[16]: dfByGender = dfmembers.pivot_table(values='guestID', index='year',
                                         columns='gender', aggfunc='count')
      dfByGender
```

gender	F	M
year		
2015	128	148
2016	101	124
2017	38	104
2018	31	97
2019	11	89

To this table, we can add the percentage of women joining each year:

```
[17]: dfByGender['pctFemale'] = round(dfByGender.F /
                                (dfByGender.F + dfByGender.M) * 100)
      dfByGender
```

gender	F	M	pctFemale
year			
2015	128	148	46.0
2016	101	124	45.0
2017	38	104	27.0
2018	31	97	24.0
2019	11	89	11.0

We can also visualize the above table with:

```
[18]: sns.barplot(data=dfByGender, x=dfByGender.index, y='pctFemale')
```

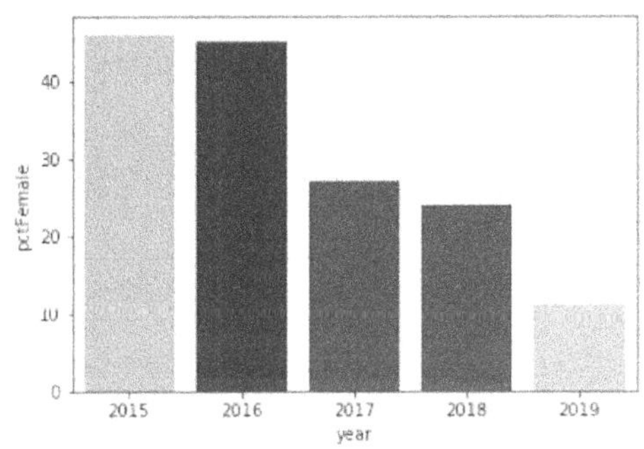

The decline of women as a percentage of new joiners is clear from the table and chart and may merit further investigation into potential reasons and areas for improvement.

Section Summary

In this section, we analyze membership data for a fictional health club to better understand the demographics and trends. In addition to using pandas for capturing the data, we introduced the seaborn visualization package, which can help us explore data with a few lines of code. In particular, for this dataset, we uncovered some membership demographics and trends that could lead to recommendations to improve membership for the health club.

Chapter Review

pandas provides a concise way to turn its DataFrame data into a variety of charts. It does this by using the Matplotlib package. For simple charts, you can use pandas alone without knowing Matplotlib. However, you can use Matplotlib statements to refine your charts if desired.

In this chapter, we visualized three datasets. For electric car data, with few statements, we generated different charts with pandas, including line, bar, and pie charts. In some cases, in order to properly visualize the data, we need to rearrange it—for example, to transpose the DataFrame or sort its rows.

In the next sections of the chapter, we looked at daily stock and index prices, and charted prices and returns. In addition to time series charts, we looked at statistical types of charts: histogram, KDE, and box plot. Finally, we generated a scatter chart to review the relationship between two variables.

In the last part of the chapter, we used visualization to analyze data from a fictional health club. Our approach here was to use exploratory data analysis to try to discover patterns and trends in the data. For this, seaborn is a good package to concisely visualize a dataset. In this chapter, we looked at bar charts, histograms, and a scatter plot as a few of the many visualizations you can create with the seaborn package.

Key Terms

adjusted closing price	For stocks, the adjusted closing price is a smoother time series than the closing prices, since it restates history to account for stock splits and dividends paid.
bin	A histogram is composed from the count of observations in each bin, which covers a range between two values.
kind	The pandas `plot()` method has a kind parameter to define what type of chart to generate.
legend	A legend indicates the color and or style within a chart to distinguish the different parts of the data.
plot()	The pandas `plot()` method calls the Matplotlib package to generate a chart.
symbol	Each stock has a unique symbol to distinguish it.
transpose	Rotate a table so that the rows become columns and the columns become rows.

Programming Problems

1) DJIA vs. a Stock

The Dow Jones Industrial Average (DJIA) is a stock market index.

a) Create a line chart that shows the DJIA adjusted price over time.
b) Create a line chart that shows the DJIA adjusted price as compared with Apple's stock price. For this chart, use dual *y* axes.
c) Compute the daily returns for the DJIA, and display these in a histogram.
d) Create a scatter chart showing the relationship of the DJIA daily returns to the daily returns of Apple's stock price.

2) Income Statements

A company has two divisions that have provided historical and project income statements. The data can be found in diva-incstmt.csv and divb-incstmt.csv. Create the following charts to visualize the data:

a) a stacked bar chart of historical and projected sales over time that includes both divisions' sales
b) a clustered bar chart of sales over time that includes both divisions' sales
c) two separate pie charts, one for each division, showing the different kinds of expenses forecasted for 2021

3) Tracking Infections

A county is tracking infections over time in order to watch for potential outbreaks. The data are included in countyhealth.csv and countypop.csv. For these data:

a) Create a line chart showing infections over time that includes all cities.
b) Create a line chart that shows each city's infections over time as a percentage of the city's population.
c) Create a stacked bar chart showing the total number of infections in each month.

4) Health Club Visits

The data file analyzed in Sections 16.7 and 16.8 related to people who became members of the health club. The club also has a data file (visitors.csv) that includes data for everyone who took a tour of the health club. This file includes data about people who joined as well as those who did not join. The health club has asked us to look further at these data to better understand the profile of their visitors as well as see if there are patterns of interest about people who visited but did not join. You may wish to review the charts created in 16.8 as a guide for potentially interesting charts to visualize the visitors data.

Chapter 17 μσ

Statistics

Chapter Objectives

In this chapter, you'll learn:

- how to calculate descriptive statistics using Python, NumPy, and pandas
- how to create pivot tables to summarize detailed data using pandas `pivot_table()`
- how to calculate and visualize time series moving averages with pandas
- selected SciPy functions for statistically comparing populations
- how to test if data appear to follow the normal distribution
- how to analyze data relationships with regression using SciPy and scikit-learn

Chapter Outline

17.1 Descriptive Statistics

Overview of descriptive statistics

In this section, we'll look at several descriptive statistics for both the center as well as the spread of numeric data. When you analyze a dataset, the most common questions of interest are how measures are centered as well as how data spread from their center. When there are no missing data, using simple Python statements is sufficient. As we'll see, pandas has some statistical methods that simplify getting descriptive measures on Series and DataFrames; these methods also simplify the handling of missing data.

For this section, we'll use a small dataset of test scores for a few students. Assume students have taken a midterm and final in a course. Their grades are:

Name	Midterm	Final
Stefanie	99	88
Adam	85	-
Becca	98	99
John	72	81
Jane	89	91

Table 17-1: Test score data

Note that there is one missing grade—Adam has not yet taken the final for the course.

Calculating the mean with Python and NumPy

Mean and median are two measures for the center of a set of data. First, let's look at using Python to calculate the means (averages) of the midterm and final scores described above. The mean is simply the total of all scores divided by the number of scores. Assume the midterms scores are in a list, as in:

```
[1]: midterms = [99, 85, 98, 72, 89]
```

Then the midterm mean can be calculated using built-in Python functions `sum()` and `len()`:

```
[2]: sum(midterms) / len(midterms)
     88.6
```

This will calculate the mean of the midterm scores as 88.6. However, if you use similar statements for the final scores:

```
[3]: finals = [88, None, 99, 81, 91]
     sum(finals) / len(finals)
     TypeError: unsupported operand type(s) for +: 'int' and 'NoneType'
```

then the second statement will generate a `TypeError`. This is because the `sum()` function doesn't handle the `None` value in the list.

As missing values are important to handle, let's look closer at this issue. Another convention for storing missing values is to use the special NumPy value `NaN` (not a number). This is in fact the approach that pandas uses to mark a numeric cell as missing a value. With this convention, we can define the final scores as:

```
[4]: import numpy as np
     finals = [88, np.nan, 99, 81, 91]
```

As with `None`, `np.nan` in the above list is a placeholder indicating there is no known value. The mean could be attempted to be calculated as:

```
[5]: sum(finals) / len(finals)
     nan
```

In this case, the above statement does not generate an error. However, the result will be a NumPy `nan`, which is still not a useful result.

Although you can work around this problem by filtering out missing values, a simpler approach is to use the NumPy function `nanmean()`, which will calculate the mean of all values that are not `nan`. In other words, the statement:

```
[6]:  np.nanmean(finals)
```

```
89.75
```

will give the desired result of 89.75.

Note that NumPy provides other `NaN`-aware functions: `nansum()`, `nanmin()`, and `nanmax()`.

Tip

About Missing Values

It is common for datasets to have missing values. In Python, you might represent these as the special Python value `None`. The approach in pandas is to represent missing values as the special NumPy value `NaN`. Further, pandas has methods like `isna()` and `dropna()`, to help find and filter out missing values. You should avoid using special values to represent missing values, like 0 or -999, as these can cause undetected incorrect errors in statistical results.

Calculating the mean and median with pandas

Previously we looked at calculating the mean using Python and also using NumPy. In this section, we'll explore a number of pandas methods for computing the mean as well as other descriptive statistics. As we'll see, the pandas descriptive statistics methods handle missing values. Further, data can be summarized across rows and/or columns of a DataFrame.

In order to explore pandas descriptive statistics, let's set up the tests scores in Table 17-1 as a DataFrame:

```
[7]:  # create a DataFrame of test scores
      import pandas as pd
      scores = pd.DataFrame({'midterm': [99, 85, 98, 72, 89],
                             'final':[88, None, 99, 81, 91]},
                             index=['Stefanie', 'Adam', 'Becca', 'John', 'Jane']
                            )
      scores
```

	midterm	final
Stefanie	99	88.0
Adam	85	NaN
Becca	98	99.0
John	72	81.0
Jane	89	91.0

The mean of the midterms can be calculated with:

```
[8]:  scores.midterm.mean()
```

```
88.6
```

The above statement works because `scores.midterm` is a pandas Series, and a pandas Series provides the `mean()` method to calculate the average of its values.

While mean is perhaps the most common measure for a set of values' central point, another measure is the median. The median is less sensitive to outliers and so sometimes preferred to mean for describing a data-set's center. The median is calculated by determining the value that has the same number of values below and above it. For example, if we order the values of the midterm scores:

```
72, 85, 89, 98, 99
```

we see 89 is the middle value, and so it is the median. This is what would be returned by the pandas method median:

```
[9]: scores.midterm.median()
```

```
89.0
```

For these data, the mean and median are close. As noted, the median is less sensitive to outlier values. For example, if the lowest midterm score were 50 instead of 72, the median would still be 89.

Let's look again at the final grades. Recall that there is one missing value. pandas represents this as a NumPy NaN, and the `mean()` and `median()` methods will ignore the missing values. So we can calculate the mean of the final grades with:

```
[10]: scores.final.mean()
```

```
89.75
```

What is the median of the final scores? In this case, pandas will omit the missing value, leaving the values:

```
88, 99, 81, 91
```

We can again find the median by ordering these values:

```
81, 88, 91, 99
```

In this case, we have an even number of values. The median is given by the average of the two middle values, in this case (91 + 88) / 2 = 89.5. This is the result given by:

```
[11]: scores.final.median()
```

```
89.5
```

To summarize, pandas can calculate the mean and median to describe the middle of a DataFrame column of numbers. For these calculations, pandas will ignore missing values.

So far, we've calculated the mean and median for each DataFrame column individually. If you'd like to calculate a mean or median for all columns, you can use the DataFrame `mean()` and `median()` methods. For example, given the scores DataFrame defined previously, the expression:

```
[12]: scores.mean()
```

```
midterm    88.60
final      89.75
dtype: float64
```

will calculate the mean score for both the midterm and final and store the results in the variable `means`. Further, the result is a pandas Series type variable, allowing the results to be labeled with the column headings midterm and final. Program 17-1 shows a complete program and output.

Program 17-1

```
1  # calculate the mean of each test across all students
2  import pandas as pd
3
4  # create a DataFrame of test scores
5  scores = pd.DataFrame({'midterm': [99, 85, 98, 72, 89],
6                         'final':[88, None, 99, 81, 91]},
7                         index=['Stefanie', 'Adam', 'Becca', 'John', 'Jane']
8                         )
9  # calculate the means for all columns
10 means = scores.mean()
11
12 # display the result
13 print(type(means))
14 print(means)
```

Program 17-1 Output

```
<class 'pandas.core.series.Series'>
midterm    88.60
final      89.75
dtype: float64
```

In Program 17-1, line 10, the result type of the `means` variable is a pandas Series, and its labels are taken from the DataFrame headings. In this way, pandas provides a labeled set of results from the `mean()` method.

Similarly, a pandas DataFrame can provide the medians for its numeric columns.

```
[13]: scores.median()

midterm    89.0
final      89.5
dtype: float64
```

As we've seen, the `mean()` and `median()` methods can summarize the columns of a DataFrame. It is also possible to use these methods to summarize each *row* of a DataFrame. In order to calculate the average of each student's midterm and final grades, the named parameter `axis` can be provided and set to the value `'columns'` (or equivalently, the value 1). Here we calculate the average of each student's midterm and final scores:

```
[14]: scores.mean(axis='columns')

Stefanie    93.5
Adam        85.0
Becca       98.5
John        76.5
Jane        90.0
dtype: float64
```

Calculating the variance and standard deviation with pandas

In addition to describing how the data are centered with mean and median, pandas also provides methods for describing how the data are spread out. In particular, the pandas method `var()` calculates the variance of a set of numbers and `std()` calculates the standard deviation.

Variance measures how spread out the data are from the mean. In particular, the steps for calculating variance for a set of numbers are:

1) Calculate the mean of the numbers.
2) Calculate the squared difference between each number and the mean.
3) Total the values in step 2.
4) Divide by the total number of values.*

For example, for the final exam scores, the variance would be:

$((88 - 89.75)^2 + (99 - 89.75)^2 + (81 - 89.75)^2 + (91 - 89.75)^2) / 4 = 41.6875$

The variance can be calculated by the pandas `var()` method. As with `mean()` and `median()`, the `axis` parameter indicates whether the calculation is across the rows or columns. Further, there is a parameter named `ddof` (delta degrees of freedom), which is set to 0 if the data represent the complete population (as in our test scores example), or we can omit this to use the default of 1 if the data are a subset of an overall population.

```
[15]: scores.final.var(ddof=0)
```

```
41.6875
```

Standard deviation is another measure of spread. It is calculated by taking the square root of variance. For the final exam scores, the standard deviation would be (to four decimal places):

sqrt(41.6875) = 6.4566

The standard deviation can be calculated using the pandas `std()` method. As with `var()`, the `std()` method has named parameters `axis` and `ddof`. For our test scores, we can compute the standard deviation of the two tests using:

```
[16]: scores.std(axis='rows', ddof=0)
```

```
midterm    9.850888
final      6.456586
dtype: float64
```

Other descriptive statistics methods available in pandas include `min()` and `max()`. To summarize the methods discussed in this section, Program 17-2 generates the mean, median, standard deviation, maximum, and minimum of the test scores, and consolidates these into a summary DataFrame.

* If the data are from a subset of a population of values, then set ddof=1. Statistical theory proves that in this case, this adjustment to the denominator provides a better estimate of the population variance.

Program 17-2

```
# calculate statistics of test scores
import pandas as pd

# create a DataFrame of test scores
scores = pd.DataFrame({'midterm': [99, 85, 98, 72, 89],
                       'final':[88, None, 99, 81, 91]},
                       index=['Stefanie', 'Adam', 'Becca', 'John', 'Jane']
                      )
# create summary matrix
summary = pd.DataFrame({'midterm':[], 'final':[]})

summary.loc['mean']   = scores.mean()
summary.loc['median'] = scores.median()
summary.loc['std']    = scores.std(ddof=0)
summary.loc['max']    = scores.max()
summary.loc['min']    = scores.min()
print(summary)
```

Program 17-2 Output

```
          midterm      final
mean    88.600000  89.750000
median  89.000000  89.500000
std      9.850888   6.456586
max     99.000000  99.000000
min     72.000000  81.000000
```

Spreadsheet Compare/Contrast
Descriptive Statistics

Several of the measures presented in this chapter are available in spreadsheets, including `average()`, `mode()`, `stdev.p()`, and `var.p()`. In general, the results between Python and spreadsheet descriptive statistics will match, but there are some differences. For example, if there is more than one most common value in a dataset, a spreadsheet will return only one of these values for the mode, while pandas will return all the most commonly occurring values (as a Series).

Categorical measures

So far in this section, we've focused on statistical measures—including mean, median, and standard deviation—that are appropriate for summarizing numerical data. For other types of data, including binary and category data, other statistics are appropriate, including counts and percentages. In this part, we'll look at a brief example using attendance data.

For the students, we also have a file with attendance for each date. We may be interested in summarizing each student's attendance, as well as the attendance on each date. We can load these data with:

```
[17]: attendance = pd.read_csv('attendance.csv', index_col='student')
      attendance
```

student	week1	week2	week3	week4
Stef	y	y	y	y
Adam	y	y	y	n
Becca	y	y	y	n
John	y	y	n	n
Jane	n	y	y	y

In order to calculate some basic attendance measures, we'll convert 'y' values to 1 and 'n' values to 0 using the pandas map() function. For example, to convert the first week's attendance, we could use:

```
[18]: attendance.week1.map({'y':1,'n':0})
```

```
student
Stef     1
Adam     1
Becca    1
John     1
Jane     0
Name: week1, dtype: int64
```

To do this conversion for all weeks, we can iterate over the columns using:

```
[19]: for i in range(1, 5):
          week = 'week' + str(i)
          attendance[week] = attendance[week].map({'y':1,'n':0})
      attendance
```

student	week1	week2	week3	week4
Stef	1	1	1	1
Adam	1	1	1	0
Becca	1	1	1	0
John	1	1	0	0
Jane	0	1	1	1

Once the data are restated in binary, we can easily get the attendance percentages using:

[20]:
```
attendance.sum(axis='columns') / len(attendance.columns) * 100
```
```
student
Stef     100.0
Adam      75.0
Becca     75.0
John      50.0
Jane      75.0
dtype: float64
```

[21]:
```
attendance.sum(axis='rows') / len(attendance) * 100
```
```
week1     80.0
week2    100.0
week3     80.0
week4     40.0
dtype: float64
```

Section Summary

In this section, we looked at several approaches for calculating descriptive statistics in Python. These include measures for the data's center (mean, median) as well as measures for the data's spread (variance, standard deviation). In general, pandas provides a unified approach for calculating descriptive statistics, as it can handle row-wise and column-wise calculations, as well as missing values.

Test Your Understanding

1) The SciPy function `nanmean()` will ignore values of `None` when calculating the mean of a set of numbers.
   a) True
   b) False
2) The pandas descriptive statistics methods (such as `mean()` and `std()`) will ignore missing values when calculating results.
   a) True
   b) False
3) For a DataFrame, pandas can only calculate column-wise statistics.
   a) True
   b) False

17.2 Summarizing Data with pandas Pivot Tables

In the previous section, we covered how to calculate descriptive statistics such as mean, median, variance, and standard deviation. In this section, we'll expand on this by seeing how you can partition and describe a detailed dataset using pandas pivot tables. Pivot tables are summary tables that can include counts, subtotals, means, and other measures based on a potentially large dataset of detailed data. As we'll see, pivot tables are easy to create in pandas and can give insight into patterns or trends in a dataset. If you've used pivot tables in spreadsheets, you'll see that the features available with pandas are similar and powerful.

For this section, we'll use sales data for a fictitious company that sells four products in four regions. Their sales system captures each transaction, including the time of purchase, the product ID, the region the customer is in, and the number of items purchased. Our data file has over 9,000 transactions for the year 2021. The dataset is multi-dimensional (time, product, region) with one quantitative measure (number of items purchased in each transaction).

Given these data, there are a number of questions we might be interested in. For example:

- How many units were sold in the past year for each product?
- What are the overall quarterly sales trends?
- What are the sales trends for each product?
- Are there regional differences or trends for product sales?

Pivot tables allow us to answer these questions by generating summary reports based on the detailed data. The pandas `pivot_table()` method takes a DataFrame table of detailed data and, based on the parameters we provide, provides as output a DataFrame summary table.

For all the examples in this section, we'll use the `pivot_table()` method. We can load the detailed sales data into a DataFrame using:

```
[1]: import pandas as pd
     sales = pd.read_csv('sales.csv', parse_dates=['time'])
```

The sales data contain time of purchase, product ID, region where purchased, and number of items purchased. We can see the first few rows with `head()`:

```
[2]: sales.head()
```

	time	product	region	count
0	2021-01-01 01:28:26	110	N	1
1	2021-01-01 04:22:06	110	N	3
2	2021-01-01 06:21:05	110	N	1
3	2021-01-01 09:10:33	110	N	2
4	2021-01-01 10.39.22	110	N	0

The first question we'll answer is, What were the total unit sales for each of the products during the year? We can use the following statement:

```
[3]: sales.pivot_table(values='count', index='product', aggfunc='sum')
```

	count
product	
110	4801
120	3305
130	2958
140	2058

In the above statement, the setting `values='count'` indicates what column of our sales data we'd like to summarize in our pivot table result. The setting `index='product'` indicates that our result rows will be based on the distinct products in the detailed data. Last, the setting `aggfunc='sum'` requests that the count data be totaled. The result is a DataFrame showing the unit sales for each product during the period the detailed data covers.

The `aggfunc` parameter can be used to request other summary measures. For example, if we were interested in the average number of units purchased at each sale, we could set `aggfunc='mean'`:

```
[4]: pd.set_option('display.precision', 1)
     pivot = sales.pivot_table(values='count', index='product', aggfunc='mean')
     pivot.columns = ['Avg # per Transaction']
     pivot
```

	Avg # per Transaction
product	
110	1.0
120	2.2
130	2.0
140	1.4

In the code above, we set the `columns` property to give the result a more descriptive column heading. Based on the table above, we see products 120 and 130 have the higher average counts per transaction as compared with products 140 and 110.

Another question of interest is, How have sales evolved over time? To answer this, we'll compute a new column, quarter, based on the time column, in order to look at our sales on a quarterly basis.

```
[5]: sales['quarter'] = sales['time'].dt.to_period('Q')
     sales.head()
```

	time	product	region	count	quarter
0	2021-01-01 01:28:26	110	N	1	2021Q1
1	2021-01-01 04:22:06	110	N	3	2021Q1
2	2021-01-01 06:21:05	110	N	1	2021Q1
3	2021-01-01 09:10:33	110	N	2	2021Q1
4	2021-01-01 10:39:22	110	N	0	2021Q1

The statement above creates a new column in the `sales` DataFrame using the `to_period()` method, which creates a value indicating in what quarter each sale took place. We could choose another time period—for example, monthly—but we decide that quarterly is a good granularity for the purposes of this analysis, and this is denoted by the use of `'Q'` as the argument to the `to_period()` method. Other frequencies are possible—for example, `'M'` for monthly or `'W'` for weekly.

Now that we've augmented our data with the quarter column, we can look at how overall sales have evolved over time using `columns='quarter'`:

```
[6]: sales.pivot_table(values='count', columns='quarter', aggfunc='sum')
```

quarter	2021Q1	2021Q2	2021Q3	2021Q4
count	3326	3240	3400	3156

In the results above, we see the lowest sales volume in the last quarter. To look closer, we can generate a two-dimensional report to review sales by product over time. The statement for this is:

```
[7]:  sales.pivot_table(values='count', index='product', columns='quarter',
                        aggfunc='sum')
```

quarter	2021Q1	2021Q2	2021Q3	2021Q4
product				
110	1357	1293	1173	978
120	824	801	833	847
130	727	690	830	711
140	418	456	564	620

Since both `index` and `columns` are specified in the statement above, the result is a table with product rows and quarterly columns.

From the above table, we see that sales in product 110 have been declining each quarter, whereas sales in product 140 have been growing, although from a smaller starting point.

As the above result is a DataFrame, we can easily visualize it as a chart, to see the trends more clearly. For example, we can create a clustered bar chart using:

```
[8]:  %matplotlib inline
      pivot = sales.pivot_table(values='count', index='product', columns='quarter',
                                aggfunc='sum')
      pivot.plot(kind='bar')
```

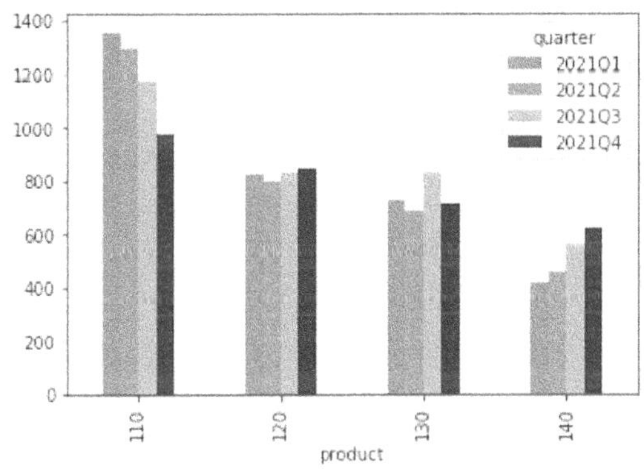

The sales dynamics over time can be reviewed further by including the regions in our report. For example, we can break out our product sales by region by including a list of index dimensions with `index=['region', 'product']`. The following statement gives us a three-dimensional summary report (flattened into two dimensions). Note that we include the totals of each row and column by using the setting `margins=True`:

```
[9]: sales.pivot_table(values='count', index=['region','product'],
                      columns='quarter', aggfunc='sum', margins=True)
```

	quarter	2021Q1	2021Q2	2021Q3	2021Q4	All
region	product					
E	110	183	167	132	189	671
	120	190	211	195	193	789
	130	180	200	210	191	781
	140	81	107	169	181	538
N	110	817	726	695	532	2770
	120	236	237	293	345	1111
	130	195	163	241	191	790
	140	127	104	146	164	541
S	110	176	231	166	127	700
	120	174	206	133	120	633
	130	177	175	216	151	719
	140	105	131	119	123	478
W	110	181	169	180	130	660
	120	224	147	212	189	772
	130	175	152	163	178	668
	140	105	114	130	152	501
All		3326	3240	3400	3156	13122

Tip
About Pivot Tables, Filtering, and Joining

In the examples in this section, we have summarized an entire DataFrame. However, if you want to summarize a subset of the detailed data, you can use a pandas filter expression to create a subset of rows of interest. Further, if you need additional data beyond your core detailed DataFrame, you can use pandas to augment the data by joining two or more datasets. That pivot tables use a DataFrame as input means you have a variety of tools in pandas to create the detailed data of interest to then summarize using pivot tables.

Section Summary

The pandas `pivot_table()` method is a flexible and concise way to create summary tables from detailed data. Using this method's `values` and `aggfunc` parameters, you can set which data to aggregate, and how to aggregate it. Further, the layout of the summary table is controlled by the `index` and `columns` parameters. As the `pivot_table()` result is a DataFrame, you can further process the results—for example, using `plot()`—to visualize the results.

Test Your Understanding

1) The `pivot_table()` method can generate summary results that total values from a detail DataFrame but does not support other summaries such as averages.
   a) True
   b) False

2) A named parameter for the `pivot_table()` method is:
   a) index
   b) columns
   c) aggfunc
   d) all of the above

17.3 Analyzing Time Series with pandas Moving Statistics

In the previous section, we looked at product sales over time. In this section, we'll look further at time series, and how pandas can summarize data series over a window of time. When the summary statistic for a window is the mean, we refer to this as the moving average of the time series. Other summary measures over windows of time such as standard deviation can be calculated as well. Statistics calculated for moving time windows are one way to see a smoothed-out picture of individual data points in order to more clearly see trends over time.

Moving averages

We'll look first at moving averages by considering a quality control question. A manufacturer would like the weight of each product to be at least 10 ounces, as stated on the product's box. If the actual product weight is below 10 ounces, this is an issue, since the customer is not getting as much as the packaging indicates. Since there is some variability in the machine that fills each product's box, the machine is set to fill 10.1 ounces. To keep watch over the product weight, the manufacturer has decided to calculate the moving average of the weights and monitor if the 10-point moving average moves below 10 ounces. The 10-point moving average is the average of the last 10 observed weights.

Example data are in the file measures2.csv, which we can load in a DataFrame:

```
[1]: import pandas as pd
     measures = pd.read_csv('measures2.csv', index_col='#')
     measures.head()
```

#	measure
1	10.11
2	9.85
3	10.09
4	10.03
5	10.51

We can see a time series plot of the data using:

```
[2]: measures.plot()
```

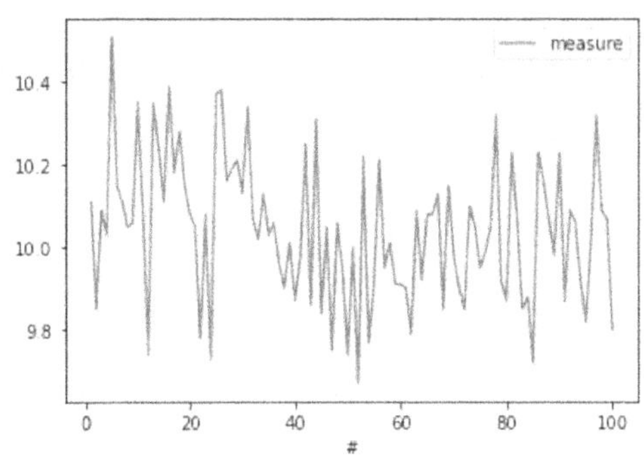

We can calculate the 10-measure moving average using the statement:

```
measures['measure'].rolling(10).mean()
```

and review the first few results using:

```
[3]:  measures['measure'].rolling(10).mean().head(12)
```

```
#
1          NaN
2          NaN
3          NaN
4          NaN
5          NaN
6          NaN
7          NaN
8          NaN
9          NaN
10      10.131
11      10.129
12      10.118
Name: measure, dtype: float64
```

Note that the first 9 values of the moving average result are NaN, since pandas requires 10 data points for a 10-item moving average.

Recall that our product's packaging indicates 10 ounces, below which we don't want our product weight to fall below. We can plot the moving average, including our 10-ounce watch level, using:

```
[4]:  measures['ten'] = 10
      measures['ten'].plot()
      measures['measure'].rolling(10).mean().plot()
```

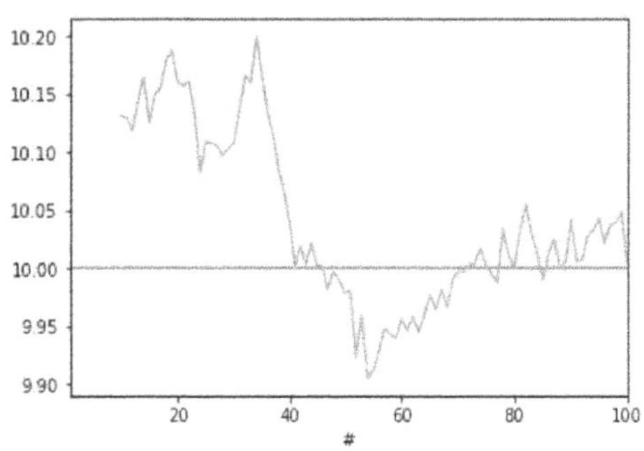

In the chart above, we can see the moving average dropped below the 10-ounce watch level soon after the 40th product measurement.

Moving standard deviation

In the previous section, we looked at moving averages by using `mean()` together with the `rolling()` method. Several other statistics can be calculated with rolling windows of data, including `sum()`, `median()`, `min()`, and `max()`, as well as `var()` for variance and `std()` for standard deviation. In this section, we'll look at the volatility of a stock market index by calculating its daily return moving standard deviation. We'll use data from the hypothetical stock market index USIX. Volatility is one measure of stock price movement that investors may watch to help with their investment decisions.

We can load the data available in the file USIX.csv using the statement:

[5]:
```
usix = pd.read_csv('USIX.csv', index_col='Date', parse_dates=True)
usix.head()
```

Date	Open	High	Low	Close	AdjustedClose	Volume
2017-01-03	450.08	451.66	447.76	450.48	412.295	182733000
2017-01-04	451.24	453.50	451.22	453.16	414.748	157488800
2017-01-05	452.54	453.16	450.96	452.80	414.418	156758000
2017-01-06	453.06	455.50	451.80	454.42	415.901	143119800
2017-01-09	453.82	454.14	452.84	452.92	414.528	93879400

For our analysis, we decide to review the daily price returns. The following statement narrows the data to adjusted closing price, and adds the daily return calculation:

[6]:
```
usix = usix[['AdjustedClose']]
usix['Return'] = usix['AdjustedClose'].pct_change()
usix.head()
```

Date	AdjustedClose	Return
2017-01-03	412.295	NaN
2017-01-04	414.748	0.005950
2017-01-05	414.418	-0.000796
2017-01-06	415.901	0.003579
2017-01-09	414.528	-0.003301

With the daily returns calculated, we can plot the 20-day rolling standard deviation using:

[7]:
```
usix['Return'].rolling(20).std().plot()
```

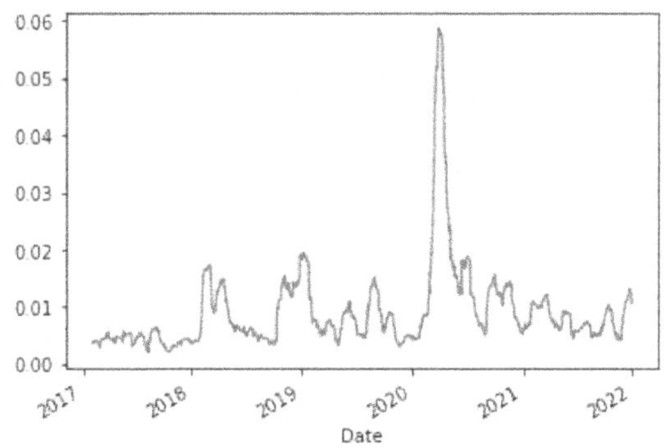

In the above chart, we can see a marked increase in volatility during the middle of 2020 followed by a return to the earlier levels of volatility.

Section Summary

Moving averages and standard deviations are easy to calculate using the pandas `rolling()` method. As these results are Series, they can be readily appended as a column in a DataFrame. In combination with charting, the rolling statistics can give a clearer picture of the dynamics of a measure over time, showing a smooth picture of measures such as mean or standard deviation.

Test Your Understanding

1) The DataFrame method for generating moving averages is called:
   a) `moving()`
   b) `rolling()`
   c) `movavg()`
   d) none of the above

2) In addition to moving averages, pandas can generate moving standard deviations.
   a) True
   b) False

17.4 Comparing Populations

In this section, we'll look at a few statistical tests that can help us check for differences between sets of observations. In certain data, you may be interested in whether there were meaningful differences between data subsets—for example, if earlier vs. later measures were materially different or if there are significant differences between different demographic groups. For example, we can use a statistical test to confirm that the mean product weights reviewed in the prior section were materially different between the early and later measurements. This could be of interest to see if a setting on the product filling machine needs to be adjusted. Further, we can check if the variances of these samples are nearly the same or materially different.

Depending on what is being compared and what assumptions are made, there are different statistical tests. For example, comparing discrete (categorical) measurements between two groups would involve a different test than comparing continuous (float) measurements. For our product weights, we have continuous measures (rounded to two decimal places), and we'll assume the observations are independent and are governed by the normal distribution.

Comparing population means

If we'd like to check if the early product weights have a mean that is statistically significantly different from the later product weights, we can use a t-test. One Python package providing this test is SciPy, which has the function `ttest_ind()`. We provide this function with two sets of values (the two sets of values whose means we want to compare), and it will return a ***p-value***. The p-value is a value calculated by a statistical test to determine the probability that the sample data are consistent with a hypothesis. In this case, if that statistic is less than a certain value (e.g., 0.05 for 95% confidence), we can conclude the two populations' means are different.

In order to perform this test, we will set up the data with the following statements, as we did in the prior section:

```
[1]: import pandas as pd
     measures = pd.read_csv('measures2.csv', index_col='#')
     measures.head()
```

#	measure
1	10.11
2	9.85
3	10.09
4	10.03
5	10.51

Based on a visual review of the moving average in the prior section, we decide to divide our sample of 100 observations into two populations, the first 35 observations and the last 60 observations, using:

```
[2]: early = measures.measure[0:35]
     later = measures.measure[40:]
```

Note that `early` and `later` are pandas Series:

```
[3]: type(early)

     pandas.core.series.Series
```

Now we can statistically compare the earlier and later measurements using the t-test:

```
[4]: import scipy.stats as stats
     result = stats.ttest_ind(early, later)
     round(result.pvalue,4)

     0.0002
```

The p-value result is 0.0002, which is less than 0.05, so we can conclude with 95% confidence that the two subsets of measures have a different mean. In other words, something seems to have happened to our product filling machine that requires adjustment.

Comparing population variances

Continuing with the product weight data, we want to determine if the measurements early on have a different variance than the later measures. If the variances are similar, we decide to make a simple adjustment to the product filling machine. However, if the variances are different, we will want to have a technician service the machine.

There are several statistical tests for comparing variances. One common test is the F-test, which we will use. First, we determine the variances of the early and later measurements using the pandas `var()` method:

```
[5]: earlyVar = early.var()
     round(earlyVar,4)
```
```
0.0323
```

```
[6]: laterVar = later.var()
     round(laterVar,4)
```
```
0.025
```

Although these values are different, we want to check if they are different enough based on their sample size and a level of confidence. In order to check these, we calculate the F statistic as the ratio of the larger variance divided by the smaller variance:

```
[7]: F = earlyVar/laterVar
     round(F,4)
```
```
1.2904
```

Next, we calculate the p-value using the cumulative density function of the F distribution:

```
[8]: pvalue = 1 - stats.f.cdf(F, len(early)-1, len(later)-1)
     round(pvalue, 4)
```
```
0.1926
```

In this case, we are performing a one-sided test: In other words, is the variance of the earlier measures greater than the variance of the later measures? For 95% confidence, we check if the p-value is greater than 0.05; since it is, we conclude the early and later variances are not statistically different.

The F-test for comparing variances assumes our data are normally distributed. There are other tests that do not assume normality—for example, the Fligner-Killeen test, which is available in the SciPy function `fligner()`.

As SciPy doesn't have a convenient function to perform an F-test for comparing variances, we provide one in Program 17-3.

Program 17-3

```
# F-test to compare 2 variances

from numpy import var
from scipy.stats import f

#define F-test function
def ftest(x, y):
    xvar = var(x, ddof=1) # variance of x
    yvar = var(y, ddof=1) # variance of y
    if xvar > yvar:
        ratio = xvar/yvar
        p = 1-f.cdf(ratio, len(x)-1, len(y)-1) #find p-value using F statistic
    else:
        ratio = yvar/xvar
        p = 1-f.cdf(ratio, len(y)-1, len(x)-1) #find p-value using F statistic
    return xvar, yvar, ratio, p

import pandas as pd
df = pd.read_csv('measures2.csv')
early = df.measure[0:35]
later = df.measure[40:]

#perform F-test
xvar, yvar, ratio, pvalue = ftest(early, later)
print('xvar:   ', round(xvar,4))
print('yvar:   ', round(yvar,4))
print('ratio:  ', round(ratio,4))
print('pvalue:', round(pvalue,4))
```

Program 17-3 Output

```
xvar:   0.0323
yvar:   0.025
ratio:  1.2904
pvalue: 0.19264
```

Section Summary

There are a variety of approaches for statistically comparing populations. In studies or analyses, you may want to check for meaningful differences between subsets of data in order to draw conclusions or make recommendations, and this can be supported with statistical tests. In this section, we looked at how to compare continuous measures to see if two populations are similar in mean and variance using functions available in the SciPy package. There are other approaches to comparing populations, which you may want to explore depending on the types of data you need to compare.

Test Your Understanding

1) A package that can perform the t-test is:
   a) pandas
   b) NumPy
   c) SciPy
   d) none of the above

2) One test for comparing population variances is:
   a) the popvar test
   b) the F-test
   c) the t-test
   d) all of the above

17.5 Confidence Intervals

In the previous section, we compared two sets of measurements in terms of their sample mean and variance. Another statistical question we may want to ask is, Given a sample statistic such as mean, what is the range around that statistic that captures the true underlying mean with some level of confidence? For example, for the weight measurements in the prior section, while we can measure the sample mean, we do not know what the true underlying mean is for product weights as related to the filling machine. But based on the sample measurements, we can calculate a ***confidence interval,*** a pair of statistics that brackets the true mean with a certain statistical confidence level. The confidence interval is based on the mean of the sample, the number of observations, the variation of the measurements, and the desired level of confidence—for example, 95% confidence.

Let's calculate a confidence interval for the true weight mean based on the first five weight observations in our data. The data and their mean are:

```
[1]:  import numpy as np
      # the first 5 weight measures
      data = [10.11, 9.85, 10.09, 10.03, 10.51]
      mean = np.mean(data)
      round(mean,3)

      10.118
```

In order to determine a confidence interval for the product weight mean, we can use SciPy's `t.interval()` function:

```
[2]:  import scipy.stats as stats
      confint = stats.t.interval(alpha=0.95, df=len(data)-1,\
                                 loc=mean, scale=stats.sem(data))
      confint

      (9.817630212189776, 10.41836978781022)
```

Above, alpha is set to 0.95, indicating we are seeking a 95% confidence level. The df (degrees of freedom) parameter is set to the number of observations minus 1. The scale parameter is set to the standard error of the mean using SciPy's `sem()` function. This measure (an abbreviation for "standard error of mean") is the sample standard deviation divided by the square root of the number of observations.

The result returned by `interval()` includes the lower and upper bounds of our confidence interval. In this case we are 95% confident the confidence interval (9.818, 10.418) brackets the true mean.

Returning to our product weight measures, we are interested in the confidence intervals of our early measurements and how this compares with the confidence intervals of our later measures. To streamline this comparison, here is a function that calls `interval()`:

```
[3]:  def confinterval(x, conf=.95, decimals=None):
          result = stats.t.interval(alpha=conf, df=len(x)-1,
                                    loc=np.mean(x), scale=stats.sem(x))
          if decimals != None:
              result = (round(result[0], decimals), round(result[1], decimals))
          return result
```

The `confinterval()` function above takes as its first parameter, `x`, the data. The named parameter `conf` is the desired confidence, set to 95% by default. The function also allows us to provide a desired number of decimals to round the result to using the `decimals` parameter.

We'll now load our measurements, as we did in the prior section, using:

```
[4]: import pandas as pd
     measures = pd.read_csv('measures2.csv', index_col='#')
     measures.head()
```

	measure
#	
1	10.11
2	9.85
3	10.09
4	10.03
5	10.51

Again, we'll divide our measurements between early and later measures and convert these into lists using:

```
[5]: early = measures.measure[0:35].tolist()
     later = measures.measure[40:].tolist()
```

With the `confinterval()` we defined earlier, we can calculate the confidence interval for the underlying mean using the early measurements with:

```
[6]: confinterval(early, decimals=3)
```

```
(10.069, 10.193)
```

This indicates that we are 95% confident that the confidence interval (10.069, 10.193) brackets the underlying population mean.

As mentioned, the confidence interval depends on the statistical level of confidence. For example, we can see different confidence levels for the later measurements using:

```
[7]: for conf in (.9, .95, .99):
         print(conf, confinterval(later, conf=conf, decimals=3))
```

```
0.9 (9.962, 10.03)
0.95 (9.955, 10.037)
0.99 (9.942, 10.051)
```

Section Summary

We can calculate a confidence interval to bracket a statistical measure with some level of statistical confidence. For example, the SciPy function `t.interval()` can be called to calculate the confidence interval for the true underlying mean. The confidence interval will depend on the data: the number of observations, mean, variance, and level of confidence.

Test Your Understanding

1) The confidence interval for a mean is based on:
   a) number of observations
   b) median of the data
   c) number of missing values
   d) all of the above

2) The SciPy `t.interval()` function has which named parameter?
   a) alpha
   b) df
   c) scale
   d) all of the above

17.6 Reviewing a Distribution and Identifying Outliers

In the previous sections, we've looked at summary measures of data, such as mean, variance, and standard deviation. In this section, we'll look at the distribution of data both visually and statistically. Further, we'll identify outliers in the data that are in the tails of the distribution. Returning to the stock market data from Section 17.3, we'll look at the daily price returns of the stock market index USIX and check if it follows a normal distribution. The normal distribution often describes data that follow a bell-shaped distribution, and many approaches have been developed to work with data that are normally distributed.

Are observations normally distributed?

We'll load the daily index price data in USIX.csv, similar to the initial steps in Section 17.3:

```
[1]: import pandas as pd
     ticker = 'USIX'
     filename = ticker + '.csv'
     df = pd.read_csv(filename, index_col='Date', parse_dates=True)
```

Note that in the above, we set the ticker variable so that it is easy to change the analysis to another index or individual stock. Related, we use the more general variable name df instead of usix (as was used in Section 17.3).

We'll focus on the adjusted closing prices and their returns:

```
[2]: df = df[['AdjustedClose']]
     df['Return'] = df['AdjustedClose'].pct_change()
     df.head()
```

	AdjustedClose	Return
Date		
2017-01-03	412.295	NaN
2017-01-04	414.748	0.005950
2017-01-05	414.418	-0.000796
2017-01-06	415.901	0.003579
2017-01-09	414.528	-0.003301

Since the first return value is missing, we'll remove the first row of our data using:

```
[3]: df = df[1:]
```

We can visually review a dataset's distribution using a histogram. In pandas, we can use `plot()` with the `kind` set to `'hist'`:

```
[4]: df['Return'].plot(kind='hist', bins=50)
```

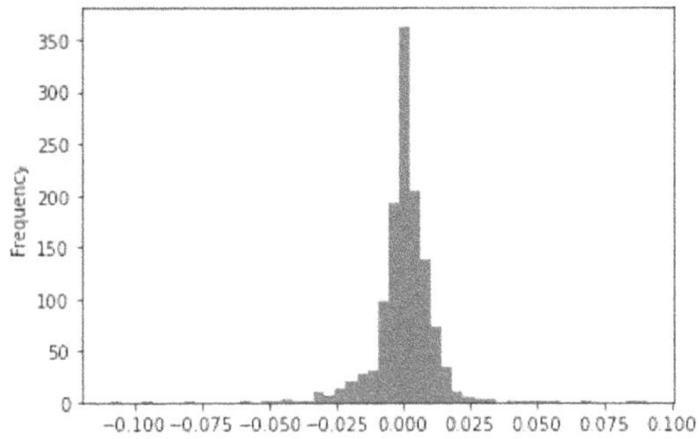

In the above plot, we use `bins=50` to override the default of 10 partitions in our chart, since we have a large number of data points. The result above has a bell-shaped curve that is similar to the shape of the normal distribution.

Beyond the visual check above, we may want to confirm whether the data appear to follow a normal distribution. This can be done by checking the ***skewness***, which is a statistic related to whether the data lean one way or another. We can also check the ***kurtosis***, which checks whether the extreme parts of the distribution, or tails, have more or less points than expected.

pandas has the `skew()` and `kurt()` methods to calculate these values, which can be found with:

```
[5]: print('Skewness=', round(df.Return.skew(),2))
     print('Kurtosis=', round(df.Return.kurt(),2))

     Skewness= -0.79
     Kurtosis= 18.24
```

These values can be compared to what is expected for a normal distribution, based on a degree of confidence and the number of observations. These checks are available in NumPy, and we can use these statements:

```
[6]: import scipy.stats as stats
     print('Skewness test p- value:',round(stats.skewtest(df.Return).pvalue, 4))
     print('Kurtosis test p- value:',round(stats.kurtosistest(df.Return).pvalue, 4))

     Skewness test p-value: 0.0
     Kurtosis test p-value: 0.0
```

For our data, the p-values to four decimal places are 0. For a 95% confidence test, since these p-values are less than 0.05, we conclude that the distribution of our returns is not following a normal distribution.

Checking for outliers

Although our data are not normally distributed, we can still use their standard deviation to identify outliers—in this case, very large daily movements in price that are in the tails of our distribution. The outliers could be related to events that we may want to understand further. For selecting outliers, we'll use the standard deviation of daily price returns and, in particular, check for those returns that exceed three standard deviations. We can do this in pandas using the `std()` method:

```
[7]: sd = df.Return.std()
     round(sd, 3)
```

```
0.012
```

We can then use this in a filter expression—for example, to find the largest price returns as:

```
[8]: df[df.Return > (sd*3)]
```

Date	AdjustedClose	Return
2018-12-26	468.112	0.050525
2020-03-02	598.865	0.043308
2020-03-04	606.169	0.042034
2020-03-10	558.816	0.051744
2020-03-13	521.810	0.085487
2020-03-17	489.802	0.053993
2020-03-24	473.875	0.090601
2020-03-26	509.053	0.058390
2020-04-06	516.186	0.067166

and the smallest returns as:

```
[9]: df[df.Return < (-sd*3)]
```

Date	AdjustedClose	Return
2018-02-05	492.508	-0.041823
2018-02-08	480.752	-0.037508
2020-02-27	576.428	-0.044912
2020-03-09	531.323	-0.078094
2020-03-11	531.575	-0.048748
2020-03-12	480.715	-0.095678
2020-03-16	464.711	-0.109425
2020-03-18	465.002	-0.050633
2020-03-20	445.909	-0.043093
2020-04-01	479.722	-0.045004
2020-06-11	585.859	-0.057650

We can combine the smallest and largest returns into one result using:

```
[10]: outliers = df[(df.Return < (-3*sd)) | (df.Return > (3*sd))]
```

The outliers DataFrame has rows corresponding to smallest and largest daily returns. We can visualize these using:

```
[11]: import matplotlib.pyplot as plt
      plt.scatter(x=outliers.index, y=outliers.Return)
      plt.xticks(rotation = 45)
      plt.show()
```

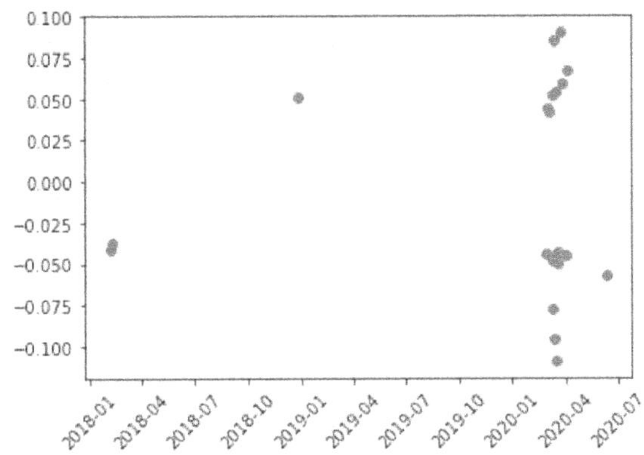

From the chart, it is clear that most large moves in the index occurred in the first quarter of 2020, with a few other large moves occurring in other periods.

Section Summary

In certain cases, we may want to review the distribution of data—for example, to check for a normal distribution. We can do this visually with a histogram. We can also use statistical tests to confirm if data seem to be governed by a particular distribution. For example, SciPy provides functions to check if data have skewness and kurtosis consistent with a normal distribution. In this section, we also looked for outliers that fell outside a number of standard deviations from the mean.

Test Your Understanding

1) Statistics that can be used to check if data follow a normal distribution are:
   a) mean and variance
   b) standard deviation and variance
   c) skewness and kurtosis
   d) all of the above

2) One approach for finding outliers in a dataset is to use:
   a) a filter based on number of standard deviations
   b) skewness and kurtosis
   c) a t-test
   d) the p-value

17.7 Analyzing a Relationship with Simple Regression

In this section, we'll look at fitting a simple regression line between two sets of data. In particular, we'll take the returns of the of a hypothetical stock market index (symbol USIX) as an independent variable, and see if they are related to the returns of a particular stock, Orange Technologies (symbol ORNG). In financial theory, the returns of the market are expected to have a relationship with the individual stocks in that market.

We can use the USIX index data described earlier in Section 17.3. Similarly, we can create a DataFrame of a stock's prices:

```
[1]: import pandas as pd
     stockSymbol = 'ORNG'
     stock = pd.read_csv(stockSymbol+'.csv', index_col='Date', parse_dates=True,
                         usecols=['Date', 'AdjustedClose'])
     stock.columns = ['Stock_Adj_Close']
     stock.head()
```

Date	Stock_Adj_Close
2017-01-03	54.595
2017-01-04	54.534
2017-01-05	54.812
2017-01-06	55.423
2017-01-09	55.930

```
[2]: marketSymbol = 'USIX'
     market = pd.read_csv(marketSymbol+'.csv', index_col='Date', parse_dates=True,
                          usecols=['Date', 'AdjustedClose'])
     market.columns = ['Market_Adj_Close']
     market.head()
```

Date	Market_Adj_Close
2017-01-03	412.295
2017-01-04	414.748
2017-01-05	414.418
2017-01-06	415.901
2017-01-09	414.528

Though not required, it will be convenient to consolidate the stock and index data into one DataFrame. We can do this with:

```
[3]: df = pd.concat([stock, market], axis=1, join='inner')
     df.head()
```

Date	Stock_Adj_Close	Market_Adj_Close
2017-01-03	54.595	412.295
2017-01-04	54.534	414.748
2017-01-05	54.812	414.418
2017-01-06	55.423	415.901
2017-01-09	55.930	414.528

Next, we calculate the daily percentage changes, and add these to the DataFrame:

```
[4]: df['Stock_Return'] = df['Stock_Adj_Close'].pct_change()
     df['Market_Return'] = df['Market_Adj_Close'].pct_change()
     df = df[1:] # drop the first row (return is NaN)
     df.head()
```

Date	Stock_Adj_Close	Market_Adj_Close	Stock_Return	Market_Return
2017-01-04	54.534	414.748	-0.001117	0.005950
2017-01-05	54.812	414.418	0.005098	-0.000796
2017-01-06	55.423	415.901	0.011147	0.003579
2017-01-09	55.930	414.528	0.009148	-0.003301
2017-01-10	55.987	414.528	0.001019	0.000000

Before fitting a regression line, we can review if there is a visual relationship between the returns:

```
[5]: df.plot.scatter('Market_Return', 'Stock_Return', marker='.')
```

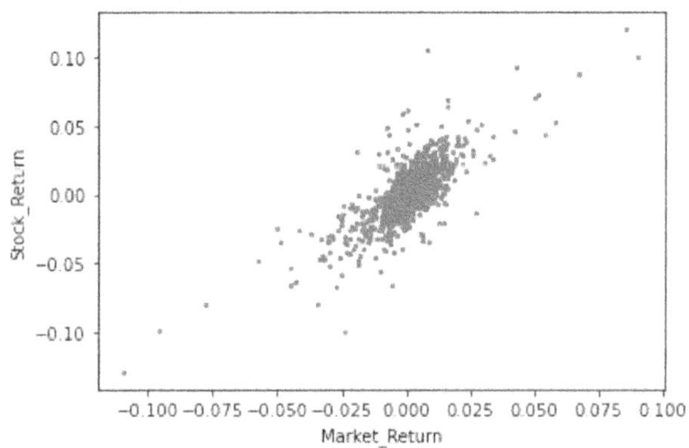

Based on the above chart, it appears there is a linear relationship between the daily market returns and the stock returns. To check this statistically, we can fit a regression line using the SciPy `linregress()` function:

```
[6]: from scipy.stats import linregress
     beta, alpha, rvalue, pvalue, stderr = \
        linregress(df.Market_Return, df.Stock_Return)
```

`linregress()` returns five results, including the slope and intercept of the fitted line as well as measures of goodness of fit. For our data, we can review these using:

```
[7]: print('Beta:      ', round(beta,2))
     print('Alpha:     ', round(alpha,2))
     print('R-squared:', round(rvalue**2 * 100), '%')
     print('pvalue:    ', round(pvalue, 4))
     print('stderr:    ', round(stderr, 4))
```

```
Beta:      1.23
Alpha:     0.0
R-squared: 58 %
pvalue:    0.0
stderr:    0.0295
```

The low p-value gives us confidence that this relationship is statistically significant. We can further visualize the fitted line over the scatter chart using:

```
[8]: import matplotlib.pyplot as plt
     plt.scatter(df.Market_Return, df.Stock_Return, marker='.')
     plt.plot(df.Market_Return, alpha + beta*df.Market_Return, color='black')
     plt.xlabel(marketSymbol)
     plt.ylabel(stockSymbol)
     plt.title('Daily Returns')
     plt.show()
```

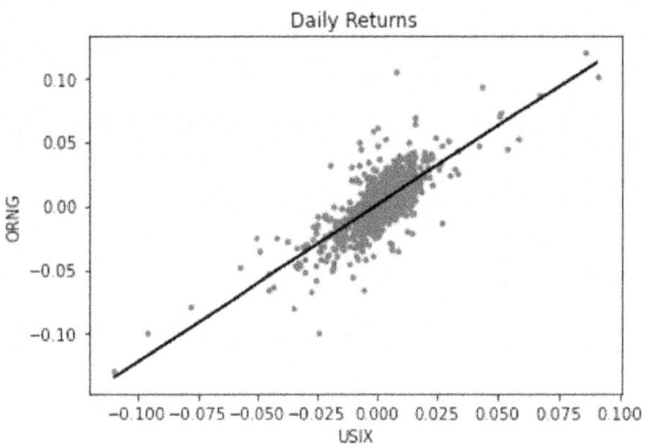

Section Summary

In this section, we looked at simple regression to model the relationship between one independent variable and one dependent variable. SciPy is one package that provides a straightforward way to fit a line for an independent and dependent variable using the `linregress()` function. Also, Matplotlib provides a straightforward way to plot a scatter chart to visually review the potential relationship between two variables.

Test Your Understanding

1) The SciPy `linregress()` function returns how many results?
   a) one
   b) two
   c) three
   d) five

2) A statistic for checking if a regression is statistically significant is:
   a) alpha
   b) beta
   c) p-value
   d) none of the above

17.8 Forecasting Using Multiple Regression

In the previous section, we looked at how simple regression can describe a relationship between two sets of data. In some cases, there is more than one factor that influences a dependent variable. In this section, we'll see how regression can relate two independent variables to a dependent variable. In particular, we'll look at a sales forecasting example for a company whose sales have been growing over time. Sales have also been strong in the last calendar quarter of recent years due to holiday sales. In this example, we'll consider sales as the dependent variable and time and holiday seasons as the independent variables. By considering these two factors, we hope to better be able to describe sales history mathematically, and potentially use this for forecasting future sales.

An initial model

Our dataset is in the file sales.csv. This has historical sales (in thousands) for a product, as well as an integer value for each time period and a binary column that is 1 if the quarter is a holiday quarter. We can load the data using:

```
[1]: import pandas as pd
     sales = pd.read_csv('sales.txt', index_col='quarter')
     sales
```

quarter	t	sales	holiday
2020Q1	1	4.2	0
2020Q2	2	4.8	0
2020Q3	3	6.1	0
2020Q4	4	8.9	1
2021Q1	5	8.2	0
2021Q2	6	8.8	0
2021Q3	7	10.1	0
2021Q4	8	13.0	1
2022Q1	9	12.2	0
2022Q2	10	12.8	0
2022Q3	11	13.1	0

We can plot the historical sales data using:

```
[2]: sales.plot('t','sales', kind='scatter', marker='.')
```

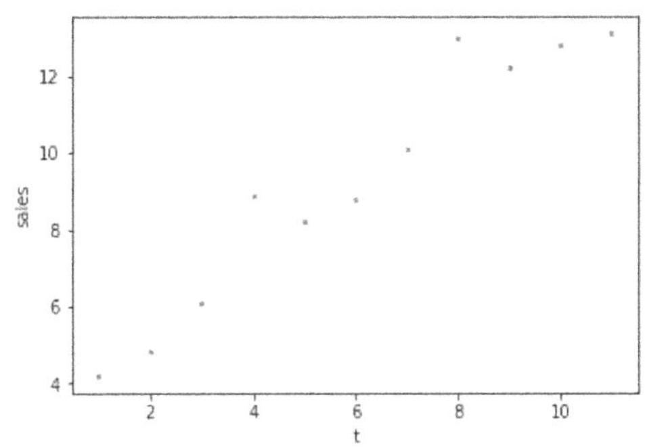

In the above chart, we can see two outliers, these correspond to the holiday sales, which are higher than the usual progression of sales. Further, we see that sales are trending upward. In order to forecast the next few quarters of sales, we can consider fitting a regression line where time (1, 2, . . . , 11) is an independent variable, and sales is the dependent variable. For this section, we'll use the scikit-learn package, since it will allow us to later use a multiple regression model. In order to fit a model between time and sales, we can use the scikit-learn LinearRegression type's `fit()` method:

```
[3]: from sklearn.linear_model import LinearRegression
     model1 = LinearRegression().fit(sales[['t']], sales['sales'])
```

The above statement determines a best fit line between time and sales. The coefficients of that line are available in the properties `intercept_` and `coef_`, which we can see using:

```
[4]: model1.intercept_.round(3) # what is the intercept of the line
     3.569
```

```
[5]: model1.coef_.round(3)[0] # what is the slope of the line
     0.954
```

We can plot how well this equation fits our data using:

```
[6]: import matplotlib.pyplot as plt
     plt.scatter(sales.t, sales.sales, marker='.')
     plt.scatter(sales.t, model1.intercept_ + model1.coef_*sales.t, marker='x',
                 color='black')
     plt.xlabel('Quarter')
     plt.ylabel('Sales')
```

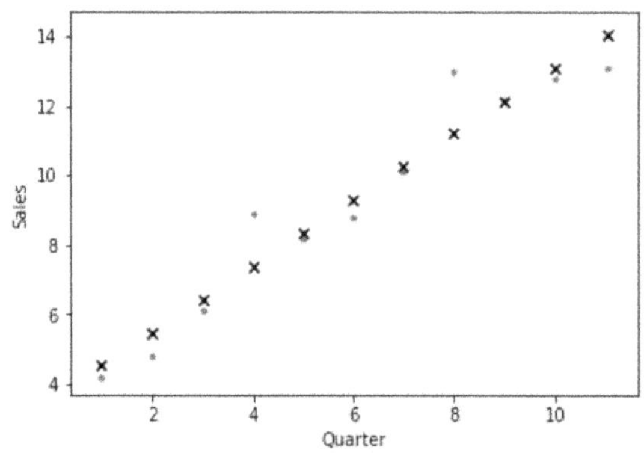

In the above chart, the *x*'s are the fitted values. These are very close to the non-holiday quarters but do not fit well for holiday sales.

A multiple regression model

We can further refine the model by adding an indicator variable set to 1 for holiday quarters and 0 for non-holiday quarters. The equation we consider fitting is:

sales = a + b1 × time + b2 × holiday

In this arrangement, the b1 coefficient captures the trend of sales increasing over time, and the b2 coefficient captures the increase we see during holiday sales.

We can fit this expanded model with the statement:

```
[7]: model2 = LinearRegression().fit(sales[['t', 'holiday']], sales['sales'])
```

and review the values for a, b1, and b2 using model2.coef_ and model2.intercept. In this case, since there are two independent variables, `coef_` will have two values corresponding to b1 and b2.

```
[8]: model2.intercept_.round(3) # what is the intercept of the model
3.2
```

```
[9]: model2.coef_.round(3) # what are the values of b1 and b2
array([0.954, 2.028])
```

We can review how well this expanded model fits the actual sales using:

```
[10]: import matplotlib.pyplot as plt
      plt.scatter(sales.t, sales.sales, marker='.')
      plt.scatter(sales.t,
        model1.intercept_ + model1.coef_*sales.t + model2.coef_[1]*sales.holiday,
        marker='x', color='black')
      plt.xlabel('Quarter')
      plt.ylabel('Sales')
```

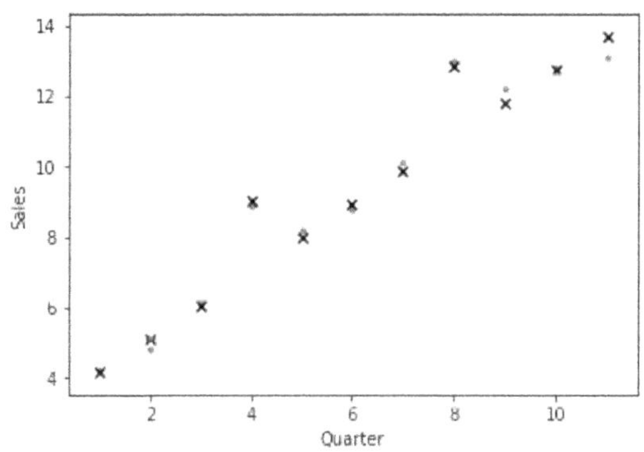

The above chart shows that our second model, which accounts for increased holiday sales, better fits the data. If we decide to adopt this model to help forecast the next two quarters, we can use LinearRegression's `predict()` method:

```
[11]: pointsToPredict = [[12,1],[13,0]]
      model2.predict(pointsToPredict).round(1)
array([16.7, 15.6])
```

The argument to `predict()` is a list of lists. In this case, each sublist contains the time we are predicting for (quarters 12 and 13) and whether each is a holiday quarter (1 for holiday, 0 if not).

Above we used charts to check how well our model fit the data. We can also check this statistically using LinearRegression's `score()` method. For example:

```
[12]: model2.score(sales[['t','holiday']], sales['sales']).round(3)*100
99.3
```

This gives the result 99.3%, which is the R^2 for the model. The R^2 is a measure of how well our model fits the data; a value of 100% indicates a perfect fit. In practice, forecasting models won't typically have such a good fit for the data—but the use of mathematical models to forecast is still a good approach to consider.

Section Summary

In this section, we looked at how scikit-learn's LinearRegression type can be used to forecast with a multiple regression model. Its methods include `fit()`, to fit an intercept and coefficients to relate independent and dependent variables. The `intercept_` and `coef_` properties provide the specific results of the regression. The `predict()` method can be used to apply a fitted model to a set of independent variables in order to predict one or more fitted values.

Test Your Understanding

1) SciPy's `linregress()` function can perform multiple regression.
   a) True
   b) False

2) scikit-learn's LinearRegression type provides which method for fitting coefficients for a regression model?
   a) `regress()`
   b) `fit()`
   c) `predict()`
   d) none of the above

Chapter Review

In this chapter, we've seen how you can use Python and certain packages for a variety of statistics. Whereas we may use visualization to get insight from data, statistics allows us to go deeper and provide additional support for our analyses. For descriptive statistics, NumPy and pandas are good alternatives.

Descriptive statistics can also be used within pivot tables to summarize detailed data using pandas. For time series data, rolling statistics can be calculated and visualized using the pandas `rolling()` method.

Selected SciPy functions are available for statistically comparing populations. To check if a distribution is normal, SciPy also provides functions to check skewness and kurtosis. For analyzing relationships between data, we reviewed how you can use SciPy and scikit-learn to perform simple and multiple regression analysis.

Key Terms

confidence interval	A pair of values that bracket an unknown population value.
dependent variable	A set of values that may depend on one or more other (independent) variables.
descriptive statistics	Statistics such as mean, median, variance, and standard deviation to describe a set of values.
F-test	A statistical test for comparing the variances of two populations.
histogram	A chart to visualize the distribution of data.
independent variable	A set of measures whose values may relate to another (dependent) variable.
kurtosis	A descriptive statistic for how heavy or light the tails of the distribution are.
mean	A descriptive statistic of the average of a set of values.
median	A descriptive statistic of the middle of a set of values.
moving average	For a time series, the average of a subset of points.
multiple regression	A technique to best fit two or more independent variables to a dependent variable.
normal distribution	A bell-shaped distribution of data.
pivot table	A summary table of descriptive statistics with results divided into rows and/or columns along certain dimensions.
p-value	A value calculated by a statistical test to determine the probability that the sample data are consistent with a hypothesis about the underlying population.
simple regression	A technique to best fit a line to describe the potential relationship between an independent and dependent variable.
skewness	A descriptive statistic for how much a distribution of data leans relative to its mean.
standard deviation	A descriptive statistic of how spread out the data are relative to the mean.
t-test	A statistical test for comparing the means of two populations.
variance	A descriptive statistic of how spread out the data are relative to the mean.

Programming Problems

1) Pivot Tables—Health Club Guests

A health club has been keeping track of people who are interested in joining, which includes all people who have visited to take a tour of the facilities. The data are in visitors.csv. Included are the following fields:

field	description
guestID	An integer assigned to each visitor
visitDate	The date the member first visited the club, before they joined
gender	M=Male, F=Female
age	Age of member when they visited the club
married	M=Married, S=Single
pool	y=interested in a pool membership, n=not interested in a pool membership
gym	y=interested in a gym membership, n=not interested in a gym membership

In order to better understand the demographics of its visitors, the gym management has asked you to create the following summary tables:

a) counts of Male vs. Female visitors
b) counts of Married vs. Single visitors
c) the count by age range (using 1 for people up to 19 years, 2 for 20–29 years, etc.)
d) the count of visitors by year of visit date
e) a table of visitor counts by year of visit (rows) and gender of visitor (columns)
f) a table of visitor counts by age range (rows) and year of visit (columns)

2) Population Tests—Health Club Members

A health club has been keeping track of key dates for its members: specifically when they visited to tour the club (before they joined) and when they stopped being a member. These dates (visitDate, endDate) are included in the data in members.csv. The club would like to see if the length of time between visit date and end date (in months) is significantly different between men and women. Use a t-test to check whether this time length (term) is statistically different between men and women. Note that you will need to filter where status is "inactive."

3) Moving Averages—Health Club Monthly Membership Terms

For the health club data described in problem 2:

a) Augment the data to include a column of the year and month when each member visited the club as the row of the table, called `'yearmonth'`. Use the format yyyymm—for example, 201501 is January 2015.
b) Augment the data to include a column for the number of months the member was active (difference between visitDate and endDate) as `'term'`.
c) Use `pivot_table()` to calculate the average term length. For this table, use `'yearmonth'` for the pivot table rows.
d) For the pivot table created in part (b), calculate the six-month moving average of membership term length.

4) Moving Average—DJIA

For the Dow Jones Industrial Average (DJIA) daily adjusted closing prices:

a) Read these data into a DataFrame.
b) Augment the DataFrame with the 10-day and 30-day moving averages of the daily adjusted closing prices.
c) Visualize the DJIA adjusted closing prices and the 10-day and 30-day moving average in a chart.

5) Distribution and Outliers—DJIA

For the Dow Jones Industrial Average (DJIA) daily adjusted closing prices:

a) Read these data into a DataFrame and augment the data with the daily returns of adjusted closing price.
b) Plot a histogram of the daily returns.
c) Calculate the skewness and kurtosis of the DJIA daily returns.
d) Determine using a statistical test if the daily returns follow a normal distribution (with 95% confidence).
e) Based on the standard deviation of the daily returns, identify those dates and returns that were smaller than -3 standard deviations or larger than +3 standard deviations.

6) Distribution and Outliers—Stock

For a stock of interest, download its daily adjusted closing price for five years of history.

a) Read these data into a DataFrame and augment the data with the daily returns of adjusted closing price.
b) Plot a histogram of the daily returns.
c) Calculate the skewness and kurtosis of the daily returns.
d) Determine using a statistical test if the daily returns follow a normal distribution (with 95% confidence).
e) Based on the standard deviation of the daily returns, identify those dates and returns that were smaller than -3 standard deviations or larger than +3 standard deviations.
f) For dates identified in part (e), were there any company-specific events?

7) Simple Regression—Stock and DJIA

For a stock of interest, download its daily adjusted closing price for 5 years of history.

a) Read these data into a DataFrame and augment the data with the daily returns of adjusted closing price.
b) Read the data for the Dow Jones Industrial Average prices into a DataFrame.
c) For both your stock as well as the DJIA, augment the data with the daily returns of adjusted closing price.
d) Plot a scattergram of DJA daily returns (x axis) vs. the stock's daily returns (y axis).
e) Perform a linear regression between the DJIA returns (independent variable) and the stock's returns (dependent variable).
f) For the regression, what is the equation of the fitted line?
g) For the regression, is the relationship significant with 95% confidence?

Chapter 18

Text Analysis

Chapter Objectives

In this chapter, you'll learn:

- several techniques for analyzing textual data, including some capabilities provided by the nltk and scikit-learn packages
- useful Python language features for text analysis, including list comprehensions and a recap of lambda functions and selected string methods
- how to perform basic and more advanced word frequency analysis, to discover the most commonly used words in text
- one approach to sentiment analysis, to determine the degree of positive or negativity in text
- an example of text classification as applied to email spam identification

Chapter Outline

18.1 About Text Analysis

In much of this book, we've focused on numerical problems and used Python and packages to work with these. In contrast, text analysis involves programmatically processing text to gain insights or provide capabilities where human language is the underlying input. In this chapter, we'll look at a few types of text analytics:

- *word frequency analysis* to determine the most common words in text. This can be useful on its own to understand the focus of a particular text. Word frequencies are also useful as an input to other text analyses such as classification.
- *sentiment analysis* to determine how positive or negative a text is. For example, this can be useful to analyze a large body of product reviews or customer feedback.
- *text classification* to tag text in certain ways. For example, email can be programmatically analyzed to determine if a message is likely to be spam. Other uses include organizing content for easier access and lookup as well as analyzing and characterizing customer feedback.

Python in combination with suitable packages is a good option for text analysis. In particular, in this chapter, we'll use the nltk package, which is the Natural Language Toolkit. We'll also use the scikit-learn package, which includes a model for text classification.

In practice, the size of the text passages of interest could be very large. For example, we may have several thousand product reviews that we want to better understand in order to consider potential product improvements or decide on a marketing approach. Though it would be hard to read and interpret all these, we can use software to read and characterize the text. Although we'll use small examples in this chapter, the techniques presented are feasible and especially valuable for much larger cases.

nltk packages

Note that when you install the nltk package, it comes with a core set of capabilities, but unlike most Python packages, it does not include all its available models and data. Instead, nltk is designed to let you subsequently download the models and data you want to use. Note that these supplemental parts of nltk are also called `'packages'`. In this chapter, we'll refer to these as ***nltk packages***, so as not to confuse these with Python packages.

In order to download nltk packages, you use nltk's `download()` function. We can do this interactively in the Python shell or a notebook cell. For example, to use the nltk tokenizer available in nltk's punkt package, you can interactively use the statement:

```
>>> nltk.download('punkt')
```

After you download an nltk package, it is ready for you to use—you don't need to download again before each run of your code. You can also use the `download()` function to upgrade to the latest version of an nltk package.

There are over 100 nltk packages, divided into several categories (also known as subdirectories). In this chapter, we'll use the few nltk packages listed in Table 18-1.

section where used	nltk package	nltk package category
18.3 Word Frequency	averaged_perceptron_tagger	taggers
	punkt	tokenizers
	stopwords	corpora
18.4 Sentiment Analysis	vader_lexicon	sentiment

Table 18-1: nltk packages used in this chapter

> **Spreadsheet Compare/Contrast**
> **Spreadsheets for Strings**
>
> Although often less used, spreadsheets have several functions for working with strings. However, these are focused on basic operations like combining strings or taking substrings.
>
> Conversely Python has not only a rich set of string-related functionality, but in combination with packages like nltk and scikit-learn, it can be used for in-depth text analysis.

18.2 Some Python Prerequisites

In this chapter, we'll use the nltk package to work with and analyze text. In combination with this, we'll also use some Python language techniques: selected useful string methods, lambda functions to concisely define a function in one statement, and list comprehensions that let us build lists using filtering and/or transformations.

Some useful string methods

In Chapter 8, we looked at a number of ways we can work with strings. Since text analysis is all about strings, we'll recap now some useful string methods and introduce a few that we haven't covered yet.

When analyzing text, we may want to filter out any items in the text that have digits or symbols. The string `isalpha()` method is a simple way to see if a string is made up only of letters. It returns `True` if this is the case and otherwise returns `False`.

Another useful string method is `lower()`, which for a given string will return the lowercase version of it. For example, if `s` is `'Hello'`, `s.lower()` will return `'hello'`.

We'll also use the method `startswith()`, which can check if the first letters of one string are a certain other string. For example, if `s = 'Noun'`, the expression `s.startswith('N')` will return `True`.

Also, we'll be using the string `join()` method. This can be used to create a string from several strings in a list. For example, if `wordlist = ['a', 'set', 'of', 'words']`, we can combine these into one string using `' '.join(wordlist)`, which will return `'a set of words'`.

Lambda functions recap

In Chapter 11, we introduced lambda functions. These let us define a function in one statement, allowing us to include function logic very concisely in our code. In particular, in Chapter 11, we used the pandas `apply()` method, which takes a function as its input argument in order to create new values for every row of a DataFrame. For example, if we have a DataFrame column where each cell has a list of words, we can replace each list with a string using:

```
df['message'] = df['message'].apply(lambda words: ' '.join(words))
```

List comprehensions

We've seen in Chapter 6 that Python lists have a variety of methods and features for concisely working with an ordered set of items. Another feature in Python, list comprehensions, allows us to generate lists in one statement. As we'll see, we can use filtering, transformations, or both to create a list from a source list or string.

Let's start with the following example. Assume you have a list of words (in a variable called `words`) and you'd like to create a new list that has the lowercase version of these words. You can do this in a few statements as:

```
lowerList = []
for word in words:
   lowerList.append(word.lower())
```

In the above code, we are explicitly iterating through the list and applying the `lower()` method to each list item, then appending this result.

Instead, we can do this in one statement with a list comprehension:

```
lowerList = [word.lower() for word in words]
```

In the above statement, the square brackets contain a transformation, `word.lower()`, followed by a `for` statement. These expressions are also used in the three-line example above, but the list comprehension allows us to combine them into one statement. Note that the `append()` method is not needed in the list comprehension, since it is implicitly done for us.

List comprehensions can also be used to create a filtered list from a source list. For example, if we want to filter out strings that are not purely made up of letters from a list called `stringList`, we could use:

```
words = [s for s in stringList if s.isalpha()]
```

In the above list comprehension, we include a filtering `if` statement after the `for` statement. In this example, there is no transformation needed, so the start of the list comprehension simply has the variable `s`.

Furthermore, in some cases we may want to filter a list as well as transform its items. Combining the above two examples, if you'd like to include only strings that have only letters and also transform these to lowercase, you could use:

```
words =[s.lower() for s in strings if s.isalpha()]
```

As a final example, note that in a list comprehension, we can use other types of variables that work with `for`. These are called iterables—lists are one example of an iterable, but strings are also iterables, since `for` can iterate over a string's characters. As an example, here we create a string that includes only letters of a source string called `sourceString`:

```
' '.join([c for c in sourceString if c.isalpha()])
```

To recap the general syntax of list comprehensions:

```
[expression for item in iterable (if logical_expression)]
```

Section Summary

In this section, we covered a few techniques that we'll use in the upcoming sections. In particular, we'll use a few string methods to help manipulate text passages, including `isalpha()`, `lower()`, and `join()`. Also in some cases, we'll use a one-statement function that can be concisely defined as a lambda function. Last, list comprehensions were introduced as a technique to filter and/or transform lists and strings.

Test Your Understanding

1) A Python lambda function can contain multiple statements.
   a) True
   b) False

2) A Python list comprehension can create a list using:
   a) filtering
   b) a transformation
   c) (a) and (b)

18.3 Word Frequency

If you have a large amount of text, one basic programmatic analysis you can perform is to determine how frequently each word appears. The most frequently occurring words may give us a sense of what the passage is about. In addition to being useful on its own, techniques for determining word frequency are useful in other types of text analysis, as we'll see in subsequent sections.

For this section and the next, we'll assume we have a set of product reviews. In particular, we are considering two types of guitars, and hope to learn about them from their text reviews. The dataset provided with this book is small—in fact, you could read these reviews to get a sense of them. But the techniques we'll cover in this chapter apply equally well to a very large set of reviews or other types of text.

The product reviews are in a file named GuitarReviews2out.txt. Each line of the file corresponds to a single product review. The individual fields of the file are separated by the pipe character | as opposed to commas, since commas commonly occur in the text.

Getting started—a basic word frequency analysis

We'll use a pandas DataFrame to hold the data. We can load the file using:

```
[1]: # DATA ACCESS: load a previously downloaded file of reviews into a DataFrame
     import pandas as pd
     df = pd.read_csv('GuitarReviews2out.txt', sep='|', index_col='#')
     df.head()
```

#	product	title	review
1	BenderCaster	My new favorite guitar	I've been playing the Wes Hall for many years,...
2	BenderCaster	Good but not Great	The Bender Caster sounds clean, the 3-position...
3	BenderCaster	Try before buying	I tried the Caster in my local music store and...
4	BenderCaster	Love It!	Everything about the Bender Caster met or exce...
5	BenderCaster	Great Guitar, Great Price	I've been playing Bender guitars for years, th...

As seen above, we have only a few fields. The product field is the name of the product being reviewed, the title field is a brief title for the review, and the review field is the main text of the review. To further check the data, we can display the first review:

```
[2]: # CHECK: the first review
     df.iloc[0].review
```

```
'I've been playing the Wes Hall for many years, but finally decided to buy this Bender
Caster. My Wes has terrific sustain, but the Caster gives me a funky sound I love! The
maple neck is terrific too. Would highly recommend this guitar!'
```

Although this looks like multiple lines, keep in mind it is a single string containing the review text. For our analysis, we'll combine all the reviews into one composite text. This can be done using:

```
[3]: # PREPARE: combine reviews
     allTextList = df.review.to_list()
     allText = ' '.join(allTextList)
```

Initially, we'll perform a simple set of operations to determine the frequency of each word in the above text. The steps will be:

1) Break up the text into words (called ***tokens***).
2) Count the occurrence of each word.
3) Plot the occurrence of the most frequently occurring words.

In order to perform these steps, we'll use the nltk package, which is the Natural Language Toolkit. In order to use this package, we need to import it:

```
[4]: import nltk
```

Further, this package has additional components that we'll need to download. This is done with the nltk `download()` function. For example, in order to tokenize the text, we'll use the nltk punkt package. We can download it (or upgrade to the latest version) using:

```
[5]: nltk.download('punkt')
```

```
[nltk_data] Downloading package punkt to /root/nltk_data …
[nltk_data]   Unzipping tokenizers/punkt.zip.
True
```

With punkt downloaded, we can use it to tokenize the text. Although you could use Python string methods to try to split up the text, nltk uses punkt in a smarter way to find distinct sentences and words—for example, handling punctuation. In particular, we'll use nltk's `word_tokenize()` function to split the text into a list of tokens and review the first few of the lists items using a slice:

```
[6]: # TOKENIZE: nltk can tokenize the text
     tokens = nltk.word_tokenize(allText)
     tokens[:8]
```

```
['I', ''', 've', 'been', 'playing', 'the', 'Wes', 'Hall']
```

Next we can calculate how often each word occurs in the above list using the nltk `FreqDist()` function:

```
[7]: # STATISTICS: count # of occurrences of each word (nltk FreqDist class)
     wordFrequency = nltk.FreqDist(tokens)
     wordFrequency
```

```
FreqDist({'.': 34, 'the': 26, 'I': 19, ',': 17, ''': 12, 'but': 11, 'The': 11, 'to': 8,
'is': 8, 'and': 8, … })
```

Last, we can generate a simple chart using the FreqDist `plot()` method:

```
[8]: # VISUALIZE: Review the summary (nltk FreqDist)
     wordFrequency.plot(30)
```

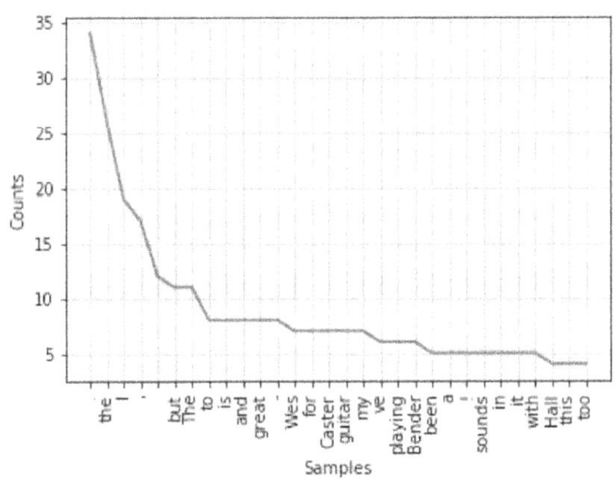

To recap our initial basic word frequency review, Program 18-1 has the steps covered above.

Program 18-1

```
import pandas as pd
import nltk

# load - read reviews into a DataFrame
df = pd.read_csv('GuitarReviews2out.txt', sep='|', index_col='#')

# combine all reviews
allTextList = df.review.to_list()
allText = ' '.join(allTextList)

# parse - tokenize the text
tokens = nltk.word_tokenize(allText)

# statistics - determine the word frequency
wordFrequency = nltk.FreqDist(tokens)

# visualize the results
wordFrequency.plot(30)
```

Some basic refinements

We can improve our word frequency with a few simple refinements. In particular, in this section, we'll look at the following:

- removing tokens that have non-letters
- ignoring the distinction between upper- and lowercase
- removing commonly occurring words (known as stop words)

The first refinement involves filtering out tokens that do not have letters only. For example, in our initial simple analysis we have punctuation like periods among the top tokens. Given a list of tokens, we can filter out tokens with non-letters using this list comprehension:

```
[9]:  # FILTER: only keep tokens that have letters
      words1 = [token for token in tokens if token.isalpha()]
      words1[:8]

      ['I', 've', 'been', 'playing', 'the', 'Wes', 'Hall', 'for']
```

Tip

Regular Expressions

In Python and other languages, regular expressions are a way to match and replace patterns of characters. Regular expression functionality is available in Python in its re package. Although we use list comprehension in this chapter for match and replace, regular expressions are more concise and may be more efficient. However, the specific syntax for regular expressions can be cryptic and involves some study to learn.

The next refinement, making our analysis case insensitive, can be thought of as harmonizing like words. For example, if one sentence starts with *Playing*, and another sentence includes the word *playing*, we'd like to count this as the same word. We can do this by transforming all uppercase letters into lowercase using a list comprehension on the list called `words1`:

```
[10]: # HARMONIZE: we can fold words to lower case using list comprehension
      words2 = [word.lower() for word in words1]
      words2[:8]

      ['i', 've', 'been', 'playing', 'the', 'wes', 'hall', 'for']
```

The above list comprehensions are easy to test interactively—you can define a list of strings and verify that these statements transform or filter as expected.

Note that in this section, we use variable names like *words1*, *words2*, and so on. This is to reflect the stepwise process for refining a set of tokens or words in order to reach our goal of a meaningful word frequency analysis.

The next refinement involves filtering out common words that do not give us much information. These are referred to as ***stop words*** and are available in lists in different languages in nltk. For example, in the initial word frequency in the prior section, among the most common words are *the*, *I*, *but*, and *to*. We'd like to filter these and other common words out of our word frequency.

We can download the stopwords package from nltk using:

```
[11]: nltk.download('stopwords')

      [nltk_data] Downloading package stopwords to /root/nltk_data …
      [nltk_data]   Unzipping corpora/stopwords.zip.
      True
```

Then, in order to get the list of stop words in English, use:

```
[12]: from nltk.corpus import stopwords
      stopWords = stopwords.words('english') # get the NLTK stopwords
      type(stopWords)

      list
```

Note that in the above, `stopWords` is a Python list. You can use the `len()` function to see how many words it contains:

```
[13]: len(stopWords)

      179
```

To see some example words, we can take a slice of the list:

```
[14]: stopWords[:8]

      ['i', 'me', 'my', 'myself', 'we', 'our', 'ours', 'ourselves']
```

Given this list of stop words, we can filter these out from our text using another list comprehension:

```
[15]: # FILTER: Remove the stop words
      words3 = [word for word in words2 if word not in stopWords]
      words3[:8]

      ['playing', 'wes', 'hall', 'many', 'years', 'finally', 'decided', 'buy']
```

Having removed stop words, we have a list of words that can be summarized with `FreqDist()` and visualized with `plot()`, similar to code used in the prior section.

```
[16]: # STATISTICS: count the # of occurrences of each word (nltk FreqDist class)
      wordFrequency3 = nltk.FreqDist(words3)
      wordFrequency3
```

```
FreqDist({'guitar': 8, 'great': 8, 'wes': 7, 'caster': 7, 'playing': 6, 'bender': 6,
'sounds': 6, 'hall': 4, 'terrific': 4, 'years': 3, … })
```

```
[17]: # VISUALIZE: display a word frequency chart
      wordFrequency3.plot(30)
```

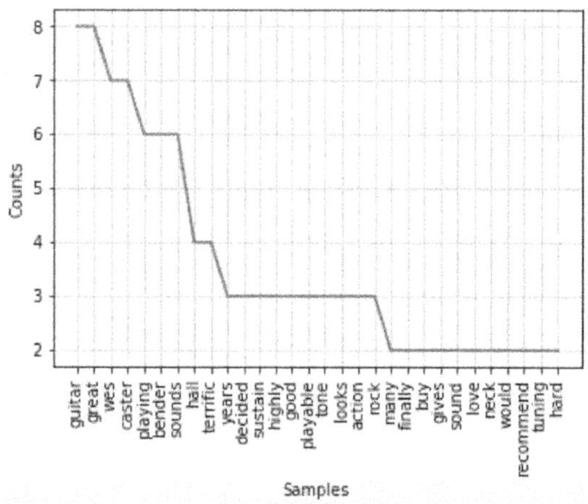

To recap, in this section, we looked at a few possible refinements to our word frequency analysis: filtering non-alphabetic tokens, folding words to lowercase, and filtering out stop words. Program 18-2 shows a revised word frequency analysis that incorporates these enhancements.

Program 18-2

```
import pandas as pd
import nltk
from nltk.corpus import stopwords

# load: read reviews into a DataFrame
df = pd.read_csv('GuitarReviews2out.txt', sep='|', index_col='#')

# prepare: combine all reviews
allTextList = df.review.to_list()
allText = ' '.join(allTextList)

# parse - tokenize the text
tokens = nltk.word_tokenize(allText)

# filter: keep tokens that have letters, using list comprehension
words1 = [token for token in tokens if token.isalpha()]

# harmonize: fold the words into lower case, using list comprehension
words2 = [word.lower() for word in words1]

# filter: Remove the stop words
stopWords = stopwords.words('english') # get the NLTK stopwords
words3 = [word for word in words2 if word not in stopWords]

# statistics: determine the word frequency
wordFrequency = nltk.FreqDist(words3)

# visualize the results
wordFrequency.plot(30)
```

Program 18-2 Output

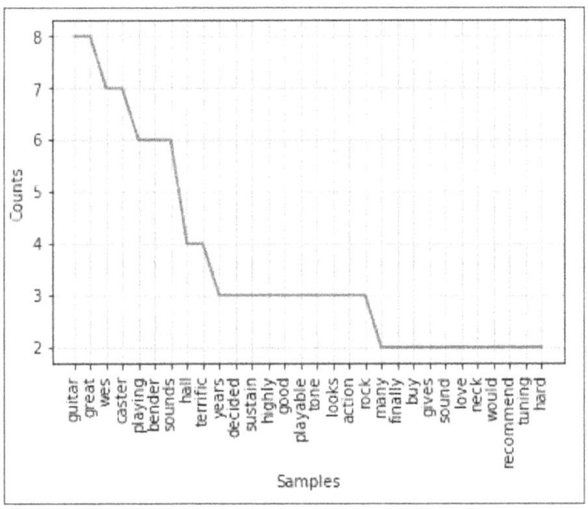

Some additional refinements

In this final section related to word frequency, we'll look at more advanced ways to refine our analysis. This includes stemming, which allows us to combine words with the same root. For example, *play* and *playing* are related and we would like to count them together in our word frequency. Another technique we'll use is part-of-speech (POS) filtering, which will allow us to count only particular types of words, such as nouns.

For stemming, nltk provides a few different approaches. We'll use the Porter stemmer, which has a `stem()` method we can use as follows:

```
[18]: # HARMONIZE: Use Porter Stemmer
      porterStemmer = nltk.PorterStemmer()
      words4 = [porterStemmer.stem(word) for word in words3]
      words4[:8]
```

```
['play', 'we', 'hall', 'mani', 'year', 'final', 'decid', 'buy']
```

Note that above, we see some words (play, we) as well as the root of some words (mani, decid).

Another technique we can use is POS filtering. For example, we may decide to analyze word frequency on nouns only. In order to do this, we can use nltk's `pos_tag()` function on the tokens. This requires the averaged_perceptron_tagger.

```
[19]: nltk.download('averaged_perceptron_tagger')
```

```
[nltk_data] Downloading package averaged_perceptron_tagger to
[nltk_data]     /root/nltk_data ...
[nltk_data]   Unzipping taggers/averaged_perceptron_tagger.zip.
True
```

The `pos_tag()` function returns a list of word/tag pairs, where the tag is a standard code for the word's part of speech. We can review a sample using:

```
[20]: # AUGMENT: Add part of speed to each token
      wordsWithTags = nltk.pos_tag(tokens)
      wordsWithTags[:8]
```

```
[('I', 'PRP'),
 (''', 'VBP'),
 ('ve', 'RB'),
 ('been', 'VBN'),
 ('playing', 'VBG'),
 ('the', 'DT'),
 ('Wes', 'NNP'),
 ('Hall', 'NNP')]
```

The POS tags are a standardized list; those tags beginning with *N* are nouns—for example, NNP refers to proper singular nouns. We can extract the list of words that are nouns using:

```
[21]: # FILTER: include only nouns (tags beginning with N)
      words5 = [word for (word,tag) in wordsWithTags if tag.startswith('N')]
      words5[:8]
```

```
['Wes', 'Hall', 'years', 'Bender', 'Caster', 'My', 'Wes', 'sustain']
```

We can use stemming and POS to further refine our guitar reviews example. Program 18-3 combines these and prior techniques into one program that determines the word frequency of all reviews.

Program 18-3

```
import pandas as pd
import nltk
from nltk.corpus import stopwords

# load: read reviews into a DataFrame
df = pd.read_csv('GuitarReviews2out.txt', sep='|', index_col='#')

# prepare: combine all reviews
allTextList = df.review.to_list()
allText = ' '.join(allTextList)

# parse: tokenize the text
tokens = nltk.word_tokenize(allText)

# filter: keep tokens that have letters, using list comprehension
words1 = [token for token in tokens if token.isalpha()]

# harmonize: fold the words into lower case, using list comprehension
words2 = [word.lower() for word in words1]

# augment: Add part of speed to each token
wordsWithTags = nltk.pos_tag(words2)

# filter: include only nouns (tags beginning with N)
words3 = [word for (word,tag) in wordsWithTags if tag.startswith('N')]

# filter: Remove the stop words
stopWords = stopwords.words('english') # get the NLTK stopwords
words4 = [word for word in words3 if word not in stopWords]

# harmonize: Use Porter Stemmer
porterStemmer = nltk.PorterStemmer()
words5 = [porterStemmer.stem(word) for word in words4]

# statistics: determine the word frequency
wordFrequency = nltk.FreqDist(words5)

# visualize the results
wordFrequency.plot(30)
```

Program 18-3 Output

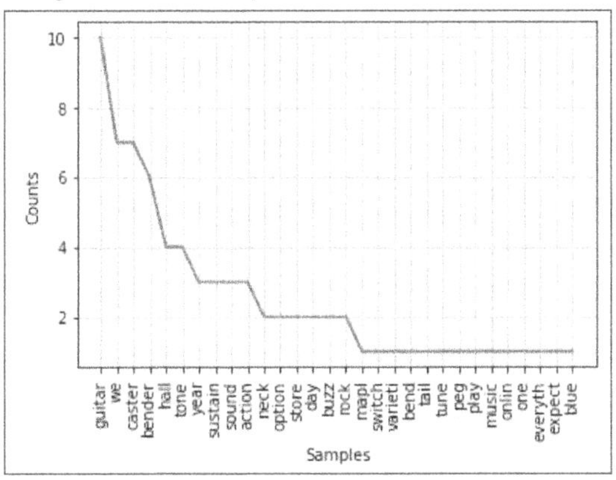

As we've covered a number of steps for word frequency, Table 18-2 lists and categorizes these.

Category	Processing	Type, Function, or Method	nltk package
Prepare	passage joining	Series `tolist()` `str join()`	
	tokenizing	nltk `word_tokenize()`	punkt
	rejoining	`str join()`	
Harmonize	ignore case	`str lower()`	
	stemming	nltk PorterStemmer	
Filtering	Letters only	`str isalpha()`	
	drop stop words	nltk `stopwords.words()`	stopwords
	part of speech focus	nltk `pos_tag()` `str startswith()`	averaged_perceptron_tagger
Statistics	counts and chart	nltk FreqDist	

Table 18-2: Word frequency processing steps

Section Summary

In this section, we introduced word frequency analysis as a way to find the most common words in text in order to gain an understanding of what is covered. Basic analysis can be done simply with nltk's `word_tokenize()` function and the `FreqDist` type. Refinements to the analysis allow us to derive a more useful word frequency—for example, by filtering out tokens with non-letters and folding words to lowercase. Common stop words can be filtered out as well. Further, related words can be combined by using a stemmer. Last, we can focus on a particular part of speech—for example, nouns—in order to get a more refined word frequency result.

Test Your Understanding

1) The process of separating a sentence into words and other strings is called:
   a) filtering
   b) tokenizing
   c) stemming
   d) harmonizing

2) The words weak and week have the same stem.
   a) True
   b) False

18.4 Sentiment Analysis

In the previous section, we analyzed a set of text reviews using word frequency, to get a sense of what the text was about. In this section, we'll look further at the same reviews using sentiment analysis, an approach for determining how positive or negative a passage is. Although we have a very small set of reviews, in practice, this can be used on a very large dataset to programmatically determine the reviewers' sentiment for a product or service.

There are a few approaches available for sentiment analysis. We'll use one that is available in nltk called the `SentimentIntensityAnalyzer`, which implements the VADER model. VADER is an acronym for Valence Aware Dictionary for Sentiment Reasoner, and it aims to determine whether a text passage is positive or negative.

For this section, we'll again use the guitar reviews covered earlier. We can load these using:

```
[1]: # DATA ACCESS: load a previously downloaded file of reviews into a DataFrame
     import pandas as pd
     df = pd.read_csv('GuitarReviews2out.txt', sep='|', index_col='#')
     df.head()
```

#	product	title	review
1	BenderCaster	My new favorite guitar	I've been playing the Wes Hall for many years,...
2	BenderCaster	Good but not Great	The Bender Caster sounds clean, the 3-position...
3	BenderCaster	Try before buying	I tried the Caster in my local music store and...
4	BenderCaster	Love It!	Everything about the Bender Caster met or exce...
5	BenderCaster	Great Guitar, Great Price	I've been playing Bender guitars for years, th...

To get started with the VADER model, we need these import statements:

```
[2]: import nltk
     from nltk.sentiment import SentimentIntensityAnalyzer
```

As with other nltk components, we need to download a required component using the following:

```
[3]: # Prereq for using the Vader analyzer
     nltk.download('vader_lexicon')

     [nltk_data] Downloading package vader_lexicon to /root/nltk_data ...
     True
```

To perform the sentiment analysis, we create an analyzer using:

```
[4]: # create a Vader analyzer
     analyzer = SentimentIntensityAnalyzer()
```

To get started with the analyzer, we'll get the first product review:

```
[5]: # for an initialize test, use the first review
     text1 = df.iloc[0].review
     text1

     'I've been playing the Wes Hall for many years, but finally decided to buy this Bender
     Caster. My Wes has terrific sustain, but the Caster gives me a funky sound I love! The
     maple neck is terrific too. Would highly recommend this guitar!'
```

The `polarity_scores()` method gives us a set of sentiment scores for the text:

```
[6]:  # calculate the polarity scores for 1 review
      analyzer.polarity_scores(text1)

      {'neg': 0.028, 'neu': 0.621, 'pos': 0.351, 'compound': 0.9646}
```

Note that the result from `polarity_scores()` is a Python dictionary with four pairs. The first three provide scores regarding the percentage of the passage that is negative, neutral, and positive. These values are in decimal and always add up to 1. The fourth item in the dictionary is a composite score between -1 and 1, indicating overall how negative or positive the passage is. In our example above, the compound score is 0.9646 indicating this is a very positive review.

We can apply `polarity_scores()` to each review individually to determine how positive or negative each review is using:

```
[7]:  compoundList = []
      for index, row in df.iterrows():
          text = row.review
          scores = analyzer.polarity_scores(text)
          compound = scores['compound']
          print(format(index,'2d'), format(compound, '6.2f'), row.title)
          compoundList.append(compound)

       1   0.96 My new favorite guitar
       2   0.81 Good but not Great
       3   0.49 Try before buying
       4   0.95 Love It!
       5   0.91 Great Guitar, Great Price
       6   0.98 Amazing Guitar
       7   0.92 None better
       8   0.97 Great Choice
       9  -0.82 Good Guitar but Pricey
      10   0.53 It's it
```

We can see above that the sixth review is calculated to be the most positive, and the ninth review is the only negative review.

More concisely, we can use the DataFrame `apply()` method to calculate and append the compound scores using:

```
[8]:  # determine polarity scores for text, and return the compound score
      def compoundScore(text):
          scores = analyzer.polarity_scores(text)
          return scores['compound']
```

```
[9]: df['compound'] = df['review'].apply(compoundScore)
     df
```

#	product	title	review	compound
1	BenderCaster	My new favorite guitar	I've been playing the Wes Hall for many years,...	0.9646
2	BenderCaster	Good but not Great	The Bender Caster sounds clean, the 3-position...	0.8074
3	BenderCaster	Try before buying	I tried the Caster in my local music store and...	0.4854
4	BenderCaster	Love It!	Everything about the Bender Caster met or exce...	0.9533
5	BenderCaster	Great Guitar, Great Price	I've been playing Bender guitars for years, th...	0.9136
6	WesHall	Amazing Guitar	I've been playing an entry level guitar for ye...	0.9751
7	WesHall	None better	The Wes Hall is THE GUITAR for rock & roll. I...	0.9184
8	WesHall	Great Choice	I've been playing guitar occasionally but now ...	0.9738
9	WesHall	Good Guitar but Pricey	I was excited to order the Wes but given the p...	-0.8225
10	WesHall	It's it	Unbelievable! It's the classic, the original....	0.5255

Program 18-4 recaps the steps presented in this section.

Program 18-4

```
# Sentiment analysis on reviews in a DataFrame
import pandas as pd
from nltk.sentiment import SentimentIntensityAnalyzer

# load - read reviews into a DataFrame
df = pd.read_csv('GuitarReviews2out.txt', sep='|', index_col='#')

# create an analyzer
analyzer = SentimentIntensityAnalyzer()

# define a function that calculates and returns the compound VADER score
def compoundScore(text):
    scores = analyzer.polarity_scores(text)
    return scores['compound']

# apply the analyzer on all the reviews in the DataFrame, and display
df['compound'] = df['review'].apply(compoundScore)
print(df[['title', 'compound']])
```

Section Summary

Sentiment analysis provides measures of the positivity or negativity of text. For example, this approach can be used to programmatically analyze a large body of product reviews. The VADER model, implemented in nltk, is one approach for this. The `SentimentIntensityAnalyzer` type will calculate the percentages of the text that are positive, neutral, and negative. It also determines a composite score that ranges from -1 to +1 to provide an indication of how negative or positive the passage seems to be.

Test Your Understanding

1) The nltk `SentimentIntensityAnalyzer` `polarity_scores()` method returns a Python:
   a) integer
   b) string
   c) list
   d) dictionary

2) The `polarity_scores()` method includes a compound score whose value is between:
   a) 0 and 1
   b) -1 and 1
   c) 0 and 100

18.5 Text Classification

In this last section, we'll look at an example of text classification, which involves classifying a set of texts into two or more groups. One use of text classification is to try to identify which emails might be spam (junk mail), as opposed to not spam. There are more spam emails than non-spam emails, so programmatic filtering of likely spam is a desired feature for email systems. We'll look at one approach for this by analyzing a small set of publicly available emails from the Enron Corporation.

There are a few approaches to text classification; in this section, we'll use a naïve Bayes classifier to evaluate the probability of each word occurring in a spam or non-spam message. The individual word probabilities can be then used to assess whether a message is more likely to be spam or not. In order to take this approach, we'll use some of the techniques introduced earlier for word frequency.

Here is the outline for applying the naïve Bayes approach with word counts for spam detection:

- Obtain of sample of messages.
- Determine by inspection which messages are spam vs. non-spam.
- Divide the message sample into a training set (to fit the model) and a test set (to evaluate the fitted model's quality).
- Process each message using tokenization and stemming to create a "bag of words."
- Determine for each message the count of each word stem.
- Fit the model on the training messages.
- Apply the fitted model to assess how well it fits the test messages.

Preparing the data

The data we have are in a text file with one line per email. The file uses the pipe character | as the delimiter between fields. We can read this into a pandas DataFrame using:

```
[1]: # Read the data file of emails - 1 line per email, pipe delimited
     import pandas as pd
     data = pd.read_csv('emails2.txt', usecols=['isSpam', 'Message'], sep='|')
```

In the above statement, the fields of interest are specified with the `usecols` parameter. The binary `isSpam` field is 1 if the message was previously manually determined to be spam; otherwise, it is 0. The string `Message` field contains the text body of emails.

Having loaded the data, we can check how many spam vs. non-spam messages we have with:

```
[2]: # check how many spam vs. non-spam emails (1=spam)
     data.isSpam.value_counts()
```

```
0    140
1    139
Name: isSpam, dtype: int64
```

Based on the above counts, we have nearly the same number of spam vs. non-spam emails in our dataset.

Note that the data have non-spam emails first, followed by spam emails. This allows you to see a few non-spam emails with `data.head()` and a few spam emails with `data.tail()`. We can also browse the text of any particular email with `iloc[]` and slicing. For example:

```
[3]: # look at an example (non-spam) message
     data.iloc[0].Message[:160]
```

```
"just to update you on this project ' s status : based on a new report that scott mills
ran for me from sitara , i have come up with the following counterparties"
```

Having loaded the data into a DataFrame, we'll now process the text to prepare for word frequency calculations. As we did in the earlier section, we can remove non-letters, tokenize the text, and apply stemming. The result of these transformations will be placed into a new column in the DataFrame, called `Message2`, so we can compare the result with the original email text. Here are statements to perform these steps:

```
[4]: # take out non-letters
     def nonletterRemover(text):
         alist = [c if c.isalpha() else ' ' for c in text]
         return ''.join(alist)
     data['Message2'] = data['Message'].apply(nonletterRemover)
```

```
[5]: # tokenize
     from nltk.tokenize import word_tokenize
     tokenizer = lambda text: word_tokenize(text)
     data['Message2'] = data['Message2'].apply(tokenizer)
```

```
[6]: # apply Porter Stemmer
     from nltk import PorterStemmer
     porterStemmer = PorterStemmer()
     stemmer = lambda words: [ porterStemmer.stem(word) for word in words ]
     data['Message2'] = data['Message2'].apply(stemmer)
```

```
[7]: # rejoin text
     rejoiner = lambda words: ' '.join(words)
     data['Message2'] = data['Message2'].apply(rejoiner)
```

We can review the stem lists for a few messages with `head()` and `tail()`:

```
[8]: # compare the initial and transformed text for a few non-spam messages
     data.head()
```

	isSpam	Message	Message2
0	0	just to update you on this project ' s status ...	just to updat you on thi project s statu base ...
1	0	the above referenced meters need to be placed ...	the abov referenc meter need to be place on a ...
2	0	(see attached file : hpll 228 . xls) - hpll ...	see attach file hpll xl hpll xl
3	0	daren , it ' s in . bob - - - - - - - - - - - ...	daren it s in bob forward by robert cotten hou...
4	0	daren , fyi . bob - - - - - - - - - - - - - - ...	daren fyi bob forward by robert cotten hou ect...

```
[9]: # compare the initial and transformed text for a few spam messages
data.tail()
```

	isSpam	Message	Message2
274	1	we have very compet itive pricing on hundreds ...	we have veri compet itiv price on hundr of the...
275	1	good day , your application has been pre - app...	good day your applic ha been pre approv on thu...
276	1	dear sir or madam , would you refinance if you...	dear sir or madam would you refin if you knew ...
277	1	size = 1 > order confirmation . your order sho...	size order confirm your order should be ship b...
278	1	if you ' re tired of traffic lights , speed ca...	if you re tire of traffic light speed camera a...

Analyzing the emails

Now that `Message2` has lists of stems, we can proceed to modeling. When developing a model, it is good practice to divide our data between a training set to fit the model, and a test set to evaluate how well the model fits. We can make this split using scikit-learn's `train_test_split()` function:

```
[10]: # split to 25% test data and 75% train data
# isSpam is the dependent variable, Message2 is the independent variable
from sklearn.model_selection import train_test_split
train_text, test_text, train_labels, test_labels = \
 train_test_split(data.Message2, data.isSpam, test_size=0.25, random_state=1)
```

The four returned results are each pandas Series: `train_text` and `test_text` have the email text, and `train_labels` and `test_labels` have binary values from `isSpam`. For example:

```
[11]: train_labels

227    1
264    1
147    1
230    1
266    1
      ..
203    1
255    1
72     0
235    1
37     0
Name: isSpam, Length: 209, dtype: int64
```

Now that we have a train/test split, we'll use the `CountVectorizer` to determine the word frequency for each message using:

```
[12]: # build bag of words features vectorizer and get features
from sklearn.feature_extraction.text import CountVectorizer
bow_vectorizer = CountVectorizer(min_df=1, ngram_range=(1,1))
bow_train_features = bow_vectorizer.fit_transform(train_text)
bow_test_features = bow_vectorizer.transform(test_text)
```

In the above, the `fit_transform()` method will count occurrences of each word in each email. Note that `min_df=1` tracks words occurring at least once; `ngram_range=(1,1)` will look at single words instead of word combinations.

Next we are ready to fit the model using the `MultinomialNB` type:

```
[13]: from sklearn.naive_bayes import MultinomialNB # import naive bayes
      model = MultinomialNB()
      model.fit(bow_train_features, train_labels)
```

Given the word frequencies and labels (spam / not spam), the `fit()` method will determine probabilities of individual words occurring in non-spam and spam messages. From this, a message's overall probability of being spam can be calculated. Using this probability, the `predict()` method can classify emails as spam or not. For example, to predict if each email in the test set is spam, we use:

```
[14]: # predict using model
      predictions = model.predict(bow_test_features)
```

Checking the model's accuracy

Since our test data have known values for the dependent variable in `isSpam`, we can check the accuracy of the model. First we check how large our test dataset is:

```
[15]: # number of emails in the test data
      len(test_labels)
```

```
70
```

Since we have a relatively small dataset, we can find the individual cases where the model incorrectly classified the emails as spam or non-spam. In order to do this, we'll combine the actual spam indicator and the model's spam prediction into a DataFrame using:

```
[16]: test_results = \
        pd.DataFrame({'actual':test_labels.tolist(), 'predict':list(predictions)})
      test_results
```

	actual	predict
0	0	1
1	1	1
2	1	1
3	1	1
4	1	1
...	...	...
65	0	0
66	0	0
67	0	0
68	0	0
69	1	1

Above we see 10 observations for the actual and predicted spam indicator. For these, the first row indicates that the model predicted the email as spam (predict=1), but it was in fact not spam (actual=0). All the other rows show the model result matches the actual spam indicator.

To see all rows where the model was incorrect, we can use this filter expression:

```
[17]: test_results[test_results.actual != test_results.predict]
```

	actual	predict
0	0	1
24	0	1
36	1	0
53	0	1

The result above shows only four cases where the model is incorrect. We can compute the ***accuracy***, which is the number of correct predictions divided by the total number of observations, as (70 – 4) / 70 = 94.3%.

Accuracy is one of several measures of model fit. For example, we may also be interested in how often the model correctly predicted that a spam email was spam. Or we may be interested in how often the model correctly predicted that an email was not spam. These counts of how often the model is right or wrong can be summarized in a ***confusion matrix***. The confusion matrix can be determined using the scikit-learn function `confusion_matrix()`. For example:

```
[18]: # check accuracy using confusion matrix
      from sklearn import metrics
      metrics.confusion_matrix(test_labels, predictions)

      array([[33,  3],
             [ 1, 33]])
```

The above result is a two-by-two matrix (as a NumPy `ndarray`). The results can be more easily interpreted by labeling the rows and columns, as in Figure 18-1.

		model	
		not spam	*spam*
actual	not spam	33	3
	spam	1	33

Figure 18-1: Confusion matrix for email spam prediction

In Figure 18-1, the shaded cells show where the model's spam prediction was incorrect.

We can characterize the model fit in a few more ways. For example, the model correctly characterized non-spam email in 33 of the 36 cases (91.7%); this is referred to as ***specificity***. Also the model correctly indicated which emails were spam in 33 of the 34 spam cases (97.1%); this is referred to as ***sensitivity***.

Section Summary

Text classification is a technique for placing text into one of a set of categories. An example presented in this section shows how to identify emails that are potentially spam. This was done by combining word frequency with a naïve Bayes model to estimate probabilities for words occurring in spam vs. non-spam email. From these probabilities, a particular email can be classified as potentially spam or not. The nltk and scikit-learn packages in combination with some Python techniques allowed us to perform the analysis with relatively few lines of code, to categorize spam emails with a high degree of accuracy.

Test Your Understanding

1) The scikit-learn `CountVectorizer` can be used to:
   a) find the count of numeric values that are outside a given range
   b) determine the count of distinct words occurring in a set of text
   c) (a) or (b)
   d) none of the above

2) The scikit-learn `MultinomialNB` type implements:
   a) logistic regression
   b) naïve Bayes classification that can be used with word counts
   c) (a) or (b)
   d) none of the above

Chapter Review

Although much of the book has covered numerical problems, text analysis is an important and growing area given the large amount of text available electronically. In this chapter, we've covered several techniques for analyzing textual data, including some capabilities provided by the nltk and scikit-learn packages. Some of the steps toward analyzing text can be done with Python language features, such as string methods, lambda functions, and list comprehensions. In word frequency, we covered how to tokenize and summarize text in order to reveal the most commonly occurring words. Beyond basic word frequency, we used stop word filtering, stemming, and part-of-speech filtering to help create more useful frequencies.

For sentiment analysis, we looked at one approach with the VADER model available in nltk. This provides a straightforward way to determine positive, neutral, negative, and compound scores for text. We also covered text classification using word frequencies in combination with a naïve Bayes classifier to try to determine which emails were potentially spam.

Key Terms

confusion matrix	A summary of counts for which a model correctly and incorrectly predicts the actual value of a dependent variable.
lambda function	A one-statement unnamed function.
list comprehension	A statement that generates a filtered and/or transformed list from a list, string, or other type of iterable.
naïve Bayes classifier	A type of probabilistic classifier that can be used in text classification.
nltk package	Natural language toolkit package with functionality for working with text.
part-of-speech tagging	Associating the part of speech (POS) for each word in text.
scikit-learn	A package with machine learning functionality, also known as sklearn.
sentiment analysis	Measuring the degree of positive or negative opinion in text.
spam email	Unsolicited email, also known as junk email.
stemming	Determining the stem (or root) of a word.
stop words	Words that can be filtered out from a sentence without losing the original meaning.
text classification	Dividing text into two or more categories.
token	A distinct word or string as determined by tokenizing a text.
tokenize	Determine the distinct words or strings from a text.
word frequency	The counts of words or tokens occurring in text.

Programming Problems

For these problems, use the following sample text:

The laptop was difficult to set up, and difficult to connect to the WIFI. I would go back to using the previous laptop except it no longer charges.

a) Tokenize the sample text and display the tokens.
b) For the sample text, determine and plot the word frequency.
c) For the tokens of the sample text, determine a filtered word list that excludes the stop words provided in nltk. Determine and plot the word frequency of this filtered list of words.
d) For the sample text, tokenize, filter stop words, filter out nonnouns, apply stemming, and determine and plot the word frequency.
e) Using the SentimentIntensityAnalyzer, determine the polarity scores of the sample text.

Chapter 19

Case Study: Customer Retention Analysis

Chapter Objectives

In this chapter, you'll learn:

- how a classification model can seek patterns in data
- how you can use the statsmodels and scikit-learn packages to fit models to data
- how to create and use training and testing data subsets for fitting and testing a model
- how you can use a confusion matrix to evaluate a model's fit to the data
- what a threshold level is and one approach to determine its setting

Chapter Outline

19.1 About the Case

Case study: Customer retention

In this chapter, we'll look at the case study of a fictional telecom company. In particular, the company is experiencing a lot of subscription turnover due to high competition from alternative telecom companies. The company would like to see if it can use its subscriber data to predict in the future which customers might be more inclined to switch to another telecom service. If a good forecast model can be developed, it could be used to reach out to these at-risk customers and offer them discounts in order to try to retain their business.

The company has specific information on its subscribers, including demographic, service, billing, and payment history. In particular, here are the fields available in the company's databases:

Category	Field	Description
Demographic	gender	f=female, m=male
	married	s=single, m=married
	age	customer age in years
Service	service	b=basic, p=premium, g=gold
	active	0=customer has canceled service, 1=customer is active
Billing and payment	contract	n=no annual contract, y=has annual contract
	billing	a=autopay, c=pay by check
	due	0=balance up-to-date, 1=payment 1 month behind, 2=payment 2 or more months behind

Table 19-1: Telecom case study—data fields

In this chapter, we'll step through the modeling process for a statistical model to forecast which customers may be likely to switch providers. The steps covered are:

- selecting the factors
- fitting the model to a subset of the historical data (the training data)
- testing the model on additional historical data (the test data)
- reviewing the model accuracy
- seeking a good decision rule for using the model for forecasting

Classification models

A classification model seeks to assign a label to each observation based on data for that observation. For example, a program might try to identify a letter or digit from a handwritten character. In this case, there are 36 possible labels (26 letters and 10 digits). This is a multi-label classification problem. There are also binary classification problems, where the model tries to determine if an item is one or another type. As an example of this, a text analysis program might seek to decide if an email message is spam or not. In this chapter, we'll focus on an example binary classification model related to customer retention.

19.2 Data Preparation

Reading the data

The telecom company data we have are in two files. The file customers.csv contains most of the data fields indicated in Table 19-1. There is an additional file, status.csv, which contains a field called churn. This field is set to 1 if the customer has terminated their service in the last three months. From a modeling perspective, we can consider the data in customers.csv as the factors (also called independent variables), which we hope to use to forecast the target variable (also called dependent variable) in status.csv.

We'll read the customer data file into a pandas DataFrame using `read_csv()`:

```
[1]: import pandas as pd
     customers = pd.read_csv('customers.csv', index_col='#')
     customers.head()
```

#	gender	married	age	service	contract	billing	due
1	f	m	49	p	n	a	0
2	f	m	19	p	n	c	0
3	m	m	53	p	y	a	1
4	f	m	64	g	y	a	1
5	f	s	47	b	n	a	0

Similarly, we can read the status file:

```
[2]: status = pd.read_csv('status.csv', index_col='#')
     status.head()
```

#	churn
1	0
2	0
3	1
4	1
5	1

and then combine the data using the `join()` method:

```
[3]: customers = customers.join(status)
     customers.head()
```

#	gender	married	age	service	contract	billing	due	churn
1	f	m	49	p	n	a	0	0
2	f	m	19	p	n	c	0	0
3	m	m	53	p	y	a	1	1
4	f	m	64	g	y	a	1	1
5	f	s	47	b	n	a	0	1

We can check how many customers are represented in the data by checking the number of rows in our DataFrame:

```
[4]: len(customers)
     10000
```

Transforming the data

In order to fit the data, we need to transform our available data from character codes to numeric values, since the packages we'll use (statsmodels, scikit-learn) require numeric data. Table 19-2 shows the mapping we plan to use for fields with two possible values.

Field	Mapping
gender	m=0, f=1
married	s=0, m=1
active	n=0, y=1
contract	n=0, y=1
billing	c=0, a=1

Table 19-2: Field mapping for two-valued fields

For fields with two possible character values, we'll define a new numeric field with the value of 0 or 1. The `map()` method is a concise way to do this. Here are statements that will augment our dataset with four binary columns:

```
[5]: customers['gender_binary']   = customers.gender.map( {'m':0, 'f':1})
     customers['married_binary']  = customers.married.map( {'s':0, 'm':1})
     customers['contract_binary'] = customers.contract.map({'n':0, 'y':1})
     customers['billing_binary']  = customers.billing.map( {'c':0, 'a':1})
     customers.head()
```

#	gender	married	age	service	contract	billing	due	churn	gender_binary	married_binary	contract_binary	billing_binary
1	f	m	49	p	n	a	0	0	1	1	0	1
2	f	m	19	p	n	c	0	0	1	1	0	0
3	m	m	53	p	y	a	1	1	0	1	1	1
4	f	m	64	g	y	a	1	1	1	1	1	1
5	f	s	47	b	n	a	0	1	1	0	0	1

There is one additional field to map, service, which can take on three values. For modeling purposes, we define two new columns, service_premium_binary and service_gold_binary. Although we could also define service_basic_binary, there is no need, as this case is covered when service_premium_binary = 0 and service_gold_binary = 0. To create the new columns, we use the `map()` method:

```
[6]: customers['service_premium_binary'] = \
         customers.service.map({'b':0, 'p':1, 'g':0})
     customers['service_gold_binary'] = \
         customers.service.map({'b':0, 'p':0, 'g':1})
     customers[['service', 'service_premium_binary', 'service_gold_binary']].head()
```

#	service	service_premium_binary	service_gold_binary
1	p	1	0
2	p	1	0
3	p	1	0
4	g	0	1
5	b	0	0

The above statements complete the data augmentation and give us several binary columns that we can use to fit models with packages in the upcoming sections.

Note that we have not focused on data quality issues, as the dataset we have has clean data. In practice, we may need to discover and repair or exclude certain data, as covered in Chapter 11.

Section Summary

In order to fit a statistical model, we need to prepare the data for the software package to be used. This includes data preparation steps such as cleaning and transforming the data. Data transformations can be done with pandas—for example, using its `map()` and `apply()` methods in order to map string data to numeric values.

Test Your Understanding

1) In order to use statsmodels or scikit-learn, factors can have string or integer values.
   a) True
   b) False

19.3 Selecting the Factors

As a next step, we'll determine which factors are worthwhile to include in our model. In general, we want to include the set of factors that truly influence the target variable and omit other factors in order to avoid overfitting our model to the data. In some types of models, the fitting algorithm will determine which factors are selected. However, in a regression model like the one we'll use, we need to select our factors. One approach to this is to fit the model with a broad candidate set of factors and see which resulting coefficients are statistically significant.

Later in this chapter, we'll use the scikit-learn package. But while it can fit a logistic regression model, it does not readily provide a check for statistical significance of the fitting coefficients. For this reason, we'll first use another package, statsmodels, to initially fit our model and help us narrow down our factors.

To use the statsmodels package, we can import it as:

```
[7]: import statsmodels as sm
```

Next, we filter in the factors that we want to include in the model. These are all the numeric columns except for churn, which is the target variable. We can create the factors DataFrame as:

```
[8]:  X = customers[['age', 'gender_binary', 'married_binary',
                     'service_premium_binary', 'service_gold_binary',
                     'contract_binary', 'billing_binary', 'due'
                    ]]
```

For modeling purposes, we also want to include a constant coefficient to reflect a baseline level for the target variable. So that statsmodel will include a constant coefficient in the fitted model, we can use the function `add_constant()` to augment our factors DataFrame by adding a column called `'const'`:

```
[9]:  from statsmodels.api import add_constant
      X = add_constant(X)
      X['const'].head()

      1    1.0
      2    1.0
      3    1.0
      4    1.0
      5    1.0
```

We'll now turn to the target variable. The target variable is the churn column, so we can create a Python Series referencing it as:

```
[10]: y = customers.churn
```

We're now ready to fit the coefficients for our logistic regression using all numeric variables. This can be done using:

```
[11]: from statsmodels.api import Logit
      logreg1 = Logit(y, X).fit()

      Optimization terminated successfully.
               Current function value: 0.621024
               Iterations 5
```

and then the results can be displayed using:

```
[12]: logreg1.summary()
```

Logit Regression Results

Dep. Variable:	churn	**No. Observations:**	10000
Model:	Logit	**Df Residuals:**	9991
Method:	MLE	**Df Model:**	8
Date:	Sat, 19 Feb 2022	**Pseudo R-squ.:**	0.04017
Time:	11:32:54	**Log-Likelihood:**	-6210.2
converged:	True	**LL-Null:**	-6470.1
Covariance Type:	nonrobust	**LLR p-value:**	4.034e-107

	coef	std err	z	P>|z|	[0.025	0.975]
const	0.1323	0.083	1.593	0.111	-0.030	0.295
age	-0.0035	0.001	-2.949	0.003	-0.006	-0.001
gender_binary	-0.0192	0.043	-0.446	0.655	-0.104	0.065
married_binary	-0.0988	0.044	-2.254	0.024	-0.185	-0.013
service_premium_binary	-0.2277	0.051	-4.452	0.000	-0.328	-0.127
service_gold_binary	-0.4073	0.056	-7.279	0.000	-0.517	-0.298
contract_binary	-0.6625	0.043	-15.301	0.000	-0.747	-0.578
billing_binary	-0.3220	0.043	-7.458	0.000	-0.407	-0.237
due	0.4204	0.032	13.154	0.000	0.358	0.483

The summary section related to the model coefficients is the second part of the output. The column labeled P>|z| allows us to find factors that are statistically significant—these correspond to factors that have meaningful impact on the target variable. In particular, for 95% confidence, we identify values that are less than 0.05. In the above output, we see that all factors are statistically significant except for gender_binary.

We can repeat the model fitting by redefining our factors matrix, excluding service_premium_binary. This can be done with the statement:

```
[13]: X.drop('gender_binary', axis=1, inplace=True)
```

With this pared-down set of factors in X, we can refit the model, again using `fit()`, and check the results with `summary()`:

```
[14]: logreg2 = Logit(y, X).fit()

Optimization terminated successfully.
         Current function value: 0.621034
         Iterations 5
```

[15]:
```
logreg2.summary()
```

Logit Regression Results

Dep. Variable:	churn	**No. Observations:**	10000
Model:	Logit	**Df Residuals:**	9992
Method:	MLE	**Df Model:**	7
Date:	Sat, 19 Feb 2022	**Pseudo R-squ.:**	0.04015
Time:	11:59:24	**Log-Likelihood:**	-6210.3
converged:	True	**LL-Null:**	-6470.1
Covariance Type:	nonrobust	**LLR p-value:**	4.976e-108

	coef	std err	z	P>|z|	[0.025	0.975]
const	0.1229	0.080	1.530	0.126	-0.035	0.280
age	-0.0035	0.001	-2.957	0.003	-0.006	-0.001
married_binary	-0.0989	0.044	-2.257	0.024	-0.185	-0.013
service_premium_binary	-0.2273	0.051	-4.446	0.000	-0.328	-0.127
service_gold_binary	-0.4074	0.056	-7.281	0.000	-0.517	-0.298
contract_binary	-0.6623	0.043	-15.298	0.000	-0.747	-0.577
billing_binary	-0.3219	0.043	-7.457	0.000	-0.407	-0.237
due	0.4206	0.032	13.162	0.000	0.358	0.483

In this refined model, except for the const coefficient, the values for $P>|z|$ are all less than 0.05. These coefficients are statistically significant at the 95% confidence level. Based on this, we consider our refined model as having meaningful factors, and we'll continue to work with these factors in the next section.

Section Summary

For our case study, we used the statsmodels package to help pare down which factors we'd like to use with our model. In particular, we reviewed the p-values of the model coefficients to filter out one factor from our model. The review of p-values for the model's coefficients allows us to identify those factors that are statistically significant in order to choose a good set of variables for our model. This is not the only approach that might be used for choosing a set of candidate factors—for example, we might examine the correlation between factors—however, this is out of scope for this chapter.

Test Your Understanding

1) In order to add a constant coefficient, statsmodels:
   a) provides the function `add_constant()`
   b) provides the named parameter add_constant, which can be set to True
   c) does this by default
   d) all of the above

2) The statsmodels type for fitting logistic regression models is:
   a) MultipleRegression
   b) LogisticRegression
   c) Logit
   d) none of the above

19.4 Fitting the Model

For the remainder of the chapter, we'll shift from using statsmodels to using the scikit-learn package. Although statsmodels provided us with the statistical test to pare down our factors of interest, scikit-learn has better functionality for modeling and forecasting. This includes a method for dividing our data between a training subset (to fit the model) vs. a testing subset (to evaluate the model). scikit-learn also has some other models for fitting classification data.

Dividing a dataset into a training subset (to fit the data) and one or more testing subsets (to evaluate the fitted model) is a common modeling approach. This helps reduce the chance our model doesn't overfit the data, which would cause the model to be less useful for forecasting. The data in a testing data subset can be used to verify that the fitted model is also a good fit for the held-back testing data.

Our first use of scikit-learn is to call its `test_train_split()` function to divide the data into training and testing subsets. Recall the `X` DataFrame includes only numeric columns, in particular the factors for our model, and the `y` Series includes our target variable for churn.

We can split our data using:

```
[16]: from sklearn.model_selection import train_test_split
      X_train, X_test, y_train, y_test = train_test_split(X, y, random_state=0)
```

By default, `train_test_split()` creates 75% of the rows in the training data, although we could change this using the `train_size` parameter. Further, the `random_state` parameter controls how the rows are assigned to the training vs. test subsets. In the above, we use `random_state=0`, so the split is reproducible, instead of randomly giving different results when rerun. Note that `X_train` and `X_test` are DataFrames; `y_train` and `y_test` are Series.

We may be interested in how our training data are split between customers staying vs. ending their service. We can see this with:

```
[17]: round(y_train.value_counts() / len(y_train) * 100)

      0    65.0
      1    35.0
      Name: churn, dtype: float64
```

From this, we see our training data show that 65% of customers kept their service over the three months covered by our data and 35% did not.

We are now ready to fit our logistic regression model with scikit-learn. Note that this step overlaps somewhat with our use of statsmodels; however, we are fitting the model with a subset of the rows (the training data). We can fit the model using:

```
[18]: from sklearn.linear_model import LogisticRegression
      logisticRegression = LogisticRegression()
      logisticRegression.fit(X_train, y_train)
```

In the above, the `fit()` method determines the coefficients of the logistic regression equation. These are stored as part of our `logisticRegression` variable. Although we do not need to explicitly see the model's coefficients for subsequent steps, you can use the `intercept_` and `coef_` properties to see them:

```
[19]: logisticRegression.intercept_
```

```
array([0.08719319])
```

```
[20]: logisticRegression.coef_
```

```
array([[ 0.08748334, -0.0042766 , -0.11523966, -0.25986953, -0.41815953,
        -0.64980243, -0.30744573,  0.43194216]])
```

In the above results, the order of the coefficients corresponds to the order of the X variable we provided earlier to the fit() method. For example, married_binary is the second factor and has a corresponding coefficient in our model of -0.00442766. Among our binary variables, this is the smallest coefficient, so gender appears to have the smallest impact on customer retention relative to the other binary variables.

Section Summary

In this section, we used the scikit-learn package to first divide the data into a training set and a testing set using the test_train_split() function. The split allows us to fit our model on one dataset and further evaluate it on another dataset. The training dataset was then used to fit a logistic regression model for customer churn using the LogisticRegression class and its fit() method.

Test Your Understanding

1) In order to divide a pandas DataFrame into a testing set and a training set, scikit-learn provides:
   a) a function test_train_split()
   b) the named parameter split, which can be set to True
   c) (a) and (b)
   d) none of the above

2) The type provided by scikit-learn to fit a logistic regression model is:
   a) MultipleRegression
   b) LogisticRegression
   c) MultipleLogisticRegression

19.5 Predicting with the Model

In the previous section, we used a subset of our historical data to fit a model describing how selected factors might relate to the likelihood of retaining customers. Ultimately we are interested in using this model as a forecasting tool to anticipate which customers we are likely to lose. In this section, we'll start looking at how the model can be used to forecast customer retention.

Having fit the model in the previous section, we can now use our testing data to make predictions about the likelihood of customers terminating their service. We'll use the scikit-learn method predict_proba() on the test data:

```
[21]: logicPredict = logisticRegression.predict_proba(X_test)
```

The result of predict_proba() is a matrix (as a NumPy array). Let's look at the first and last few rows of this matrix:

```
[22]: logicPredict
```

```
array([[0.72596709, 0.27403291],
       [0.78930094, 0.21069906],
       [0.37359844, 0.62640156],
       ...,
       [0.64044751, 0.35955249],
       [0.49482045, 0.50517955],
       [0.58254648, 0.41745352]])
```

Note that each pair of numbers in the inner lists sums to 1. The first value of each inner list is the predicted probability the customer will be retained; the second value is the probability the customer will not be retained. Since our target variable is a binary variable, scikit-learn provides these two values. If we had a target variable with three or more outcomes, the result from `logicPredict` would correspond to the multiple outcomes.

We are interested in the probability of losing each customer as forecasted by the model. In order to extract this, we can use the following syntax from NumPy:

```
[23]: y_test_proba = logicPredict[:,1]
```

In the above, the variable `predict` has the model-forecasted probability of losing each customer. We can review it as a histogram using:

```
[24]: series = pd.Series(y_test_proba)
      series.hist(bins=20)
```

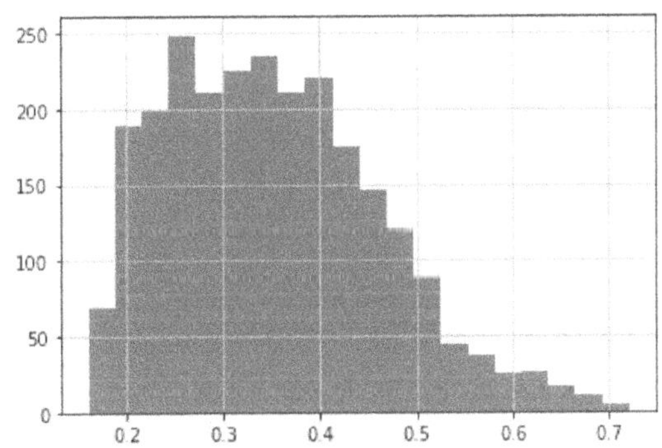

The above histogram indicates that the model forecast is that the majority of customers have less than a 0.5 probability of discontinuing service. The tail of the distribution on the right side (above 0.5) represents the customers we are at most risk of losing. Note that since this is historical data, we actually know which customers discontinued service, and we can compare this with the probabilities forecasted by the model. We will do this in the next section.

Section Summary

We used our logistic regression model, which was fit in the prior section, to predict churn probabilities for the test dataset. We were able to visualize these probabilities using a histogram. In the next section, we'll look at how well the model fits our data.

Test Your Understanding

1) Which scikit-learn `LogisticRegression` method can be used to return the predicted probability for each row of the dataset?
   a) `prob()`
   b) `predict_proba()`
   c) `fit()`
   d) all of the above

2) The result type of the LogisticRegression `predict_proba()` method is a:
   a) Python list
   b) pandas Series
   c) pandas DataFrame
   d) NumPy array

19.6 Reviewing Model Accuracy

In the previous section, we used the fitted model to predict the probability of losing each particular customer. In this section, we'll look at different measures that show how well the model fits the data. In particular, we'll use a classification model tool called the confusion matrix. This is a concise summary of how often our model (along with a chosen threshold) is right or wrong. The confusion matrix is a two-by-two table that compares what we know about our sample (in this case, whether customers were retained or not) vs. what our model forecasts for our customers' retention. Here is the general form for the confusion matrix:

		Predicted Value	
		1	0
Actual Value	1	TP	FN
	0	FP	TN

In the matrix above, the abbreviations are:

Abbreviation	Label	Meaning
TP	True Positive	Count of target variable values that are actually 1 that the model correctly predicts as 1
FN	False Negative	Count of target variable values that are actually 1 that the model predicts as 0
FP	False Positive	Count of target variable values that are actually 0 that the model predicts as 1
TN	True Negative	Count of target variable values that are actually 0 that the model predicts as 0

A perfect model would have FN and FP as 0—in other words, the model is always right. This is almost never the case, as models are imperfect, though hopefully useful. There are several measures of model accuracy. For example, ***accuracy*** is calculated as the percentage of time the model is right: (TP + TN) / (TP + TN + FN + FP).

Accuracy is the simplest measure of how well the model fits the data. We may also be interested in how well the model fits for a subset of the observations. For example, the measure for how often the model correctly predicts actual 1 cases is called *sensitivity*, calculated as TP / (TP + FN). This focuses on the first row of the confusion matrix. This and other measures are summarized in Table 19-3.

Measure	Matrix part	calculation
sensitivity	Top row	TP / (TP + FN)
specificity	Bottom row	TN / (TN + FP)
precision	First column	TP / (TP + FP)
negative predictive value	Second column	TN / (TN + FN)

Table 19-3: Confusion matrix measures

For our subscriber data, we can calculate a confusion matrix for the fitted model at a given probability cutoff. scikit-learn provides the function `confusion_matrix()` to generate a confusion matrix from a parallel set of actual values and predicted values.

In order to check the confusion matrix for the test data, we need to apply the model at a given probability threshold to predict if each customer will be retained or lost. We can do this with:

```
[25]: # for a cutoff, determine if prediction would be 0 or 1 for each observation
      cutoff = .5 # sample cutoff, can set to any value between 0 and 1
      y_test_proba_series = pd.Series(y_test_proba)
      y_test_pred = y_test_proba_series.apply(lambda x: 1 if x>cutoff else 0)
      y_test_pred.head(5)
```

```
0    0
1    0
2    1
3    0
4    0
dtype: int64
```

Given the prediction in y_test_pred, we can now determine the confusion matrix for the test data using:

```
[26]: # given truth and the model+cutoff prediction, generate a confusion matrix
      from sklearn.metrics import confusion_matrix
      cm = confusion_matrix(y_test, y_test_pred)
      cm
```

```
array([[1516,  121],
       [ 741,  122]])
```

Note that the result type of `confusion_matrix()` is a NumPy `ndarray`—in this case, a two-by-two matrix.

To display the confusion matrix as earlier shown (with TP in the upper left), we can write utility functions that access matrix elements returned from `confusion_matrix()`. We may also want to see the various measures of fit and can cover these in our function as follows:

```
[27]: # convenience function for calculating any confusion matrix ratio
      def cmratio(correct, incorrect):
          if correct+incorrect == 0: return 0
          else: return correct / (correct+incorrect) * 100

      # display confusion matrix
      def displayCM(cm, w=4):
          accuracy    = cmratio(cm[0][0]+cm[1][1], cm[0][1]+cm[1][0])
          sensitivity = cmratio(cm[1][1], cm[1][0])
          specificity = cmratio(cm[0][0], cm[0][1])
          precision   = cmratio(cm[1][1], cm[0][1])
          npv         = cmratio(cm[0][0], cm[1][0])

          fmt = str(w)+"d"
          fmt1 = ".1f"
          print(' TP:', format(cm[1][1],fmt), 'FN:', format(cm[1][0],fmt),
                        format(sensitivity,fmt1))
          print(' FP:', format(cm[0][1],fmt), 'TN:', format(cm[0][0],fmt),
                        format(specificity,fmt1))
          print('     ', format(precision,fmt1), '     ', format(npv, fmt1))
          print('Accuracy:', format(accuracy, fmt1))
```

With these functions defined, we can now display our confusion matrix, calculated above, annotated with goodness of fit measures:

```
[28]: displayCM(cm)

 TP:  122 FN:  741 14.1
 FP:  121 TN: 1516 92.6
       50.2     67.2
Accuracy: 65.5
```

According to the above confusion matrix, we see that our fitted model with a threshold of 0.5 has an accuracy of 65.5% when applied to the test data. Notably, our sensitivity is only 14.1%, which means at a 0.5 cutoff, we are incorrectly predicting many customers will leave who are actually staying. In the next section, we'll look at varying the cutoff level to see how the model can be more accurately and usefully applied.

Section Summary

We introduced the confusion matrix as a tool for reviewing how well the model fit the data. Within the confusion matrix are counts of actual classification as compared with the model's classification. For example, TP indicates the number of true positive observations—those customers who we have lost that the model predicted (above a certain probability cutoff) that we would lose. Other counts in the confusion matrix (TN, FP, FN) allow us to compute a number of measures such as accuracy, and still others to see how well the model matches the actual data.

Test Your Understanding

1) A confusion matrix size for a binary classification model:
   a) is two rows and two columns
   b) is three rows and three columns
   c) depends on the number of rows in the dataset

19.7 Setting the Model Threshold

In the previous section, we introduced the confusion matrix and its associated ratios as a way to evaluate how well the model fits the data. In this next section, we look at how we might use this information. In particular, we want a threshold probability value above which we want to reach out to customers to seek to retain them. For example, we might decide to offer a discount on an annual contract as an incentive. In this section, we'll use a cost-benefit approach to compare different thresholds in order to select one to use.

Let's assume we'll apply our model as follows. For each customer who has a probability of terminating their service above a certain level, *p*, we will offer them an annual subscription at a discounted rate. Some of these customers may take the deal. What is the value to our company with this approach?

Note that because our model is not perfect, some of the customers we offer the discount to would have continued with us without the discount. In terms of the confusion matrix, these are the FP—that is, false positives. For these lucky customers, we will lose some revenue, since we've given them a discount we didn't need to. Conversely, those customers we would have lost but who decide to take a discount contract are additional revenue we can attribute to the strategy. In essence, we want to set a threshold probability to balance the cost of unneeded discounts with the benefit of retained customers.

The approach we'll take is to consider a range of probability thresholds and, for each threshold, estimate the potential profit for offering discounts to selected customers whose probability of leaving is above the threshold. For this approach, we'll to make a number of assumptions:

For customers in the test sample who had switched service (TP):
  Percentage who would have taken the one-year discounted contract: 25%
  Profit for retaining the customer: $500
For customers in the test sample who did not switch service (FP):
  Percentage who would have taken the one-year discounted contract: 65%
  Loss due to provided discount: $120

Based on these assumptions, for any given threshold level, we can calculate for the test sample the profit for selectively offering a discounted one-year plan as:

Overall value of discount offer = TP × $500 × 25% - FP × $120 × 65%

We can generate a table of thresholds and values for the discount offer with the following code:

```
[29]: # parametric analysis of cutoff vs. profit&loss
      from numpy import arange
      pls_x = []
      pls_y = []
      print('cutoff', 'P&L ($)')
      for cutoff in arange(.05, .99, .05):
          y_test_pred = y_test_proba_series.apply(lambda x: 1 if x>cutoff else 0)

          cm = confusion_matrix(y_test, y_test_pred)

          # example P&L function based on model predict=1 column
          # true + (retention): $500/customer, assume 25% re-subscribe (take offer)
          # false + (unneeded discount): $120/customer, assume 60% take offer
          pl = cm[1][1] * 500 * .25 - cm[0][1] * 120 * .65
          print(format(cutoff,'.2f'), format(pl,'9,.0f'))

          # store values for subsequent plotting
          pls_x.append(cutoff)
          pls_y.append(pl)
```

```
cutoff P&L ($)
0.05   -19,811
0.10   -19,811
0.15   -19,811
0.20   -11,873
0.25    -1,830
0.30     9,133
0.35    13,567
0.40    13,600
0.45     9,108
0.50     5,812
0.55     4,663
0.60     3,003
0.65       970
0.70       250
0.75         0
0.80         0
0.85         0
0.90         0
0.95         0
```

The code above creates two parallel lists, which allows us to plot the profit-and-loss curve as follows:

```
[30]: pl_df = pd.DataFrame({'cutoff':pls_x, 'pl':pls_y})
      pl_df.plot(x='cutoff', y='pl', kind='scatter')
```

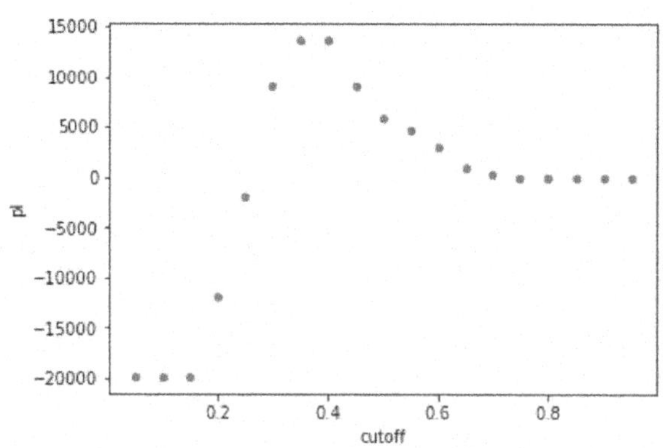

There are a few insights from the profit-and-loss table and chart above. The maximum profit for this model with the test data is at or near the threshold of 0.40. In other words, given the assumptions for customer discounted plan adoption, we should offer the discount to customers whose estimated churn probability is 40% or more. Note that the models in scikit-learn often use 0.5 as the default threshold, but in this case, that is a suboptimal decision level.

Further, note that for low thresholds, the plan generates a loss. This is because we give the discount to many customers who would not have left our service. Conversely, for very high cutoff levels, there is no value for the discount program, since no customers have an estimated churn probability that high.

Section Summary

In order to use a binary classification model, we need to decide on a probability threshold. In this section, we presented a profit-and-loss approach to decide on a threshold level that maximizes the forecasted profit of employing the model.

Test Your Understanding

1) The probability threshold:
   a) is a value between 0 and 1
   b) is used to determine if a predicted target variable is 0 or 1 when applying a binary classification model
   c) should be set before fitting a logistic regression model
   d) (a) and (b)

Chapter Review

In this chapter, we introduced an example of a statistical model applied to a case study on customer retention. Here we focused on a binary classification model to try to forecast the probability of losing a given customer based on their demographic, service, and billing data. The specific model type we used was logistic regression—although note that there are several other types of classification models that are also implemented in scikit-learn. In order to apply this approach, we first augmented the data to include numeric values representing string labels. With these data, we could determine which factors were statistically significant.

With the factors chosen, we divided the data into a training dataset for fitting the model and a testing dataset for evaluating the fitted model. The fitted model was used with different cutoff probabilities to determine a profit-maximizing strategy for offering customers service discounts in the hopes of retaining customers we are at risk of losing and ultimately improving our overall profitability.

Key Terms

accuracy	The percentage of observations that are correctly predicted by a model.
binary classification model	A type of model that predicts the probability of a target variable being 0 or 1.
classification model	A type of model that predicts the classification of an observation.
confusion matrix	A summary of counts for which a model correctly and incorrectly predicts the actual value of a dataset.
logistic regression	A type of binary classification model that fits coefficients for a set of factors to best fit a target variable.
precision	A measure from a confusion matrix for how often a model is correct for all observations that the model predicts as positive (i.e., target variable is 1).
sensitivity	A measure from a confusion matrix for how often a model is correct for the set of observations whose outcome is positive (i.e., target variable is 1).
specificity	A measure from a confusion matrix for how often a model is correct for the observations whose outcome is negative (i.e., target variable is 0).
testing dataset	A subset of data used to test a model previously fit with a training dataset.
training dataset	A subset of data used to fit a model.

Programming Problems

1) A Decision Tree Model

The scikit-learn package includes several classification models, including the DecisionTreeClassifier class.

a) Divide the data used in this chapter between training and test subsets.
b) For the training data created in part (a), use the DecisionTreeClassifier model to fit a model.
c) Use the model fit in part (b) to predict churn for each customer in the test data subset.
d) Decision trees can be limited to a certain depth. Fit alternative decision tree models to different maximum depths and compare their accuracy on the test subset data.

2) Receiver Operating Characteristic (ROC) Curve

The receiver operating characteristic (ROC) curve is an approach to reviewing model accuracy over a variety of thresholds. In particular, this curve is plotted with the false positive rate on the *x* axis vs. the true positive rate on the *y* axis. Each point corresponds to a different threshold. Using the logistic regression model developed in this chapter, determine the ROC curve points and visualize these data.

APPENDIXES

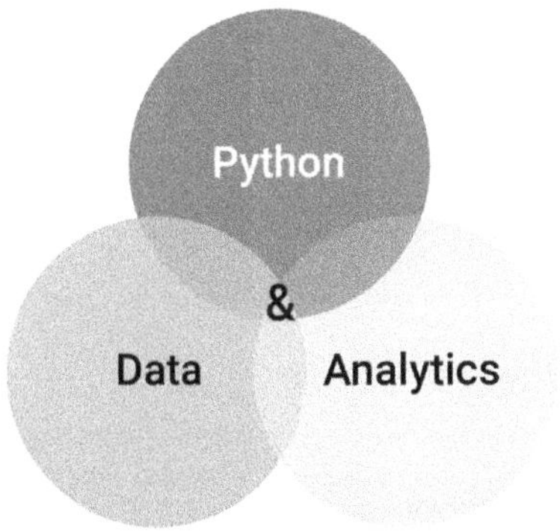

APPENDIX A

Getting Started With . . .

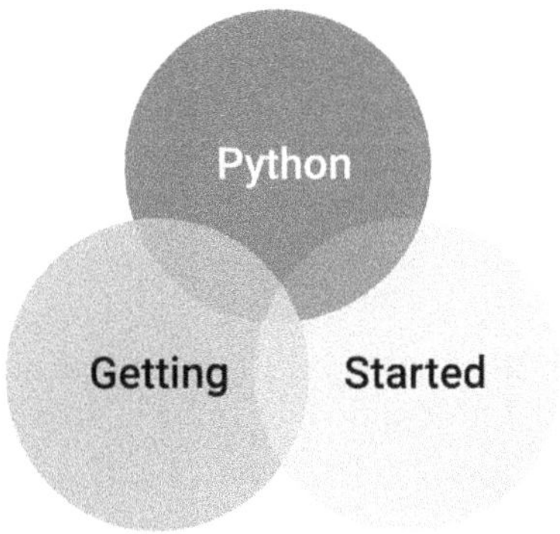

Appendixes A1 through A7 provide background on getting started with Python and Python packages. There are different development tools for using Python. Two major approaches are *notebook tools*, which allow for interactive cumulative development, and *integrated development environments* (IDEs), which allow you to edit, run, and debug programs.

Appendixes A1 and A2 introduce popular notebook tools: Jupyter Notebook and Google Colab. Certain IDEs are introduced in Appendixes A3 through A5. Appendix A3 introduces Thonny, an IDE for learning Python. The professional IDEs Spyder and PyCharm are covered in Appendixes A4 and A5.

Python packages are used throughout Parts II and III of the book. One way to get many packages is to install the Anaconda distribution, which is covered in Appendix A6. Further background about getting and using packages is covered in Appendix A7.

Appendix A1

Getting Started with Jupyter Notebook

A1.1 Overview

Jupyter Notebook is an excellent tool for learning Python, as well as creating and documenting professional Python-based analyses. It allows you to write one or more lines of code in its cells and to run one, several, or all cells at a time. In addition to code cells, you can also intersperse documentation cells that allow you to describe your code with formatted comments. These features make the notebook style of development well suited to coding as well as exploratory data analysis. In this appendix, we'll look at how to get started with Jupyter Notebook.

A1.2 Getting and Installing

You can get Jupyter Notebook in one of a number of ways. It comes with the Anaconda distribution (www.anaconda.com), which also includes Python and a large number of packages. Other options for getting Jupyter Notebook are described at jupyter.org/install.

A1.3 Starting

After you've installed Jupyter Notebook, you can start it by opening a command line window and entering:

```
jupyter notebook
```

The command line is available in Windows using the Command Prompt application, which you can start from the Windows Start menu. On the Mac, you can use the command line using the Terminal application, which you can start from the Mac Spotlight Search magnifying glass icon.

As an alternative to using the command line, if you installed Anaconda, you can use the Anaconda Navigator to start Jupyter Notebook. This is described at https://docs.anaconda.com/anaconda/user-guide/getting-started/.

Jupyter Notebook's interface runs in a web browser. After Jupyter Notebook starts, the following is displayed in your web browser (though your list of directory folders will be different):

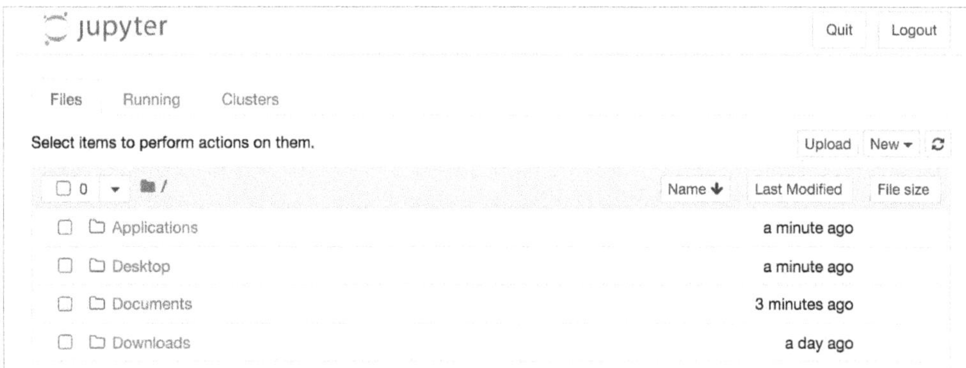

Note that the Jupyter Notebook start-up screen is related to what your current directory was when you entered the `jupyter notebook` command.

A1.4 Navigating and Running Notebooks

At the initial Jupyter Notebook screen, you can click on a directory name next to the folder icon to open it and navigate down your computer's file folder structure. For example, having clicked on Documents and then John, we have the following screen:

The above shows an empty directory. We can create a new notebook file by selecting New in the upper right and then Python 3:

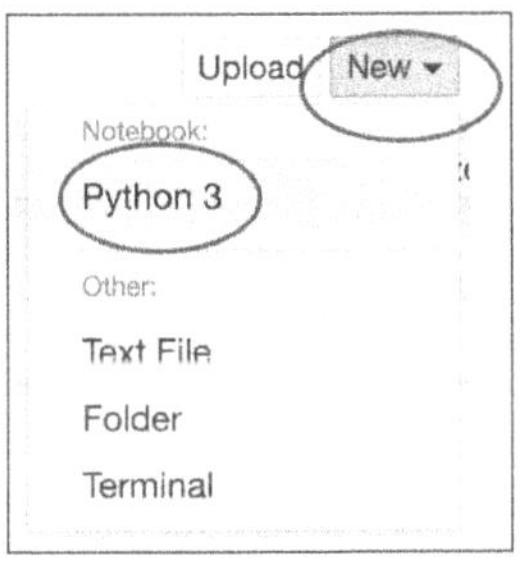

This opens a new tab or window in your browser, displaying a new empty notebook:

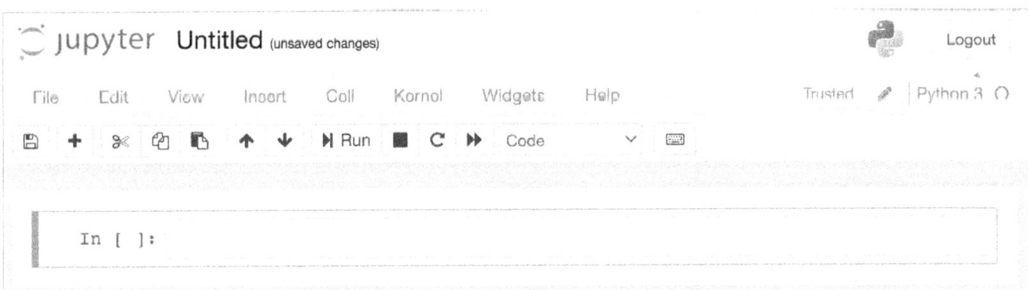

In the cell next to the `In [ ]:` prompt, you can enter one or more lines of code and then click on Run in the toolbar. For example:

Above we see that the output of our statement (in this case, `Hello`) is displayed below our input cell.

Also above we see Jupyter Notebook provided another input cell where we can further enter one or more statements. Alternatively we can go back to cell 1, edit, and rerun that cell.

If we define a variable and run that cell, that variable is available in subsequent cells. For example:

```
In [2]: sales = 124.5

In [3]: sales

Out[3]: 124.5

In [4]: print('Sales=', sales)

        Sales= 124.5
```

Jupyter Notebook periodically automatically saves your notebook in a file with the filetype .ipynb (the full path and filename is displayed in the browser URL). A new notebook is given an initial name like Untitled, but you can change this by clicking on the filename, and entering a new name.

To save your notebook, from the File main menu, choose Save and Checkpoint or Save as . . .

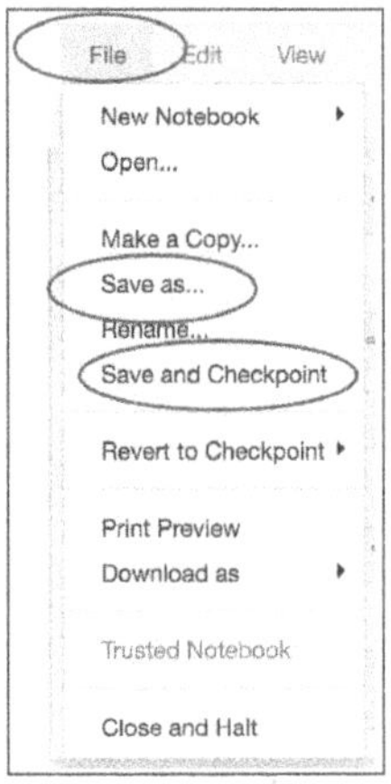

Save and Checkpoint will save your notebook with the current notebook name. Alternatively you can use Save as . . . to save your notebook with a different name.

A1.5 Markdown

Notebook cells can also contain documentation, with support for highly formatted notes using the **Markdown** language.* A cell can be indicated as Markdown using the menu near the top by changing the choice from Code to Markdown:

* Information about Markdown can be found online. See "Markdown Cells," Jupyter Notebook, accessed April 20, 2022, https://jupyter-notebook.readthedocs.io/en/stable/examples/Notebook/Working%20With%20Markdown%20Cells.html.

The simplest formatting can be done by leading a Markdown cell with one or more # signs corresponding to different size headings. Below, ## Demonstration Notebook has been entered in the Markdown cell and, once Run, appears as:

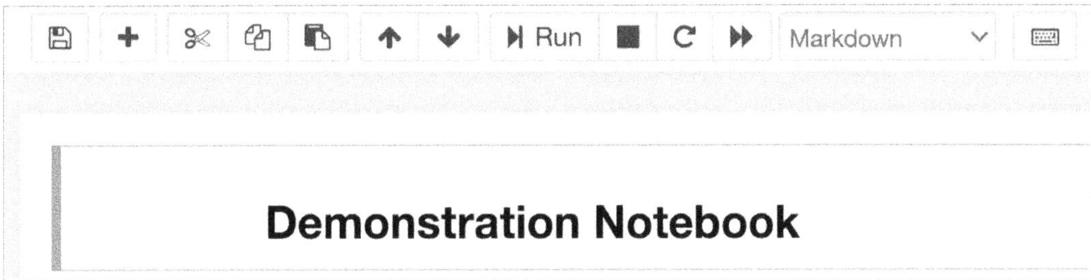

The ability to intersperse formatted notes between code cells is a distinguishing and useful feature of notebooks.

A1.6 Editing and Inserting Cells

You can evolve your notebook by working with cells. This includes:

- deleting one or more cells (Edit > Delete Cells)
- splitting a multi-line cell into two cells (Edit > Split Cell)
- merging two cells together (Edit > Merge Cell Above *or* Merge Cell Below)
- moving one or more cells up or down (up and down arrows in toolbar)

You can also insert cells using:

- Insert > Insert Cell Above
- Insert > Insert Cell Below

Jupyter Notebook also has an auto-complete feature that you can use with the tab key. For example, if you enter part of a variable name and hit tab, Jupyter notebook may complete the name of the variable for you or provide a choice list with defined variable names from which you can choose.

A1.7 Running and Clearing All

When developing a notebook, you typically run one cell at a time. As cells take shape, you may want to rerun several or all cells.

- To run all cells, use Cell > Run All.
- To run all cells, resetting the cell numbers, use Kernel > Restart & Run All.

In order to keep your cell input but clear all output results:

- Run Cell > All Output > Clear.

A1.8 Creating and Renaming Files and Directories

From within Jupyter Notebook, you can also create new directory folders as well as text files.

To create a directory folder:

1) Right-click the window and select New > Folder. This will create a folder called Untitled Folder (or similar).
2) Check the box to the left of Untitled Folder.
3) Select Rename near top, then give the folder a new name.

Similarly, you can create a text file:

1) Select New > Text File. This will create a file called untitled.txt (or similar).
2) Select File > Rename, then give the file a new name.

A1.9 Exiting Notebook Screens

In order to stop working with a notebook, from the File menu, choose Close and Halt:

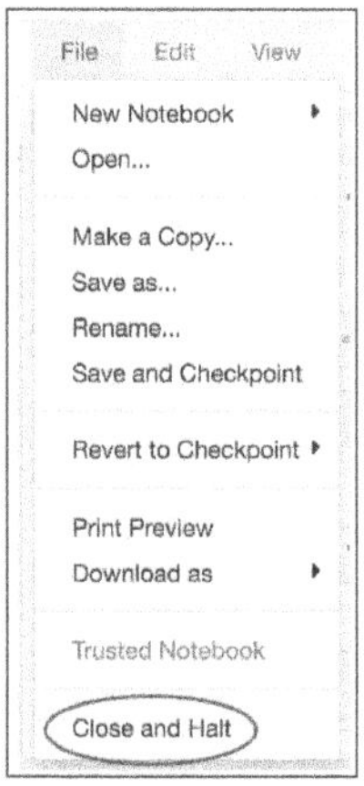

In order to exit from a Jupyter Notebook directory screen, simply close the tab from your browser.

A1.10 Opening a Previously Saved Notebook

To return to a previously saved notebook, use Jupyter Notebook Files to navigate to the folder where your notebook is saved. Then simply click on the name of the notebook (a .ipynb file) and Jupyter Notebook will open it in a new tab in your browser.

Appendix A2

Getting Started with Google Colab

A2.1 Overview

Google Colab is an excellent tool for learning Python, as well as creating and documenting professional Python-based analyses. It allows you to write one or more lines of code in its cells and to run one or several cells at a time, similar to Jupyter Notebook. In addition to code cells, you can also intersperse documentation cells. These features make the notebook style of development well suited to coding, as well as exploratory data analysis. Google Colab is cloud-based, so you don't need to install anything to get started. In this appendix, we'll look at how to get started with Google Colab.

A2.2 Getting Started

Since Google Colab runs in your browser over the internet, there is nothing you need to have installed on your computer other than a web browser. However, Colab stores your work in Google Drive, which is included with a Google account. If you don't yet have a Google account, you can create one at google.com (choose Sign in, then Create account).

A2.3 Starting

To use Colab, you'll need to sign into your Google account. You can do this from google.com.

Colab can be started from a browser by entering:

https://colab.research.google.com/

or searching for Google Colab in a web browser.

For your first time using Colab, you will see a menu frame within your browser like this:

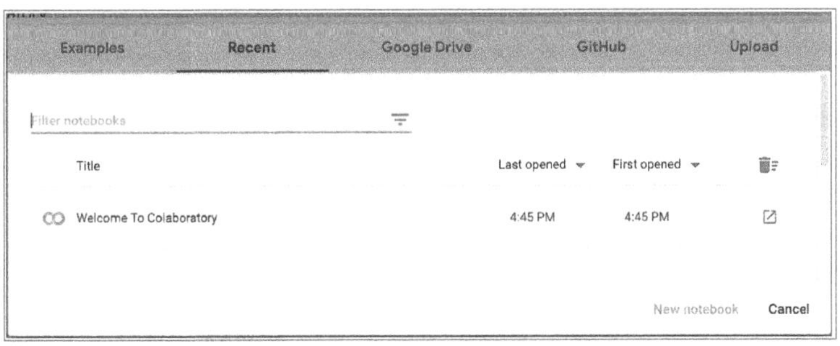

From this menu, select New notebook from the lower right. A new empty notebook screen will be displayed:

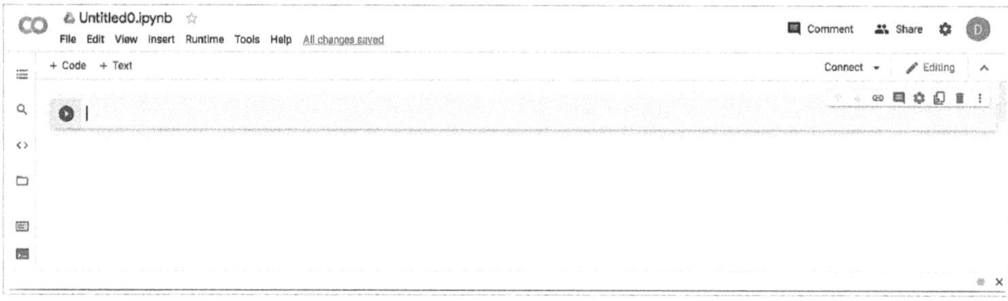

A2.4 Entering and Running

Once you have a notebook in view, you can enter code statements in its cells, then click on the triangle next to the cell to run that code. For example:

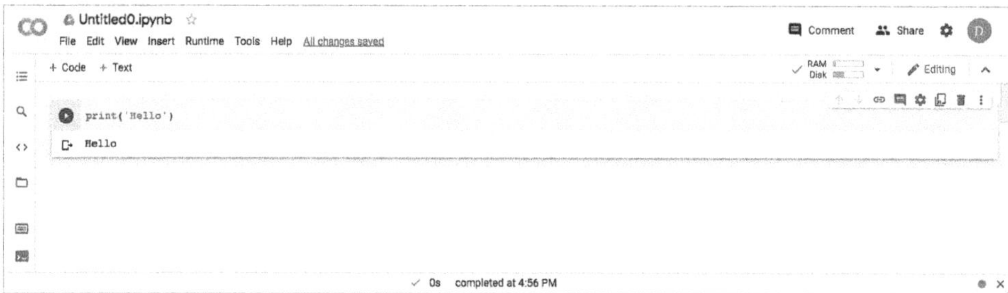

Above we see the result of our statement. We can then revise our cell code and rerun with the triangle. Or we can add a new cell by clicking on + Code at the top left.

If we define a variable and run that cell, that variable is available in subsequent cells. For example:

```
[2] sales = 124.5

[3] sales

    124.5

[4] print('Sales=', sales)

    Sales= 124.5
```

As you enter statements and run code, Colab automatically saves your notebook in your Google Drive in a file with the filetype .ipynb in the folder called Colab Notebooks. Your first new notebook is given the name Untitled0.ipynb, but you can change this by clicking on the filename and entering a new name.

Though the above examples show one statement in each cell, you can enter and run multiple statements in a single cell. The organization of code in cells is up to you, but in general, you should place statements for a logically related set of work in one cell.

A2.5 Text

Notebook cells can also contain documentation with support for formatted notes. In order to add a text cell, select the + Text choice from the toolbar. Text cells can be formatted using the toolbar available above that cell:

The ability to intersperse formatted notes between code cells is a distinguishing and useful feature of notebooks.

Once you've completed entering a text cell, you can continue working with your notebook by clicking anywhere else in your notebook, such as another cell, + Code or + Text, or by choosing a menu selection.

A2.6 Editing and Inserting Cells

You can evolve your notebook by working with cells. This includes:

- deleting one or more cells (Edit > Delete Selected Cells)
- moving one or more cells up or down (up and down arrows in cell's own toolbar at its right)

You can also insert cells anywhere in your notebook by first moving your cursor to a particular cell, and then selecting + Code or + Text.

A2.7 Running and Clearing All

When developing a notebook, you typically run one cell at a time. As cells take shape, you may want to rerun several or all cells.

- To run all cells, use Runtime > Run All.

In order to keep your cell inputs, but clear all output results:

- Select Edit > Clear All Outputs.

Clearing all outputs is useful if you'd like to demonstrate the steps of your notebook to someone, by running one cell at a time.

A2.8 Exporting and Importing Notebooks

Colab notebooks are stored as .ipynb files in Google Drive. If you'd like to download a notebook file to your computer, you can use:

File > Download > Download .ipynb

Conversely, if you have a .ipynb on your computer, and you'd like to load it into Colab, use:

File > Upload notebook

The .ipynb format is the same used by Jupyter Notebook. This allows you to use both Google Colab and Jupyter Notebook with your notebooks.

A2.9 Loading Data Files to Your Colab Session

Since Google Colab runs in the cloud, by default, it doesn't have access to data files on your computer. However, you can interactively load your local data files into your Colab session. In order to do this, click on the folder icon in the left toolbar:

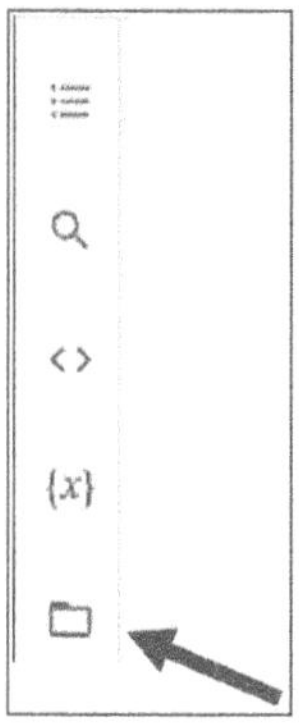

This will open the Files choices (below), which includes an icon (pointed to below) for selecting a data file to upload to your Colab session:

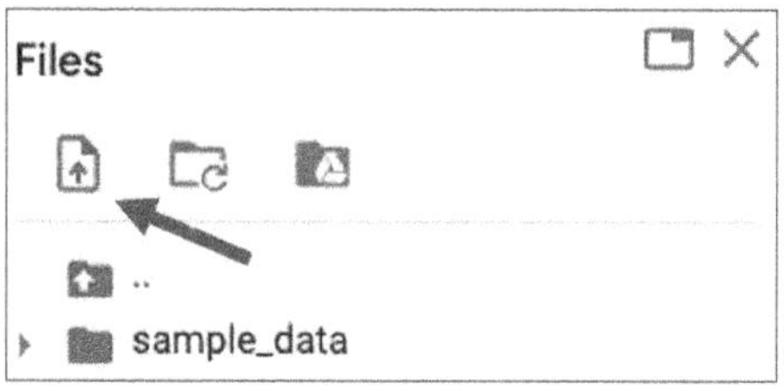

Click on this "Upload to session storage" icon and your computer's file chooser will open. Then navigate to and select the data file you'd like to load into the Colab session. Once loaded, your code in Colab will be able to read the data in this file.

Note that your data file is available to Colab for the duration of your session only. Once you leave Colab, the data file is no longer remembered for your next Colab session, so you will need to upload your data again when you return to Colab.

A2.10 Opening a Previously Saved Notebook

To return to a previously saved notebook, use File > Open notebook to open Colab's file chooser. From here, locate and select the notebook you'd like to open and Colab will open it in your browser.

Note that the Colab file chooser includes an Upload tab; from here, you can select a .ipynb saved on your computer to be loaded into Colab. You can also use File > Upload notebook to go to the Upload tab.

Appendix A3

Getting Started with Thonny

A3.1 Overview

An integrated development environment (IDE) lets you enter, run, and debug your programs. There are several IDEs available for Python. The Thonny IDE is designed with new Python programmers in mind, and is a great tool for getting started with Python. Like other Python IDEs, Thonny lets you enter and run one statement at a time, or you can create a program with multiple statements. In this appendix, we'll cover how to get started with Thonny.

Thonny is available at thonny.org. There are versions for Windows, Mac, and Linux. This appendix covers installing and using Thonny with Windows 10, but the other versions are very similar.

A3.2 Installing Thonny

In order to install Thonny on your computer, visit thonny.org in a browser. In the Thonny home page there are links to download installers for Windows, Mac, and Linux. Click on the link for your computer's operating system.

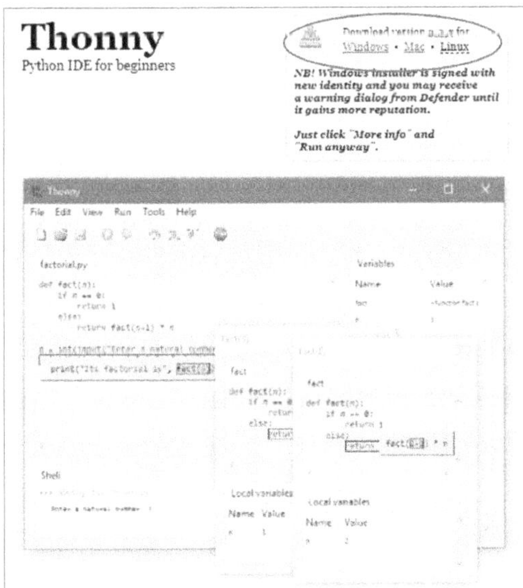

After clicking on one of the download links, your browser should download an installer file. In Windows, locate and run this file in order to install Thonny. You will be presented with a sequence of screens by the installer; you can generally select Next without changing any of the default settings. After several screens, a screen will include an Install button. After selecting this, the installer will take several minutes to install the Thonny program. Once it completes successfully, you should see this screen:

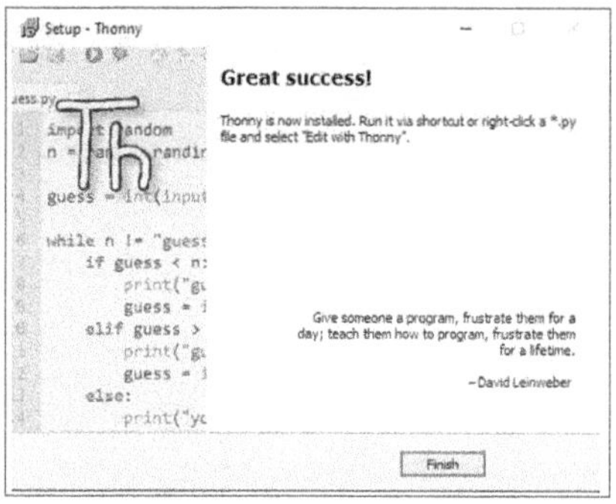

The Windows Start menu will have this entry:

For easier access to Thonny, you can place it in the Windows taskbar. To do this, right-click on the Thonny menu entry, then select More . . . , then Pin to the taskbar.

A3.3 Using Python Interactively in Thonny

When you run Thonny, the application will look as follows:

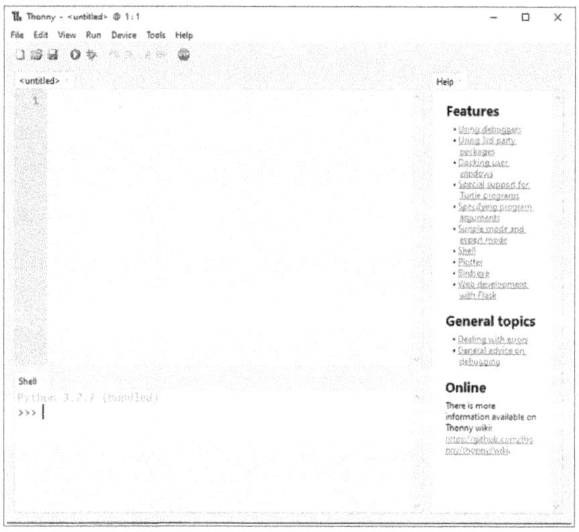

At the bottom of the screen is the interactive Python shell. This is where you can enter one statement at a time and immediately see the result. The >>> prompt indicates that you can enter a statement there. For example, below, a print statement is entered, and Python displays the output result.

```
>>> print('Hello')
 Hello
```

In order to clear the shell area, right-click in this area and select Clear from the menu.

A variable can be set in the interactive shell. For example, below, the variable *i* is set to 1. Unlike the `print()` example above, setting a variable does not result in output being displayed. In a subsequent statement, the variable can be displayed by simply typing the name of the variable at the >>> prompt.

```
>>> i = 1
>>> i
1

>>>
```

A3.4 Running Python Programs in Thonny

In the previous section, we reviewed how single statements can be run interactively in the Thonny shell. You can also enter multiple lines in a file, and run the file as a program. This is done in the top half of the Thonny screen. For example, below, a two-line program is entered:

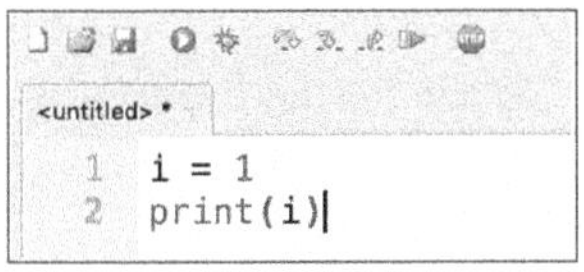

If you click on the run icon (white triangle within the green circle), Thonny will ask you to save the program in a file with a given name. This will create (or replace) a file on your computer with the name you choose and a .py file extension. Once the file is saved, Python will run it and display the results (or error messages, if any) in the bottom section.

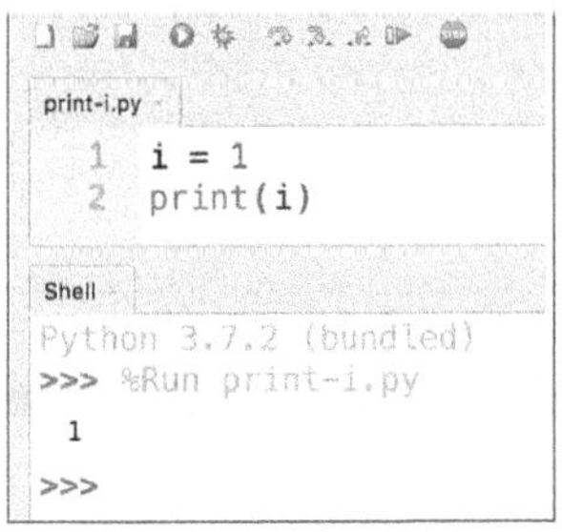

Appendix A4

Getting Started with Spyder

A4.1 Overview

An integrated development environment (IDE) lets you enter, run, and debug your programs. There are several IDEs available for Python. The Spyder IDE is a professional development tool and is also an option for getting started with Python. Like other Python IDEs, Spyder lets you enter and run one statement at a time, or you can create a program with multiple statements. In this appendix, we'll cover how to get started with Spyder.

Spyder is available at spyder-ide.org. (It is also included as part of the Anaconda distribution, which includes several tools and packages.) There are versions of Spyder for Windows, Mac, and Linux. This appendix covers installing and using Spyder with Windows 10, but the other versions are very similar.

A4.2 Installing Spyder

In order to install Spyder on your computer, visit spyder-ide.org in a web browser. On the Spyder home page, there is a choice for Download:

Clicking on Download will move you to the instructions for downloading Spyder.

After clicking on the Download button, your browser downloads an installer file. In Windows, locate and run this file in order to install Spyder. You will be presented with a sequence of screens by the installer; you can generally select Next without changing any of the default settings. After several screens, a screen will include an Install button. After selecting this, the installer will take several minutes to install the Spyder program.

Once Spyder is installed, the Windows Start menu will have an entry for Spyder. Alternatively, if you installed the Anaconda distribution, there will be an entry to Spyder within the Anaconda folder with the Windows Start menu.

For easier access to Spyder, you can place it in the Windows taskbar. To do this, right-click on the Spyder menu entry, then select More . . . , then Pin to the taskbar.

A4.3 Using Python Interactively in Spyder

When you run Spyder, the application will look as follows:

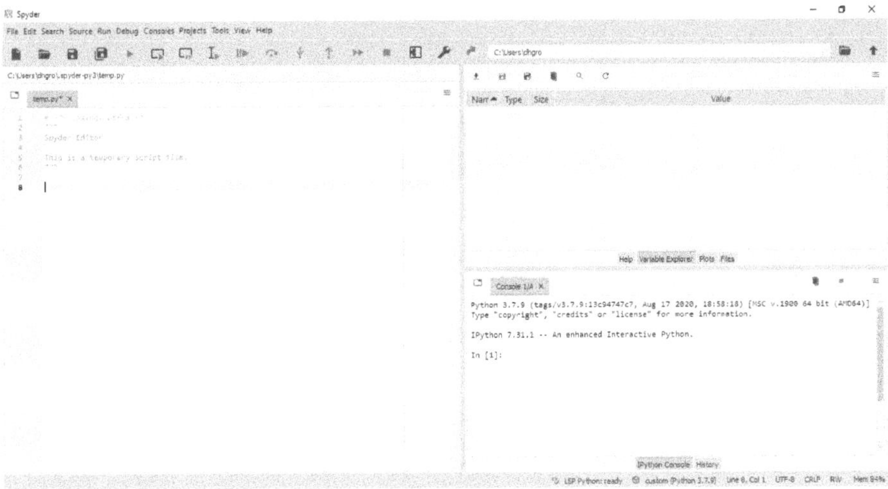

At the lower right of the screen is the interactive Python shell. This is where you can enter one statement at a time and immediately see the result. The In [1]: prompt indicates that you can enter a statement there. For example, in the image below, a print statement is entered, and Python displays the output result.

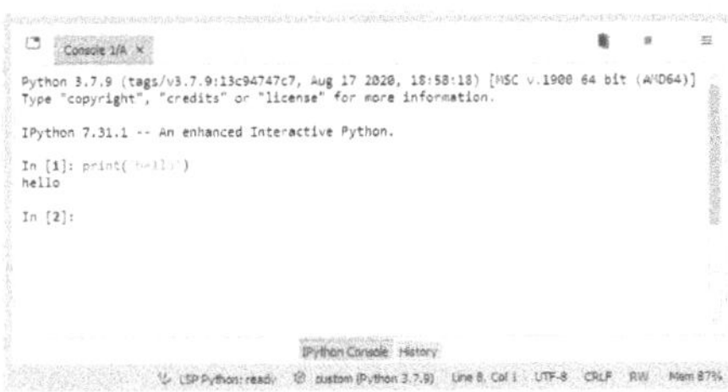

In order to clear the shell area, right-click in this area and select Clear console from the menu.

A variable can be set in the interactive shell. For example, below, the variable *sales* is set to 124.5. Unlike the `print()` example above, setting a variable does not result in output being displayed. In a subsequent statement, the variable can be displayed by simply typing the name of the variable at the [#]: prompt.

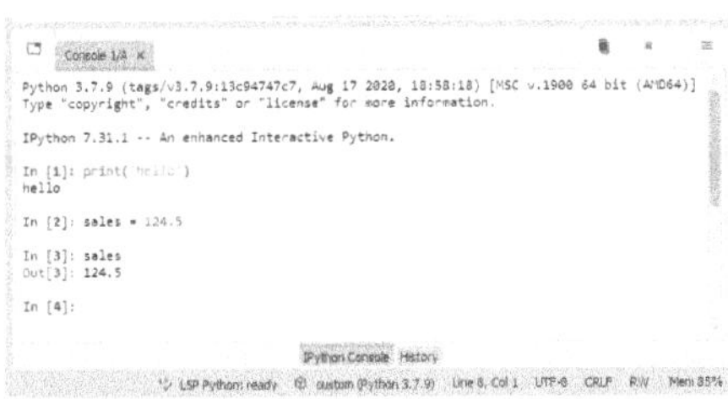

A4.4 Running Python Programs in Spyder

In the previous section, we reviewed how single statements can be run interactively in the Spyder shell. You can also enter multiple lines in a file and run the file as a program. This is done in the left half of the Spyder screen. For example, below, a three-line program is entered:

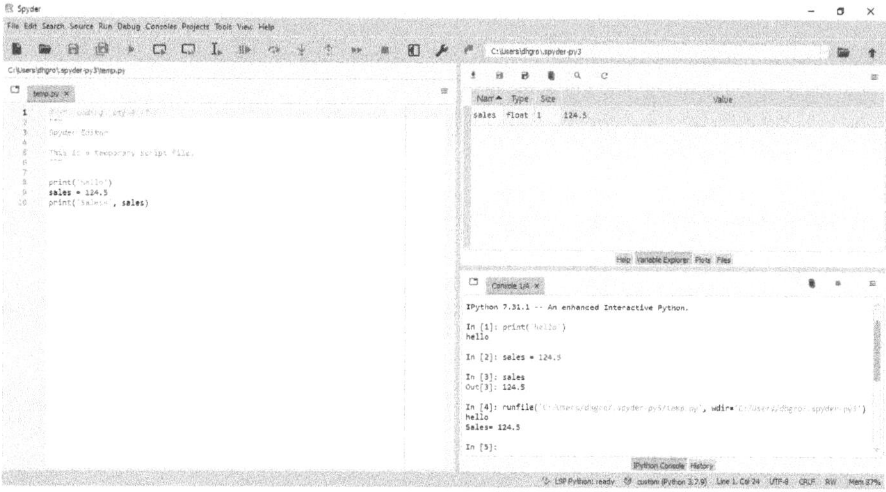

To run your program, click on the run icon (green triangle in toolbar). Spyder will save your .py file, run it, and display the results (or error messages, if any) at the console in the bottom right of your screen.

Appendix A5

Getting Started with PyCharm

A5.1 Overview

An integrated development environment (IDE) lets you enter, run, and debug your programs. There are several IDEs available for Python. The PyCharm IDE is a professional development tool and is also an option for getting started with Python. Like other Python IDEs, PyCharm lets you enter and run one statement at a time or you can create a program with multiple statements. In this appendix, we'll cover how to get started with PyCharm.

PyCharm is available at jetbrains.com/pycharm. There are two editions: Community is free, and Professional offers a free trial period. There are versions for Windows, Mac, and Linux. This appendix covers installing and using PyCharm Community with Windows 10, but the other versions are very similar.

A5.2 Installing PyCharm

In order to install PyCharm on your computer, visit jetbrains.com/pycharm in a browser. On this page, there is a choice for Download. Once you select this, the download page displays the follow choices:

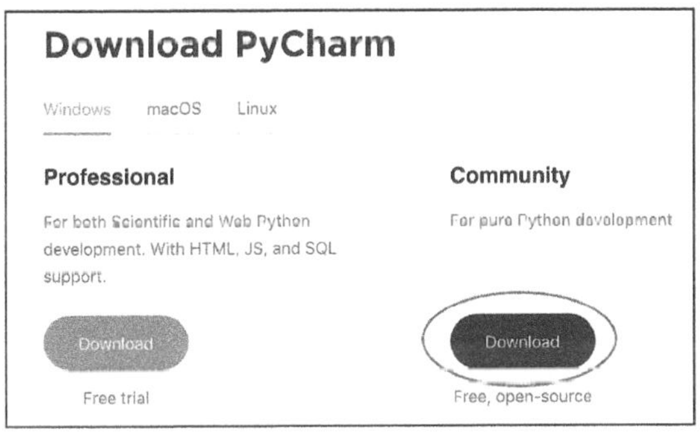

After clicking on the Download button, your browser should download an installer file. In Windows, locate and run this file in order to install PyCharm. You will be presented with a sequence of screens by the installer, you can generally select Next without changing any of the default settings. After several screens, a screen will include an Install button. After selecting this, the installer will take several minutes to install the PyCharm program.

After installing PyCharm, the Windows Start menu will have a folder entry for JetBrains; within this is a choice for PyCharm.

For easier access to PyCharm, you can place it in the Windows taskbar. To do this, right-click on the PyCharm menu entry, then select More . . . , then Pin to the taskbar.

A5.3 Using Python Interactively in PyCharm

When you run PyCharm, the application will look as follows:

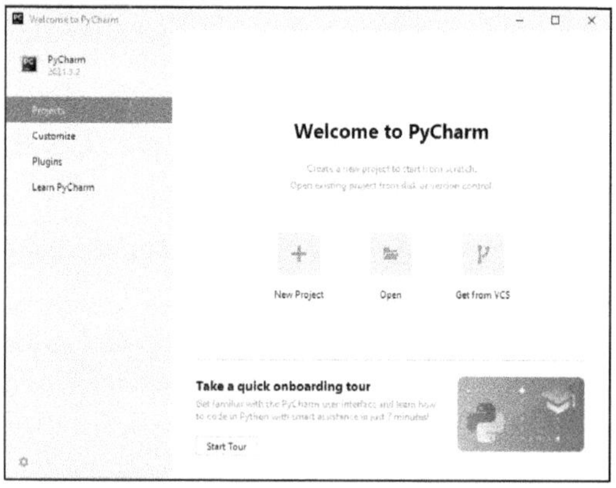

From here, you can choose New Project to get started. The next screen will prompt for the settings of your new project:

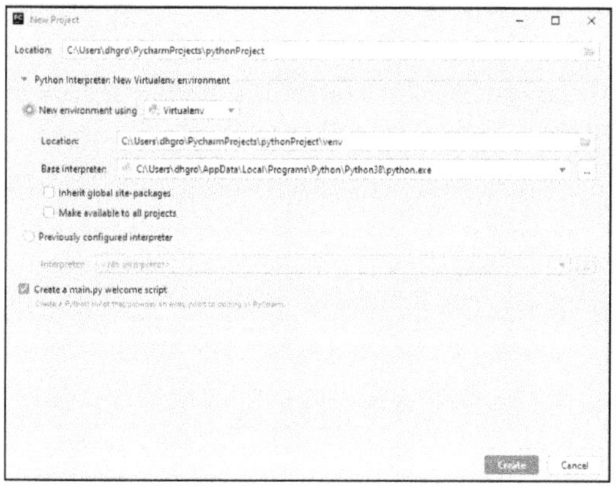

To get started you can leave these as is, but typically you may want to adjust these—for example, you may want to change the Location of your project (where a collection of files will be saved on your computer).

Once you choose Create on the above screen, you will see the PyCharm main screen:

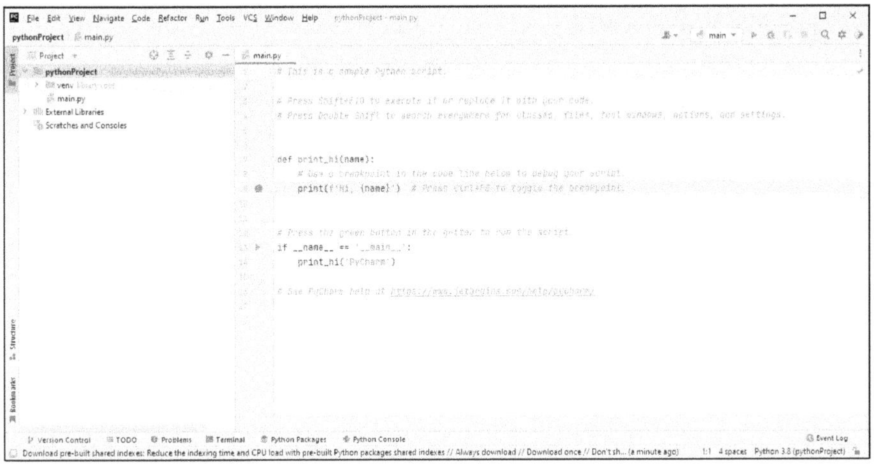

At the bottom center of the screen is a tab for opening the interactive Python shell (also known as the Python Console in PyCharm). If you select this, this shell will be displayed at the bottom left:

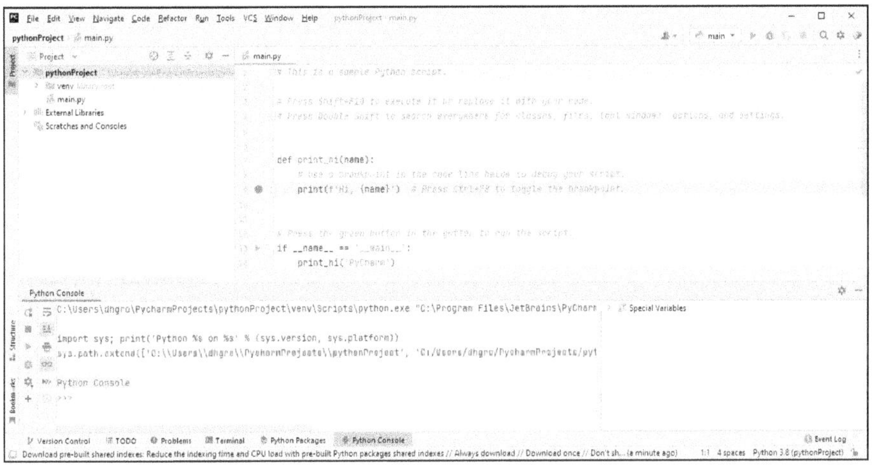

This Python Console is where you can enter one statement at a time and immediately see the result. The >>> prompt indicates that you can enter a statement there. For example, below, a print statement is entered, and Python displays the output result.

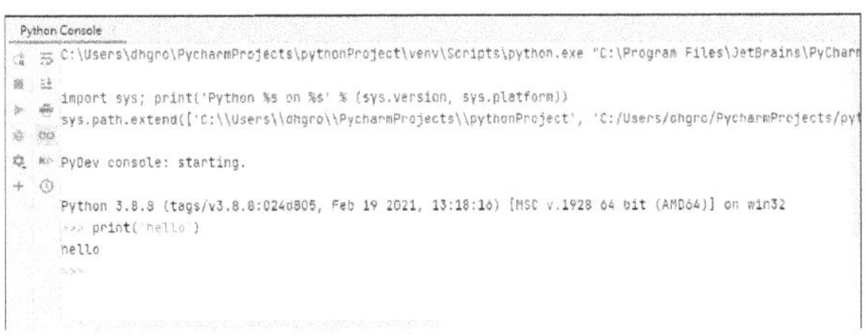

In order to clear the shell area, right-click in this area and select Clear All from the menu.

A variable can be set in the interactive shell. For example, below, the variable *sales* is set to 124.5. Unlike the `print()` example above, setting a variable does not result in output being displayed. In a subsequent statement, the variable can be displayed by simply typing the name of the variable at the >>> prompt.

A5.4 Running Python Programs in PyCharm

In the previous section, we reviewed how single statements can be run interactively in the PyCharm shell. You can also enter multiple lines in a file and run the file as a program. This is done in the top half of the PyCharm screen. You can create a new program file by right-clicking on the top folder at left (with the name of your project) and selecting New > Python File. For example, below, a three-line program is entered:

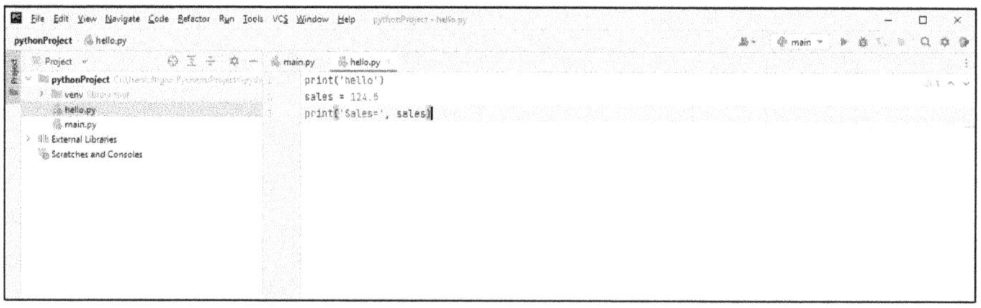

If you right-click on the program and then select `Run 'hello'`, PyCharm will run the program and its results will be displayed in the Python Console at bottom:

Appendix A6

Getting Started with Anaconda

A6.1 Overview

Anaconda is a software distribution that bundles together Python as well as many Python packages. Anaconda also includes popular Python development tools such as Jupyter Notebook and Spyder as well as a link to install PyCharm. These tools are covered in earlier appendixes.

To begin installing the Anaconda distribution, go to www.anaconda.com and click on the Download button. Note that Anaconda requires at least 3 GB of disk space and takes several minutes to install.

Anaconda includes a Navigator application (shown below), which provides a simple way to start certain tools as well as install others. The Navigator also provides a Learning page with links to documentation and tutorials.

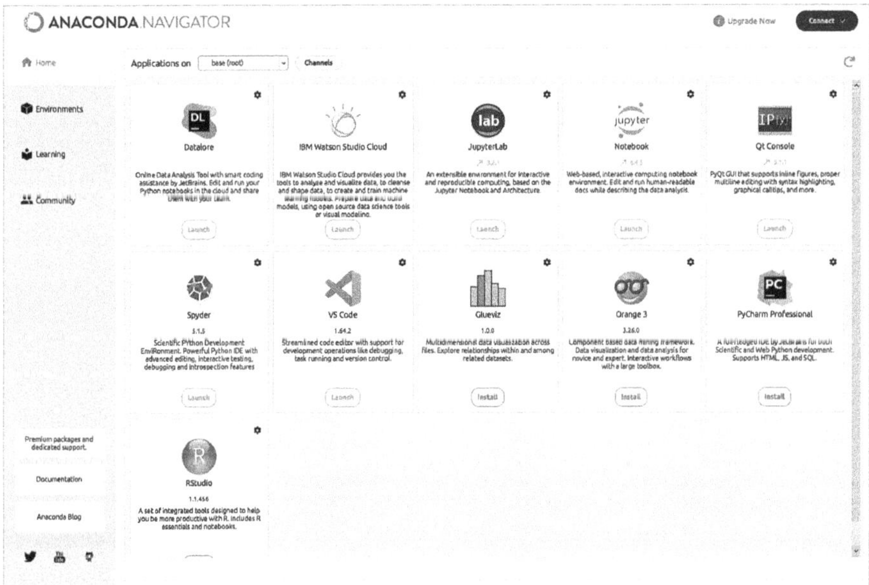

Appendix A7

Getting Started with Third-Party Packages

A7.1 Overview

In this book, we use several third-party packages in Parts II and III. A third-party package is a software library that is not part of the core Python distribution. However, software distributions (e.g., Anaconda) and environments (e.g., Google Colab) come with many packages pre-installed.

There are thousands of third-party packages available for Python. These are available in a central repository called PyPi (Python Package Index), which makes it easy to find and install packages. In this book, we've used several third-party packages; the complete list is indicated later in this appendix.

A7.2 Installing Packages

The pip utility works in conjunction with PyPi to let you install packages in your Python environment. With pip, you can: see what packages you have installed (including their versions), install a package, upgrade a package's version, and remove a package from your installation.

If you have installed Anaconda or are using Google Colab, you already have all the packages used in this book except for the python-dotenv package. Further, if you decide to work with another package that is not yet available to you, you can install that package.

To install a package on your computer using pip, you can use your computer's command line prompt, available from the Windows Command Prompt or Mac Terminal application. At the command line, use:

```
pip install <package>
```

where <package> is the name of the package to install. For example:

```
pip install python-dotenv
```

will install the python-dotenv package on your computer (if it is not already installed).

If you're using a notebook tool (e.g., Jupyter Notebook or Google Colab), you can run pip in a notebook cell by using a preceding exclamation point. For example:

```
[1]: !pip install python-dotenv

Collecting python_dotenv
  Downloading python_dotenv-0.19.2-py2.py3-none-any.whl (17 kB)
Installing collected packages: python-dotenv
Successfully installed python-dotenv-0.19.2
```

An integrated development environment (IDE) may also provide an interactive way to install packages. This varies depending on which IDE you use. See the corresponding appendix in this book or your IDE's documentation.

A7.3 Packages Used in This Book

Table A7-1 lists all the third-party packages used in this book. In order to use a package, in addition to having the package installed in your Python environment, you need to use the import statement in your program. The import statement is described in Appendix B.

The name of the package is what you use to install, upgrade, or remove a package from your Python environment. In most cases, the package name is the same name you use in an import statement. In some cases, the import name differs from the package name; this is noted in Table A7-1.

As mentioned above, if you installed Python using Anaconda or use Python with Google Colab, you will have all the packages used in the book except for python-dotenv.

Area	Package	Import name	Chapters
Core	pandas		10–19
	numpy		10, 11, 15, 17–19
	scipy		17
Web	beautifulsoup4	bs4	12
	requests		12–13
Database	sqlite3f*		14
Excel	openpyxl		15
Visualization	matplotlib		16–17
	seaborn		16
Statistics	scikit-learn	sklearn	18
	statsmodels		19
Text Analytics	nltk		18
Environment	python-dotenv	dotenv	13–14

Table A7-1: Packages used in the book

* The sqlite3 package is included with the core Python distribution.

Appendix B

Importing Packages and Modules

B.1 About Modules, Packages, and Importing

A large part of Python's power comes from the ability for developers to add new functionality by developing and making available ***packages,*** which are code libraries that you can readily use. Several of these third-party packages are covered in this book. Further, Python also comes with many supplemental ***modules*** (files of code) that are standard but not automatically loaded and available when you run interactively or when a script runs.

In order to use third-party packages, as well as certain Python modules, you need to explicitly indicate to Python that these need to be loaded. The keywords `import` and `from` let you do this, and this appendix describes some variations for using these statements. These statements work in both .py scripts as well as notebooks (e.g., Jupyter Notebook, Google Colab)

Relatedly, as your programs grow, you may decide to split code into multiple files. The import statement covers this case too, as it makes a set of code from one file available to your code defined in another file. This is covered in the last section of this appendix.

B.2 `import`

For this appendix, we'll use Python's standard module `datetime`. Similar syntax is used for importing third-party packages. The `datetime` module includes some types such as `date` for representing a year, month, and day. As `datetime` types are defined in Python in a supplemental module (i.e., these are not ***built-in types***), you need to import these types in order to use them interactively or in your program. The simplest way to do this is:

```
import datetime
```

The import statements are often placed at the top of the .py program or in an initial notebook cell, although it is possible to place these further down in your program or notebook. You can also use import statements interactively at the Python command line prompt.

After this `import` statement, the `datetime` types are available for you to use. However, you need to prefix them with `datetime`. For example:

```
d = datetime.date(2022, 6, 30)
```

The above statement creates a `date` object representing the date June 6, 2022. The Python documentation describes methods available for `datetime.date`. Full documentation for the datetime module, including all its types and methods, is available at https://docs.python.org/3/library/datetime.html.

If you'd like to reference a module with a shorter prefix, you can use a statement like:

```
import datetime as dt
```

This makes the `datetime` types available for use, but lets you prefix them with `dt` instead of `datetime`. For example, you can create a date using:

```
d = dt.date(2022, 6, 30)
```

In addition to types, functions and constants defined in a module are similarly accessed with the prefix, depending on how you use `import` described above.

B.3 Third-Party Packages

Third-party packages such as pandas and others covered in this book must be installed in order for you to use them. If you've installed the Anaconda distribution, then you have many Python packages installed on your computer. Similarly, if you use Google Colab, many packages are pre-installed for you to use. Conversely, if you do not already have a particular package you'd like to use, you need to install it.

The most basic way to install a Python package is to use the pip command. To install a package on your computer, from Command Prompt (Windows) or Terminal (Mac), you can simply enter:

```
pip install <package>
```

where `<package>` is the name of the package to install. If the package depends on other packages, and you don't yet have those installed, pip will install those as well.

The pip install command can be issued from a notebook (Jupyter Notebook or Google Colab) by prefixing it with an exclamation point:

```
!pip install <package>
```

Once a package is installed, it can be imported with the same syntax as described in this appendix. For example, if the pandas package is installed on your computer, you can load it for use in your code as:

```
import pandas as pd
```

Certain integrated development environments (IDEs) have their own interactive way to install and manage the packages available. For example, the Thonny IDE has a menu item Tools > Manage packages . . . that lets you see what packages are installed as well as find and install packages. PyCharm has a Python Packages tab where you can see install packages as well as install more.

B.4 `from/import`

In a previous section, we used the `import` statement to make available all types, functions, and constants defined in a module. As an alternative, you can load selected items from a module by using the `from` statement. For example:

```
from datetime import date
```

In this case, only the `date` type is loaded from the `datetime` module and available. Further, you do not need to use a prefix when referring to a `date`. For example:

```
from datetime import date
d = date(2022, 6, 30)
```

If there are several items you'd like to import, you can indicate them together. For example:

```
from datetime import date, time
```

B.5 Multi-file Applications

In the previous sections, `import` and `from/import` were described for loading code from modules and packages. These same statements are also used if you split up your application's code into multiple files. You may decide to divide your code into multiple files as your programs get larger. The general approach is that one file of code will import the types/functions/constants that you define in another file. Conceptually you use the same approach and syntax as previously described in this appendix—however, instead of importing others' code, you are importing your own code.

To see this, let's look at a simple example. You may decide to write a set of functions and place them in a file called bmifuncs.py. Then you can use these functions in another file by using the import statement. In this example, we assume both .py files are in the same directory (folder) on your computer. Programs D-1 and D-2 show this arrangement.

Program B-1 bmifuncs.py

```
# file bmifuncs.py with some functions defined
def bmiCalc(height, weight):
    return weight / height**2 * 703

def weightLoss(height, weight, bmiGoal=25):
    bmiNow = bmiCalc(height, weight)
    if bmiNow < bmiGoal:
        return 0
    else:
        targetWeight = bmiGoal * height**2 / 703
        return weight - targetWeight
```

Program B-2 myapp.py

```
1  # file bmiapp.py that uses some functions defined in bmifuncs.py
2  import bmifuncs # bring the code from bmifuncs.py in (note .py not included)
3
4  height = float(input('Enter height (inches): '))
5  weight = float(input('Enter weight (pounds): '))
6
7  bmi = bmifuncs.bmiCalc(height, weight)
8  print('Current BMI:', round(bmi,1))
9
10 toLose = bmifuncs.weightLoss(height, weight)
11 if toLose > 0:
12     print('Suggested weight loss (pounds):', round(toLose))
```

In the program myapp.py, line 2 shows the `import`. Also, lines 7 and 10 show the prefix used in order to access the functions defined in the other file.

Appendix C

Formatting

C.1 About Formatting

Unlike spreadsheets, by default, Python doesn't limit the number of decimals displayed when displaying a numeric result. This may be OK for interactive or personal use, but often we want to control the decimals displayed and other formatting aspects. In this appendix, we give a brief overview of two Python approaches to formatting: the `format()` function and f-strings. Date formatting is also covered in this appendix.

A third Python formatting approach using the percent operator is also available but has fewer features and is not covered here. This is also known as printf-style string formatting or old string formatting. Details about it are covered here:

https://docs.python.org/3/library/stdtypes.html#old-string-formatting

C.2 Python `format()` Function

The simplest way to limit decimal places in displayed numbers is the `round()` function. However, this does not always give the desired number of decimal places. For example, if we want two decimal places displayed, `round(10.0, 2)` will give the result `10.0` as opposed to `10.00`. Instead, the `format()` function can be used to request a desired number of decimal places.

The `format()` function takes two arguments. The first is the value, variable, or expression that we want displayed—that is, *what* we want displayed. The second argument is *how* we want it displayed, referred to as the ***format spec***. The format spec is a string that tells `format()` how to create its result. Note that the result of `format()` is always a string, even though its first argument is often a number.

The format spec is a little cryptic and less friendly than interactively formatting cells in a spreadsheet. In this section, assume the variable `x` has the value `1234.567`. If we want to display it to two decimal places, we can use: `format(x, '.2f')`

The `'.2f'` above is the format spec. In it, the `f` refers to floating point notation. If you omit the `f`, the result will be in scientific notation. The `.2` indicates two decimal points in the result. The result is rounded; in this case, the result is `'1234.57'`. Again note the result of `format()` is a string. We can assign the `format()` result to variable, or we can use it within a `print()` function:

```
print('The value of x is', format(x, '.2f'))
```

If we'd like commas in `format()`'s result, we can use a format spec like `',.2f'`. In this case, the result is `'1,234.57'`.

For multi-line output, we may want to line up numbers in a column. With `format()`, we can do this by indicating the number of characters the result should have—for example, `format(x,'10,.2f')`. The format spec `'10,.2f'` indicates:

- 10 characters
- two decimal places
- include commas
- floating point output

For `format(x, '10,.2f')`, the result is `'  1,234.57'`. Note the two spaces to the left of the results. These are added so the total number of characters in `format()`'s result is 10. By requesting the number of characters in this way, we can line up a column of results. For example:

```
print('Gross price:', format(gross, '10,.2f'))
print('Discount:   ', format(discount, '10,.2f'))
print('Net price:  ', format(gross-discount, '10,.2f'))
```

There are few other options that can be set in the format spec. The justification can be set to left using <, or centered using ^, or right using >. So for example, `format(x,'^10,.2f')` returns `' 1,234.57 '`; in this case, a space is added at the beginning and end of the result in order to provide 10 characters with the number centered.

Another format spec option is to request a plus sign for positive numbers. For example, `format(x,'^+11,.2f')` gives the result `' +1,234.57 '`.

The examples so far have focused on floating point values. `format()` can handle `int` values, in which case the format spec can have a *d* suffix, like: `format(intvar,'10d')`. Also, format can handle string values—for example, `format(strvar,'^10s')`.

Lastly, in order to return the requested number of characters, `format()` will by default use blanks to pad the result if needed. If you'd like to use another padding character, you can include that at the start of the format spec—for example, `format(x, '=^10,.2f')` will return `'=1,234.57='`.

To recap the options you can specify in the format string, here is its general form covered in this appendix:

```
[[fill]align][sign][width][,][.precision][type]
```

Table C-1 shows some example uses of the format function with numbers and strings.

Type / Name	Expression	Result (`str`)	Notes
`float`, two decimal places	`format(10.123,'.2f')`	`'10.12'`	Format number to 2 decimal places
`float`, two decimal places, 10 positions, commas	`format(1234.123,'10,.2f')`	`'  1,234.12'`	Format number to 2 decimal places, 10 positions (pad with leading spaces as needed), commas every 3 positions
`int`, add comma(s) and leading spaces	`format(1234,'10,d')`	`'     1,234'`	Format `int` by padding output to 10 spaces, and adding a comma (# of decimal places is not allowed for `int`)
String, center	`format('Sales','^8s')`	`' Sales  '`	
String, right-justified	`format('Sales','>8s')`	`'   Sales'`	

Table C-1: Example `format()` expressions

C.3 Python f-strings

Python 3.6 introduced a new way of formatting using ***f-strings***. An f-string (formatted string) is like a Python string but with two key differences:

- The string is introduced with the letter *f*.
- Within the string, you can place variables or expressions within curly braces { } to indicate something to include as well as how to format it.

As in the prior section, we'll continue to use the variable *x* with a value of `1234.567`. The f-string `f'{x}'` will evaluate to `'1234.567'`, since `{x}` is replaced by the value of `x`, and then formatted to a string. We can introduce a format spec with the curly braces after a semicolon—for example, `f'{x:.2f}'` evaluates to `'1234.57'`. To include commas, we can use `f'{x:,.2f}'`. The format specs you can use in an f-string are the same as described in the prior section of this appendix.

An f-string can contain characters outside of the curly braces; so for example, `f'x={x:,.2f}'` evaluates to `'x=1,234.57'`. And you can indicate multiple `{ }` inside the f-string—for example, if *y* is another variable, we can use `f'x={x},y={y}'`.

The format spec with an f-string can include variables or expressions. For example, if the width of a result was calculated and stored in the variable *w*, it can be used in an f-string like: `f'{x:{w}.2f}'`.

C.4 Python Dates

The Python datetime package contains several types for representing dates and times. In this section, we'll look at some variations for formatting dates. Note that to use the Python date type, you need to import the datetime package or specific types from it. For example, if you have:

```
from datetime import date
```

you can define a particular date as:

```
d = date(2021, 6, 9) # June 9, 2021
```

In order to get the string representation of a date in yyyy-mm-dd format, you can use the `isoformat()` method, which for `d.isoformat()` would return `'2021-06-09'`. This is the simplest way to get a string for a date.

For other date strings, you can use the `strftime()` method. This method takes a string format spec that lets you access the year, month, and day in particular ways. For example, `d.strftime('%m/%d/%Y')` returns `'06/09/2021'`. Month name abbreviations are available with `%b`; `d.strftime(%b %d, %Y)` and returns `'Jun 09, 2021'`. Leading zeros can be omitted; for example, `d.strftime(%b %-d, %Y)` returns `'Jun 9, 2021'`. The Python documentation describes other `%` options available; see:

https://docs.python.org/3/library/datetime.html#strftime-strptime-behavior

Dates can also be used in f-strings. For example, `f'{d}'` evaluates to `'2021-06-09'`. The `%` options can also be used—for example, `f'{d:%b %-d, %Y}'` evaluates to `'Jun 9, 2021'`.

Appendix D

Handling Errors with Exceptions

D.1 About Exceptions

Although a Python program may have valid syntax, it may encounter an error when running. In practice, we want our programs to be robust, and so in this appendix, we look at some keywords and approaches for anticipating and handling error cases.

For example, the following code is expecting the user to enter a number:

Program D-1

```
1 userInput = input('Please enter a number: ')
2 value = float(userInput)
3 print('Your value was', value)
```

If the user enters a non-numeric value, the second line will generate an error, and the program will end. For example, if we enter `'hello'` in response to the prompt from line 1, we'll get an error like:

```
Please enter a number: hello
Traceback (most recent call last):
  File "/Users/jsmith/Documents/John/program/EnterNumber.py", line 2, in <module>
    value = float(userInput)
ValueError: could not convert string to float: 'hello'
```

The error message indicates an issue on line 2—this is because the `float()` function doesn't know how to convert `'hello'` into a number. In this small example, there is no code for how to handle this error, so Python stops running the program after the `float()` function hits this error on line 2.

In this appendix, we'll look at Python's Exception framework, which generates errors in certain cases and allows you to write code to handle these errors so your programs can continue instead of stopping immediately after hitting an error.

D.2 `try` / `except`

Python provides the keywords `try` and `except`, which you can use to anticipate a ***runtime error*** and provide logic if that error occurs. This allows you to place statements indented under the `try`—these are statements to run in a normal program flow. If any of these statements encounters an error, the program will jump to the statements indented under a corresponding `except`.

For our simple number prompt example, we can anticipate the user entering something other than a number and provide a backup plan if that happens. For example:

Program D-2

```
userInput = input('Please enter a number: ')
try:
    value = float(userInput)
    print('Your value was', value)
except:
    print('Error: non-numeric value entered')
```

In the code above, if the user enters a non-numeric value, execution during line 3 will cause a jump out of the `try` block and into the `except` section. In this way, the program does not stop when the Exception happens; instead, our code handles the Exception with the `except` block, and the program ends normally.

D.3 Exception Types and Multiple Error Handling

An Exception has a specific type—there are over 50 different types of built-in Exceptions defined in Python.* For example, `ValueError` was indicated in the initial output above, and this is one type of Exception. If you want your code to distinguish between different errors for a given `try` section, you can indicate the type of error with the `except` statement. For example, we can modify the above to:

Program D-3

```
userInput = input('Please enter a number: ')
try:
    value = float(userInput)
    print('Your value was', value)
except ValueError:
    print('Error: non-numeric value entered')
except Exception as err:
    print('Error:', err)
```

In the above code, we distinguish ValueError from any other type of error (for example, an OverflowError). Also shown above is how to reference more information about the error by defining a variable as done in:

```
except Exception as err:
```

Here, the variable `err` receives additional information about the error and is used to display this in the last line of the program.

D.4 `finally`

In some cases, you want to provide logic to run whether or not an error occurred in the `try` section. This can be done using the `finally` keyword. A common use is to indicate the program is done using a file or database connection. This cleanup may be needed whether the program ran successfully or not.

To show the use of `finally`, we can enhance our program as:

Program D-4

```
userInput = input('Please enter a number: ')
try:
    value = float(userInput)
    print('Your value was', value)
except:
    print('Error: non-numeric value entered')
finally:
    print('Program ends')
```

D.5 Exceptions and Functions

In our code examples in this appendix, the `float()` function is the fragile part. In particular, internally `float()` will use the `raise` keyword to indicate that an Exception has occurred. We have seen that `try/except` can be used to handle the error. In more involved code, the call to `float()` might be in a function you write. In this

* The list of built-in Python Exceptions is here: "Built-in Exceptions," Python, last modified April 20, 2022, https://docs.python.org/3/library/exceptions.html.

case, if you do not handle the error, your function will raise the exception to its caller. In other words, Exceptions are propagated up the calling chain until (or if) they are handled by an except.

You can also use the raise keyword in your function to handle specific error cases in a particular way. For example, if your function is expecting numeric arguments, you could check this and raise an exception back to the caller of your function. For example:

```
def mathFunc(numParam):
   if not isinstance(numParam, (int, float)):
      raise ValueError("mathFunc parameter must be a number")
   # additional processing below (not shown)
```

Here we use the isinstance() built-in function to check the type of numParam. If it is not an int or float, the function raises a ValueError with more specific text.

The use of raise highlights that functions have two types of outcomes. If the function completes normally, the return statement can provide the regular result. However, if an error is hit, raise provides the error result. The code that calls a function can use the regular result in the standard way, and it can also optionally anticipate the error result within the try/except pair.

Appendix E

Defining Classes: Object-Oriented Programming

E.1 About Object-Oriented Programming

Types are fundamental to the Python language. For example, some basic types are `float`, `int` (integer), and `str` (string). Composite types include `list` and `dict` (dictionaries).

Types arise in three main ways:

- Types are included as part of Python; these are built-in types.
- Types are included in third-party packages—for example, pandas defines the `DataFrame` and `Series` types.
- You can define your own types in cases where other types are not suitable to solve your problem.

What is a type? In essence, it's a set of data (properties) and actions (methods) that can be performed on that data. For example, a string's properties are a sequence of characters, and the actions available are the many string methods such as `lower()`, `split()`, and others.

In Python, all variables have a type, even though the type is not explicitly indicated. For example, consider the statement:

```
s = 'hello'
```

Python infers that `s` is an `str` type (because it determines `'hello'` is an `str`), and keeps track of that. You can check this by using the `type()` function:

```
type(s)
```

which indicates:

```
<class 'str'>
```

In the result above, ***class*** is a synonym for type.

The combining of related properties and methods in a class is the essence of object-oriented programming. In this approach, types can model technical items—such as strings, files, or tables—as well as items in the real world, such as investments, products, and patients. Well-defined types can be concise to program with, and can help make you more productive.

In this appendix, we'll build an example class, representing a mortgage. As we'll see, our class will have properties and methods that we can use to concisely create and analyze a mortgage.

E.2 An Example with Properties and Methods: Mortgage

Before defining a class, it is important to understand the area being modeled, as well as the potentially useful actions we may want to do with it. There are many kinds of mortgages; for this appendix, we'll assume that a mortgage is a loan that has an initial amount, a fixed annual interest rate, a term in years, and monthly payments. For example, a particular mortgage might be for $100,000 (the initial amount) at 6% (the interest rate) over 30 years (the term). The amount, rate, and term are the properties we'd like our Mortgage type to keep track of.

What methods might be of interest? Our mortgages have a fixed payment due each month. Assuming an initial amount *B* (in dollars), monthly interest rate *r* (in decimal), and term *n* (in months), the monthly payment *M* for a mortgage is:

```
M = B [r(1 + r)ⁿ] / [(1 + r)ⁿ - 1]
```

The calculation of the monthly payment is one method we'd like our Mortgage type to have.

Another calculation that may be useful is how much interest (*I*) is paid over the life of the mortgage. This is given by:

I = M n - B

Where to define the class

A class can be defined in a .py file. For this appendix, we'll assume we're defining our Mortgage class in a file called Mortgage.py. As we'll see, in addition to defining the class, we'll also have statements that use the class. Also note that we're defining one class in our file; in other cases, you could define multiple classes in the same .py file.

A class can also be defined in a cell of a notebook and used in another cell in the same notebook. For example, for this appendix, you can save your code in Mortgage.ipynb. Further, you could define multiple classes in a certain notebook—for example, each class could be defined in its own cell.

Keyword `class` and the `__init__()` method

In order to define our Mortgage class, we use the `class` keyword:

```
class Mortgage:
```

The next step is to decide how each Mortgage will be created. For example, we may want to create a Mortgage using a statement like:

```
mortgage1 = Mortgage(100000, .06, 30) # $100,000 at 6% for 30 years
```

In order for this to work, we need to define an `__init__()` method. The double underscores are a convention in Python for special methods. Here `__init__()` is called whenever a new instance of a Mortgage is created. Our `__init__()` method can be defined as:

```
class Mortgage():
   def __init__(self, amount, rate, term):
      self.amount = amount # initial amount of the mortgage
      self.rate = rate     # annual interest rate in decimal
      self.term = term     # term of mortgage in years
```

The above code is a valid Python class, containing one method and three properties. The `self` variable is a Python convention that allows each Mortgage variable to keep track of its three properties. So for example, we could define two separate Mortgages:

```
mortgage1 = Mortgage(100000, .06, 30)
mortgage2 = Mortgage(110000, .05, 15)
```

In the above, `mortgage1` keeps track of its own properties separately from `mortgage2`. Also note that while Mortgage is referred to as a class (or type), the variables `mortgage1` and `mortgage2` are referred to as ***objects***. This highlights the relationship between a class and an object. A class defines the properties and methods available; an object is a specific instance with specific values for the properties.

We can now add a method to calculate the monthly payment for a given mortgage, as shown in program E-1.

Program E-1

```
class Mortgage():
   def __init__(self, amount, rate, term):
      self.amount = amount # amount of the loan
      self.rate = rate     # interest rate in decimal
      self.term = term     # term in years

   def monthlyPayment(self):
      # B [r(1+r)^n] / [(1+r)^n-1]
      r = self.rate / 12
      n = self.term * 12
      pay = self.amount * r*(1+r)**n / ((1+r)**n-1)
      return pay

mortgage1 = Mortgage(100000, .06, 30)
payment = mortgage1.monthlyPayment()
print(payment)
```

Program E-1 Output

```
599.5505251527569
```

We can round this in the `print()` statement, or build rounding into our class, which we'll do later.

In the above, we see the `self` variable as the parameter for the `monthlyPayment()` method. The Python runtime mechanics are that the variable prefix (in this case, `mortgage1`) is set to `self` when the `monthlyPayment()` method is called. In effect, the prefix to the method is "jumped" into the parentheses in order to set the `self` parameter.

Also in the above, note that `monthlyPayment()` only has the one `self` parameter—the properties (amount, rate, and term) are not included in the method's signature. This is because the `__init__()` method has previously retained these properties for subsequent use by other methods. The property values are remembered by the variable and can be accessed by the `self.` prefix. This is why methods are typically more concise than functions—the property values are available without repeating them as arguments to the method.

Methods vs. functions

In program E-1 we see that the methods `__init__()` and `monthlyPayment()` look similar to Python functions. The key difference is that a method is defined and indented within a class and a function is not defined within a class. Also, when we call a function, we don't prefix it with the name of a variable—for example, to call the `input()` function we can write:

```
name = input('Enter your name: ')
```

In contrast, in order to call a method such as `monthlyPayment()`, we prefix it with the name of the variable, such as:

```
payment = mortgage1.monthlyPayment()
```

E.3 Adding `__str__()` and `__repr__()` Methods

Our `Mortgage` class in the prior section is sufficient—it lets us create a `Mortgage` instance in a variable and then use that variable with the `monthlyPayment()` method. In this and subsequent sections, we'll augment the Mortgage class further.

If we try to print out a Mortgage variable:

```
print(mortgage1)
```

we'll get a result like:

```
<__main__.Mortgage at 0x10566a828>
```

This tells us `mortgage1` is of type `Mortgage` as well as its memory location. Typically we'd like to have a better representation of an object than this. For this, we can define the `__repr__()` method. Relatedly we can also define a string representation for the class using `__str__()`. For simple classes, these are often the same and can include the property values of the object. To our `Mortgage` class, we can add:

```
def __str__(self):
    return str(self.amount)+','+str(self.rate)+','+str(self.term)
__repr__ = __str__ # define the __repr__() method to be same as __str__()
```

The above code adds an `__str__()` method that creates a string from the Mortgage's properties. It also defines an `__repr__()` method as the same logic as `__str__()`.

With these additional methods, the statement:

```
print(mortgage1)
```

will give the result:

```
100000,0.06,30
```

__str__() vs. __repr__()

The `__str__()` and `__repr__()` methods are similar in that they provide a string representation of an object. However, `__str__()` is intended to provide a person-readable string of the object, whereas `__repr()__` is intended to provide a complete computer-readable string of the object. In particular, `__repr__()` is intended to generate a string that includes all the critical properties of the class, whereas for `__str__()`, we can decide to include some or all of the properties of our class. In our Mortgage class, since we have a few properties, we decided these two methods will give the same result.

Python will call `__str__()` or `__repr__()` depending on the specific case. For example, as described above, the `print()` function will call the `__str__()` method to get a string representation of the variable. In contrast, if we have a list of Mortgage variables, and we use `print()` to display that list, the `__repr__()` method will be called to get the string representation of each Mortgage in the list.

E.4 Adding a Class Variable: Rounding

As we saw earlier, our `monthlyPayment()` method returns a `float` that may have many decimal places. In practice, the monthly payment may be calculated to the nearest dollar or nearest cent. As well, there may be other calculations that our Mortgage will perform with rounding. Rather than hardcode a number of decimals to round to, we can introduce a variable to our class called, for example, `decimals`. Then all relevant methods can use this variable to round their results to some number of decimal places.

The `decimals` variable can be placed within the `Mortgage` class but outside of any particular method. This is referred to as a ***class variable***. For example:

```
class Mortgage:
    decimals = 2
```

Here the default setting for `decimals` is `2`, but we can easily change that using a statement like:

```
Mortgage.decimals = 0
```

The `Mortgage` class is enhanced to use the `decimals` class variable as shown below in program E-2.

Program E-2

```
class Mortgage:
    decimals = 2

    def __init__(self, amount, rate, term):
        self.amount = amount # amount of loan
        self.rate = rate     # interest rate in decimal
        self.term = term     # term in years

    def __str__(self):
        return str(self.amount)+','+str(self.rate)+','+str(self.term)

    __repr__ = __str__

    def monthlyPayment(self):
        # P [r(1+r)^n] / [(1+r)^n-1]
        r = self.rate / 12
        n = self.term * 12
        pay = self.amount * r*(1+r)**n / ((1+r)**n-1)
        return round(pay, Mortgage.decimals)
```

E.5 Adding a Method That Calls Another Method

We can augment our Mortgage class with additional methods as needed. For example, if we'd like a method that calculates how much interest is paid over the life of the mortgage, we can add it as:

```
def totalInterest(self):
    pay = self.monthlyPayment()
    totalPay = self.term * 12 * pay
    totalInterest = totalPay - self.amount
    return round(totalInterest, Mortgage.decimals)
```

Note that the above method is calling another Mortgage method, `monthlyPayment()`, as well as using the class variable `Mortgage.decimals`.

E.6 Adding a Static Method: `validate()`

One last note about classes relates to ***static methods***. These methods are defined within a class but have no sense of self. In other words, they are like functions. However, if the logic is related to a particular class, it is better to place it within that class as opposed to defining a stand-alone function.

For example, we may want a function to validate if our mortgage-related properties are valid before we try to create a `Mortgage` variable. This can be done by defining a static method, `validate()`, which takes three parameters, as follows:

```
@staticmethod
def validate(amount, rate, term):
    if (not isinstance(amount, (float, int))) or amount <= 0:
        raise ValueError('Invalid amount')
    if (not isinstance(rate, float)) or rate <= 0:
        raise ValueError('Invalid rate')
    if (not isinstance(term, int)) or term <= 0:
        raise ValueError('Invalid term')
```

The method is introduced by the decorator* `@staticmethod` to indicate it does not relate to a particular `Mortgage` variable—that is, it is similar to a function. In this case, the code checks for valid values for amount, rate, and term and, if any issues are found, will raise a `ValueError`. If the values are OK, a result of `None` is returned, indicating there is no error in the values.

E.7 The Complete Mortgage Class

In this appendix, we've covered the basics of writing your own class. Program E-3 combines all the aspects we've covered into an interactive program that prompts for a fixed rate mortgage's amount, rate, and term and then calculates and displays the monthly payment amount and total interest paid over the life of the mortgage.

* A description of decorators is beyond the scope of the book; information about decorators is available in Python documentation. For example, see "PEP 318—Decorators for Functions and Methods," Python, last modified March 9, 2022, https://www.python.org/dev/peps/pep-0318/.

Program E-3

```
class Mortgage:
    decimals = 2

    def __init__(self, amount, rate, term):
        self.amount = amount # amount of loan
        self.rate = rate     # interest rate in decimal
        self.term = term     # term in years

    def __str__(self):
        return str(self.amount)+','+str(self.rate)+','+str(self.term)

    __repr__ = __str__

    def monthlyPayment(self):
        # P [r(1+r)^n] / [(1+r)^n-1]
        r = self.rate / 12
        n = self.term * 12
        pay = self.amount * r*(1+r)**n / ((1+r)**n-1)
        return round(pay, Mortgage.decimals)

    def totalInterest(self):
        pay = self.monthlyPayment()
        totalPay = self.term * 12 * pay
        totalInterest = totalPay - self.amount
        return round(totalInterest, Mortgage.decimals)

    @staticmethod
    def validate(amount, rate, term):
        if (not isinstance(amount, (float, int))) or amount <= 0:
            raise ValueError('Invalid amount')
        if (not isinstance(rate, float)) or rate <= 0:
            raise ValueError('Invalid rate')
        if (not isinstance(term, int)) or term <= 0:
            raise ValueError('Invalid term')

amount = float(input('amount ($): '))
rate   = float(input('rate (%): ')) / 100.0
term   = int(input('term (years): '))

try:
   Mortgage.validate(amount, rate, term)
   mortgage = Mortgage(amount, rate, term)
   Mortgage.decimals = 0
   monthlyPayment = mortgage.monthlyPayment()
   totalInterest = mortgage.totalInterest()
   print(mortgage)
   print('Monthly payment= $', format(monthlyPayment, ',.0f'))
   print('Total interest= $', format(totalInterest, ',.0f'))
except ValueError as e:
   print('Error:', e)
```

Program E-3 Example Run

```
amount ($): 100000
rate (%): 5
term (years): 30
100000.0,0.05,30
Monthly payment= $ 537
Total interest= $ 93,320
```

Appendix F

Reading Files: Recap

Chapter 7 covers reading and writing files in detail. Since subsequent chapters use file reading in their examples, this appendix is a recap of the key aspects of reading a text file.

F.1 Reading Lines of a File

In order to read a text file with Python, the file must be opened. Opening a file doesn't read its contents; instead, it associates the file with a Python variable we use in subsequent statements. For example, if you have a file named data.txt that you'd like to read, you can use the statement:

```
filein = open('data.txt', 'r')
```

In the above statement, the second argument `'r'` indicates that the file will be read (instead of written to). The `filein` variable could be any valid Python variable name—for example, `f`, `fin`, or `fdata`.

We can then read each line of the file using the for statement:

```
for aline in filein:
```

The above statement uses repetition to read each line of the file, setting the variable `aline` to a string containing the contents of the file line, one line at a time. By convention, the returned string (in this case, in `aline`) will have the special newline character, referred to as `'\n'`. We typically do not want this character, so we can use the string `rstrip()` method to remove it:

```
aline = aline.rstrip('\n')
```

Program F-1 combines the above statements to read a file and display its lines to the screen.

Program F-1

```
1  # read each line of a file and display these to the screen
2
3  # open the file, connecting it to the variable filein
4  filename = 'data.txt'
5  filein = open(filename, 'r')
6
7  # using for repetition, read each line of the file in the variable aline
8  for aline in filein:
9     aline = aline.rstrip('\n') # remove the trailing newline
10    print(aline)
```

Note that the above program can be used to read a different text file by using a different filename on line 4. If the file does not exist, `open()` will raise a `FileNotFoundError`.

Where to place the data file

The simplest arrangement when working with data files is to place the data file in the same folder (subdirectory) as your .py code or .ipynb code and refer to the file without any absolute or relative path name in your `open()` statement. This is the approach we follow in the book.

However, in professional practice, it's often a good idea to have programs and data in separate subdirectories. If you place the data file(s) in a different folder from the code, the filename argument provided to the `open()`

function can have an absolute or relative path name. Alternatively, you can use the `chdir()` function in the os module to point to the folder where your data file is.

F.2 Reading Comma-Separated Value (csv) Files

For text files with comma-separated values, you can use the statements in Section F.1. In order to access each of the data items, the string `split()` method is a good choice. Further, if the file has a header line that you'd like to read but ignore, the `readline()` method can be used. For example, if you'd like to read a file named products.csv, which has two values on each line—a product ID and its price—you could use the code in Program F-2.

Here is an example for products.csv, which could be read and processed by Program F-2:

```
product-id,price
P-001,5.50
P-001x,5.95
P-002,6.50
P-002x,6.95
P-010,10.25
P-010x,10.75
Q-011,11.50
Q-011x,11.95
```

Program F-2

```
# Read a csv file with two fields, and display them to the screen

# open the file, connecting it to the variable filein
filename = 'products.csv'
filein = open(filename, 'r')

# assume there is a header record we'd like to skip over
filein.readline()

# using for repetition, read each line of the file
for aline in filein:
   aline = aline.rstrip('\n') # remove the trailing newline
   fields = aline.split(',')  # split the data into a list of strings
   productid = fields[0]      # assign the first item in the list
   price = float(fields[1])   # convert and assign the second item
   print(productid, price)    # display the data fields
```

Program F-2 Output

```
P-001 5.5
P-001x 5.95
P-002 6.5
P-002x 6.95
P-010 10.25
P-010x 10.75
Q-011 11.5
Q-011x 11.95
```

References

Air Quality Index Levels (Chapters 6, 10)
https://www.airnow.gov/aqi/aqi-basics/

S&P 500 Component Stocks (Chapters 10, 12)
Wikipedia contributors, "List of S&P 500 Companies," Wikipedia, the Free Encyclopedia
https://en.wikipedia.org/wiki/List_of_S%26P_500_companies

Dow Jones Industrial Average Component Stocks (Chapter 14)
Wikipedia contributors, "Dow Jones Industrial Average," Wikipedia, the Free Encyclopedia
https://en.wikipedia.org/wiki/Dow_Jones_Industrial_Average

Electric Car Sales by Country (Chapter 16)
Wikipedia contributors, "Electric Car Use by Country," Wikipedia, the Free Encyclopedia
https://en.wikipedia.org/wiki/Electric_car_use_by_country

Enron Emails (Chapter 18)
https://www2.aueb.gr/users/ion/data/enron-spam/index.html

Index

www.ingramcontent.com/pod-product-compliance
Ingram Content Group UK Ltd.
Pitfield, Milton Keynes, MK11 3LW, UK
UKHW051453290926
15471UKWH00008B/1189

9 781943 153992